Springer Proceedings in Geography and Sustainability

Springer Proceedings in Geography and Sustainability features state-of-the-art contributions from conferences, workshops, and similar meetings. The series presents a snapshot of timely developments in fields including but not limited to physical, human, and integrated geography; cross-disciplinary approaches to environmental challenges; and sustainability writ large. It aims to foreground trends emerging from meetings around the world, especially regions that are traditionally underrepresented.

Proposals must include the following:

- A description of the meeting's aims and theme
- The name, place, date, and website of the meeting
- The committee(s) and/or organization(s) responsible for planning the meeting
- A description of the organizer's review process for submissions

Please reach out to the publishing editor, Andrea Brody-Barre (andrea.brody-barre@springer.com), with enquiries.

Rose Ramkat · Silas Leting · Nixon Ronoh ·
Charles Rono
Editors

Towards the Sustainability of our Ecosystems

Proceedings of the International Conference on Renewable Energy, Climate Change and Ecosystem Restoration 2025

Springer

Editors
Rose Ramkat
Moi University
Eldoret, Kenya

Silas Leting
County Government of Uasin Gishu
Eldoret, Kenya

Nixon Ronoh
School of Sciences and Aerospace Studies
Moi University
Eldoret, Kenya

Charles Rono
Department of Chemistry
Kisii University
Kisii, Kenya

ISSN 3059-3344 ISSN 3059-3352 (electronic)
Springer Proceedings in Geography and Sustainability
ISBN 978-3-032-30075-1 ISBN 978-3-032-30076-8 (eBook)
https://doi.org/10.1007/978-3-032-30076-8

This Springer imprint is published by the registered company Springer Nature Switzerland AG
The registered company address is: Gewerbestrasse 11, 6330 Cham, Switzerland

This proceedings publication is dedicated to all stakeholders in climate action, which include the County Governments in the Rift Valley Ecosystem with the stewardship of North Rift Economic Bloc (NOREB)-Kenya, Intergovernmental agencies in Climate Change Management, local communities in the Rift Valley ecosystem, enablers, policy makers in the various county governments in Kenya, academic and research institutions and partners in the global earth challenge.

Foreword

Climate change and its downstream impacts on the Earth systems is a contemporary global challenge, with anthropogenic activities and industrial processes being the major contributors. Given the breadth, depth and complexity of this global challenge, strategic approaches to counter the impacts on the ecosystems and livelihoods require strategic partnerships in climate actions, collaborative initiatives and concerted efforts from all the stakeholders.

The International Conference on Renewable Energy, Climate Change and Ecosystem Restoration (ICRECC), co-hosted by Moi University and The County Government of Uasin Gishu, provides an annual platform for practitioners, policy makers, industry players, research institutions, indigenous communities and all other non-state actors in climate action to share their initiatives and action outcomes for improved climate actions.

The ICRECC-2025 theme "*Towards the Sustainability of our Ecosystems*" focussed on climate action and ecosystem restoration model that fosters mitigation strategies through energy transitions to renewable energy and nature-based mitigation strategies, adaptation of sustainable climate actions and mobilization of resource for sustainable restoration of the ecosystems in the Great Rift Valley and beyond.

Participants discussed challenges, shared knowledge, and explored actionable strategies towards adoption of renewable energy, climate smart policies and practices, and restoration of ecosystems. Through well-structured engagements with the County Government of Uasin Gishu as the model county, this conference discerned the gaps impeding progress in the county governments in sectors relevant to the overall themes of the conference. This outcome illustrates a significant leap in collective efforts to mitigate contemporary challenges retarding progress in sustainability of community initiatives. For instance, the ICRECC discussions highlighted the limited availability of data in various sectors in the county governments and converged on the importance of real-time sectoral data for data-supported decisions, policy initiatives and sustainable community-based action. It also showcased the critical role of big data technologies in accelerating progress in the restoration of ecosystems as demonstrated by some of the deployed artificial intelligence technologies in the management and restoration of wetlands such as Yala Wetlands, a wetland in the Great Rift Valley ecosystem.

ICRECC is founded on the need to document positive and productive community and stakeholder engagements at diverse levels. Together with the publishing partnership of Springer Nature, ICRECC aims to disseminate presented research findings and stakeholder policy recommendations informing cross-disciplinary issues cutting across climate change, ecosystem management and sustainability.

A collection of the selected ICRECC-2025 conference proceedings are presented by the volume's authors, namely Dr. Rose Ramkat and Dr. Nixon Ronoh who are members of The Faculty of Science and Aerospace Studies of Moi University (Kenya), Dr. Silas Kering, who is the Director of Energy at The County Government of Uasin Gishu (Kenya) and Dr. Charles Rono, who is a faculty member in the School of Pure and Applied Sciences of Kisii University (Kenya), respectively. The collaborative effort of these authors and the larger editorial team in the documentation and dissemination of the conference proceedings is greatly appreciated.

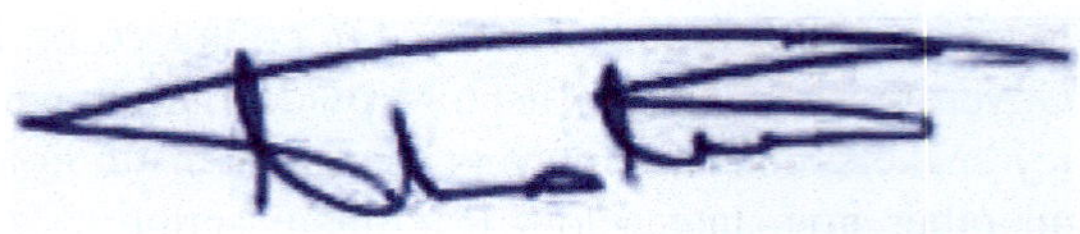

Prof. Kiplagat Kotut
Ag. Vice Chancellor
Moi University
Eldoret, Kenya

Preface

The International Conference on Renewable Energy and Climate Change (ICRECC) is an initiative to infuse theory and practice by bringing together researchers and climate action practitioners to a point of convergence. ICCREC is in itself a collaborative platform that hosts ideas that resonate, motivate, and spur climate action strategies.

ICRECC-2025 is a distinctive conference because it's founded on an atypical cooperation and partnership between a university and the county government, which is recognized as a key government entity in mainstreaming and implementation of climate action plans within the country as defined in the Kenya's Climate Change Act 2016. It further broadens the engagement to include the local community, research institutions, banking institutions, community-based organizations, state and non-state actors active in the implementation of climate action plans. The thematic focus of ICRECC-2025 aligns well with the priority thematic areas of the updated nationally determined contributions 2020–2030 of the region with emphasis on United Nations Sustainable Development Goals #6 on clean water and sanitation by considering pollution and waste management thematic focus, SDG #7 on clean and affordable energy by considering Renewable Energy thematic area, SDG #11 on sustainable cities and communities, SDG #13 on climate action, SDG #14 and SDG #15 through restoration of ecosystems and SDG #17 on partnerships by engaging several stakeholders and actors. For example, keynote presentations highlighted work on tech-based innovations in action on integrated wetlands management and land use in the Lake Victoria basin, specifically the Yala wetlands, innovations in action utilizing agricultural wastes such as banana fibres as a bioresource for textile engineering, available innovations on renewable energy technologies for large-scale use and other prospecting areas of renewable energies and the impact of artificial intelligence technologies on health, food security, livestock and livelihoods.

Publications from ICRECC-2025 border on climate change solutions, sustainability initiatives and cross-cutting climate smart initiatives and practices. The conference proceedings include work presented by a diverse set of stakeholders, including climate action advocates and financiers, climate change researchers and innovators and communities and groups living in and beyond the Great Rift region.

The present volume builds on the previous proceedings, a volume titled "Re-thinking Our Ecosystems" and recently published by Springer Nature. It broadens engagement and impact through components of renewable energy, pollution and waste management, and sustainable smart technologies essential for sustainability of ecosystems. It also recognizes the need for sustainable low-carbon urban ecosystems which are liveable and resilient by focusing on applicable and implementable renewable energy solutions in the Great Rift Valley and beyond. It further supports the need for green and clean technologies through incorporation of clean energy, pollution and waste management strategies in emerging cities.

In summary, the volume brings to fore real-time community climate change issues, actionable mitigation strategies and solutions geared towards adaptation to climate change. It includes topical discussions on the impact of climate change, mitigation strategies, discussions on restoration of ecosystems, adaptation strategies and aspects of resources and policy initiatives with significance in achieving sustainability of ecosystems.

Eldoret, Kenya — Nixon Ronoh
Kisii, Kenya — Charles Rono
Eldoret, Kenya — Silas Leting
Eldoret, Kenya — Rose Ramkat

Acknowledgements

The volume corresponding editors acknowledge the valuable input of the editorial team members, namely:

1. Prof. Frederick O. Nyamwala, Professor, School of Sciences and Aerospace Studies, Moi University, Kenya.
2. Dr. Charles Mutai, Faculty Member, School of Sciences and Aerospace Studies, Moi University, Kenya.
3. Dr. Njira Pili, Faculty Member, School of Sciences and Aerospace Studies, Moi University, Kenya.
4. Dr. Sarah Cherono Chepkwony, Faculty Member, School of Sciences and Aerospace Studies, Moi University, Kenya.
5. Dr. Denyse Snelder, Researcher, Amsterdam Sustainability Institute, Centre for International Cooperation.
6. Dr. Elvire Line Sossa, Researcher, University of Abomey-Calavi.
7. Prof. Antoine Affokpon, Professor and Researcher, University of Abomey-Calavi.
8. Dr. Michelle Eichinger, Food Systems Researcher, Clemson University.
9. Dr. Korir Kiptiemoi, Faculty Member, School of Sciences and Aerospace Studies, Moi University, Kenya.
10. Dr. Viola Kosgey, Faculty Member, School of Sciences and Aerospace Studies, Moi University, Kenya.
11. Dr. John Biwott, Faculty Member, School of Sciences and Aerospace Studies, Moi University, Kenya.

The editors further appreciate the efforts of the conference logistic team, whose membership cut across partner institutions, comprised of

1. Naomi Nkonge, Administrator, Center of Excellence, Moi University, Kenya.
2. Susan Kaluma, ICT Administrator, Moi University, Kenya.
3. Dr. Pius Chumba, CoS-Governor's Office, County Government of UasinGishu
4. Dr. Sammy Kotut, CECM—Agriculture, County Government of UasinGishu
5. Phillip Meli, County Secretary, County Government of UasinGishu

6. Mark Some, CO—Energy, County Government of UasinGishu
7. Philip Lagat, Deputy Director—Environment Department, County Government of UasinGishu
8. Sonia Sanga, Public Relations and Communications Officer, County Government of UasinGishu
9. Francis Karemeri, Statistician—Energy Department, County Government of UasinGishu

Introduction

Ecosystem-Based Approaches for Sustainable Great Rift Valley Ecosystems and Climate Resilient Communities

Ecosystem health plays a fundamental role in achieving climate-resilient ecosystems, sustaining biodiversity and resilient communities with improved livelihoods. With contemporary challenges of climate change, declining ecosystems and loss of biodiversity, diverse sectoral initiatives and investments that promote ecosystem health are essential for sustainable ecosystems and improved livelihoods of the communities dependent on such ecosystems. However, there is slow uptake on holistic and integrated approaches to ecosystem sustainability. This collection aims to describe the nested nature of some of the ecosystem components worth the attention of sectoral initiatives for sustained success in the management, governance, and sustainability of our ecosystems.

The present volume builds on the previous proceedings, a volume titled "Rethinking Our Ecosystems" and recently published by Springer Nature [1]. It broadens engagement and impact through components of renewable energy, pollution and waste management, and sustainable smart technologies essential for sustainability of ecosystems. This volume also recognizes the need for sustainable low-carbon urban ecosystems which are livable and resilient by focusing on applicable and implementable renewable energy solutions in the Great Rift Valley and beyond. It also supports the need for green and clean technologies through incorporation of pollution and waste management strategies in emerging cities.

Keywords: Ecosystem, Ecosystem health, Environmental science, Climate change, Sustainability.

Nixon Ronoh
Charles Rono
Silas Leting
Rose Ramkat

Reference

1. Ronoh N, Rono C, editors. Rethinking Our Ecosystems. Springer Proceedings in Geography and Sustainability [Internet]. Cham: Springer Nature Switzerland; 2025 [cited 2025 Oct 14]. p. 1–292. (Springer Proceedings in Geography and Sustainability; vol. 1). Available from: https://link.springer.com/10.1007/978-3-031-95781-9

Contents

About the Editors

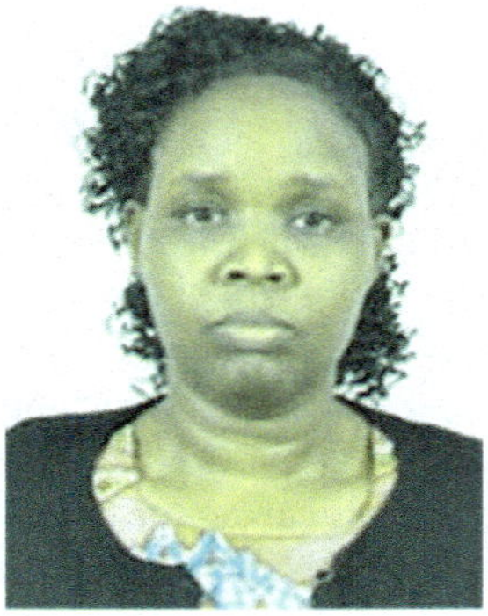

Rose Ramkat has Ph.D. in Natural Resources and Applied Life Sciences from BOKU University, Austria; M.Sc. and B.Sc. degrees from Egerton University, Kenya. She worked previously for the Ministry of Agriculture, Kenya in many projects including: National Agriculture and Livestock Extension Project, Njaa Marufuku Kenya, International Fund for Agricultural Development, and Kenya Agricultural Productivity Project. She is currently involved in collaborative projects on biotechnology, food systems, and bio-circularity. She is Deputy Center leader and Principal Investigator for Africa Center of Excellence in Phytochemicals, Textile and Renewable Energy (ACE II—PTRE, Kenya) and Dean School of Science and Aerospace, Moi University. She is representative of the International Association of Universities Cluster for UN SDG 12 and serves in several boards including Scientific Sub-Committee of DSTI—NRF Center of Excellence in Food security.

Silas Leting is a nuclear scientist, energy expert, and policy strategist. He holds a Doctor of Philosophy (Ph.D.) in Physics (Neutronics) from Moi University. He serves as the Director of Energy for the County Government of Uasin Gishu, Kenya. His leadership spans investor coordination, regulatory structuring, financial engineering, and implementation oversight, establishing the project as a benchmark for PPP-driven sustainable infrastructure. His influence extends regionally through the North Rift Economic Bloc, where he contributes to harmonized energy planning and regional investment frameworks.

Nixon Ronoh is a faculty member, School of Science and Aerospace studies of Moi University, Kenya. Nixon obtained his Doctor of Sciences degree from Vrije Universiteit Brussel (VUB), and has published research papers cutting across the fields of Computer Science and Applied Mathematics. Nixon currently works with a research team studying the applications of Artificial Intelligence technologies in innovative and sustainable strategies towards renewable energy solutions, and the adaptation and mitigation of climate change.

Charles Rono is a researcher and a faculty member in the Department of Chemistry, Kisii University. He is actively involved in collaborative interdisciplinary research focused on sustainable science solutions to climate change and ecosystem health. Some of his contributions include harnessing data chemistry and artificial intelligence for sustainability in product and process development, and phytochemicals in the development of materials for renewable energy applications.

Impact of Climate Change

Forecasting Drought Events in Kenya ASAL Region: A CNN-Based Framework Using CHIRPS Precipitation

Bernice Mugure Mathenge, Jennifer Batamuliza, and Charles Mutai

Abstract

Aim: Food insecurity and economic instability have resulted from the severe effects of droughts on agricultural output in Kenya's Arid and Semi-Arid Lands (ASALs). These regions have faced increasing vulnerability due to erratic rainfall patterns, resulting in severe crop failures and livestock losses. Despite efforts to forecast droughts, existing techniques have frequently been imprecise, thereby hindering timely intervention. By developing a CNN-based framework with CHIRPS precipitation data and the Standardized Precipitation Index (SPI), this study aimed to improve the accuracy of drought forecasts. Furthermore, the study aimed to increase the precision of drought forecasts and provide actionable insights on managing the effects of drought in ASALs by describing past drought occurrences, analyzing precipitation trends, and utilizing machine learning. **Method**: The research involved preprocessing CHIRPS data, spanning from 1981 to 2024, to calculate SPI values and address missing data using the Inverse Distance Weighting (IDW) method. A CNN-based model that integrated precipitation's temporal and spatial characteristics was developed to forecast drought situations. The model was trained to identify spatial hotspots and forecast drought events across the 14 semi-arid counties in Kenya's ASAL region. **Results**: To support timely drought mitigation, this study implemented

B. M. Mathenge (✉)
African Center of Excellence in Data Science, University of Rwanda, Kigali, Rwanda
e-mail: bernicemugure29@gmail.com

J. Batamuliza
University of Rwanda, Kigali, Rwanda
e-mail: j.batamuliza@ur.ac.rw; bajelo2000@yahoo.co.uk

C. Mutai
Department of Mathematics, Physics and Computing, Moi University, Eldoret, Kenya
e-mail: charlimtai@mu.ac.ke

R. Ramkat et al. (eds.), *Towards the Sustainability of our Ecosystems*, Springer Proceedings in Geography and Sustainability,
https://doi.org/10.1007/978-3-032-30076-8_1

a CNN-based framework that significantly improved predictive performance. The model outperformed traditional classifiers, achieving 94% accuracy and high precision and recall for both drought and non-drought events. Spatial analysis across Kenya's ASALs showed strong alignment between predicted and actual droughts, with localized patterns of under- and overestimation. These results underscore the CNN's potential for operational early warning in drought-prone regions. **Conclusion**: This study offers a reliable CNN-based drought forecasting model that strengthens resilience in Kenya's ASALs by supporting ecosystem sustainability, proactive planning, and evidence-based climate adaptation at national and county levels.

Keywords

CNN-based • Chirps precipitation data • Forecasting drought • SPI

1 Background Literature

Climate change has emerged as a fundamental driver of hydrometeorological extremes, including increased frequency and intensity of droughts, especially in climate-sensitive regions like Kenya. These shifts are largely attributed to anthropogenic factors such as fossil fuel emissions, deforestation, and land-use changes [1]. The Horn of Africa (HoA), where Kenya is located, has seen significant changes in precipitation variability and reliability, resulting in longer dry spells and erratic rainfall [2]. These climate-induced anomalies undermine ecological stability and threaten livelihoods, particularly for communities in arid and semi-arid regions.

Drought is a complex climatic hazard marked by a prolonged deficit in precipitation relative to normal conditions, often leading to agricultural losses, water scarcity, and ecological stress [3]. In Kenya, where rain-fed agriculture supports most rural livelihoods, droughts disrupt food security and economic stability [4]. Meteorological droughts, defined as seasonal or annual rainfall deficits often intensified by high temperatures, trigger cascading effects on agriculture, water resources, and livelihoods [3].

Over 80% of Kenya is classified as Arid and Semi-Arid Lands (ASALs), mainly in the northern regions bordering Somalia and Ethiopia. These areas support most rural, livestock-dependent communities and hold over 75% of the national herd [5–8]. Droughts in ASALs frequently trigger agricultural, hydrological, and socio-economic disruptions, severely impacting food security and household resilience [8]. Understanding seasonal drought patterns, especially during Kenya's two key rainy seasons (March–May and October–December), is critical for informing early warning systems and guiding policy responses [9].

Drought remains one of Africa's most pressing environmental challenges, especially in East Africa. Since October 2020, over 40 million people have been affected, with Somalia near famine and Ethiopia's drought impacting 31 million people. Kenya and Ethiopia faced six consecutive below-average rainy seasons [10], resulting in food insecurity, livestock deaths, and economic losses for farmers and pastoralists [11]. Historical droughts in 2005 and 2010–2011 emphasize the region's vulnerability [12–14]. Climate projections indicate increasing drought frequency and intensity [11], underscoring the need for accurate characterization. Tools such as the Standardized Precipitation Index (SPI) and Standardized Precipitation-Evapotranspiration Index (SPEI) offer valuable insights into the severity and temporal dynamics of droughts in Kenya [11, 15] supporting more targeted mitigation strategies.

Accurate drought forecasting is crucial to reduce socio-economic impacts, especially in Kenya's Arid and Semi-Arid Lands (ASALs). The Climate Hazards Group InfraRed Precipitation with Station data (CHIRPS) merges satellite and station data to provide high-resolution precipitation estimates, ideal for areas with sparse measurements [16]. It has proven more reliable than datasets like ARC2 and TAMSAT2 in Eastern Africa at monthly and 10-day scales [17]. With its extensive temporal coverage and sensitivity to spatial variability, CHIRPS offers a strong foundation for monitoring and analyzing rainfall trends in ASALs.

Despite the availability of improved datasets like CHIRPS, drought conditions in Kenya continue to pose serious challenges to the agricultural sector, causing inconsistent yields, crop failures, and economic disruption. Farmers and pastoralists remain highly vulnerable due to rainfall variability and the limited accuracy and timeliness of current drought prediction systems, which exacerbate these challenges and hinder effective responses.

Meteorological droughts are of particular concern as they often represent the initial stage in a cascading series of drought impacts. They are characterized by persistent deficits in precipitation over time [2, 18, 19]. Early detection and accurate forecasting of meteorological droughts are critical for pre-emptive planning in water resource management, disaster preparedness, and food security strategies. However, modeling drought occurrence and evolution remains challenging due to the inherently non-linear, non-stationary, and spatially heterogeneous nature of climate data. These complexities necessitate the use of advanced forecasting methods capable of capturing such dynamics.

Traditional models globally have struggled to capture the evolving complexity of drought patterns influenced by climate change [18]. However, advances in machine learning, particularly Convolutional Neural Networks (CNNs), offer new opportunities [18]. By leveraging CHIRPS data, CNNs can model spatial rainfall patterns more effectively, enhancing the precision and responsiveness of drought forecasts in Kenya's ASALs and contributing to more informed mitigation efforts. While tools like CHIRPS and indices such as SPI have improved drought characterization, current early warning systems remain limited in their spatial sensitivity and forecasting accuracy, particularly in the ASAL regions. Moreover, traditional statistical often fall short in capturing the non-linear spatial patterns driven by climate variability [20]. This creates a critical need for more robust, data-driven approaches to drought forecasting.

Recent research, including Gyaneshwar et al. [18], has highlighted the effectiveness of deep learning techniques, particularly Convolutional Neural Networks (CNNs), for handling complex spatiotemporal data in drought prediction. These models outperform traditional statistical approaches by learning patterns directly from gridded climate inputs, enabling more reliable predictions over larger spatial domains and longer lead times. Given the heterogeneity of drought expression across Kenya's varied landscapes, these capabilities offer significant potential for improving early warning systems and adaptation planning [21].

Kenya's arid and Semi-Arid Lands (ASALs) are increasingly vulnerable to the effects of climate change, particularly in relation to temperature and precipitation variability. Fluctuating rainfall patterns significantly influence the frequency and intensity of drought events, which in turn pose substantial threats to agricultural productivity, especially in regions where livelihoods rely heavily on livestock farming [8]. Analyses of rainfall trends reveal that focusing solely on annual averages may obscure critical short-term variations; thus, assessing intra-seasonal variability is essential to fully understand the agricultural implications of changing precipitation patterns [22].

Despite the urgency of the issue, localized studies specifically targeting drought forecasting and precipitation dynamics in Kenya's ASALs remain sparse [23, 24]. This lack of research tailored to the unique geographical, environmental, and socioeconomic conditions of the region limits the effectiveness of existing drought response strategies. It underscores the need for tools that are specifically designed to support proactive and context-appropriate drought mitigation efforts in Kenya.

In response to these challenges, the study proposes a CNN-based drought forecasting framework tailored for Kenya's ASAL regions. The approach leverages SPI values derived from CHIRPS precipitation data to achieve three core aims: to characterize drought events in Kenya using CHIRPS precipitation data and SPI analysis, to develop a deep learning model capable of forecasting droughts with improved spatial and temporal resolution, and to assess the spatial distribution of predicted droughts in the Kenya semi-arid regions to enhance the accuracy and responsiveness of early warning systems. This framework not only addresses the shortcomings of conventional methods but also supports more timely and localized mitigation planning.

Several indices have been used to monitor and analyze drought occurrence in Kenya, notably the Standardized Precipitation Index (SPI) and the Standardized Precipitation Evapotranspiration Index (SPEI) [15, 25–27]. These indices offer insights into the timing, duration, and severity of drought conditions across multiple spatial and temporal scales. For instance, Mutsotso et al. employed both the SPEI and the Normalized Difference Vegetation Index (NDVI) to examine drought frequency and its ecological implications, establishing the relationship between precipitation deficits and vegetation response in Kenyan landscapes [27]. Similarly, Frank et al. supported the application of the Effective Drought Index (EDI) for drought assessments within the Tana River Basin, citing its responsiveness to short-term meteorological changes [28]. Karanja et al. applied the SPI to evaluate seasonal and annual drought variations in Laikipia County, demonstrating its utility for regional planning and resource allocation [25]. In another notable study, Wambua et al. used the SPI to assess the geographical, temporal, and trend characteristics of meteorological droughts in the Upper Tana River Basin. Their findings confirmed the value of SPI in identifying persistent drought trends, thereby informing early warning systems and drought preparedness strategies [15].

However, these studies also reveal notable limitations. Regional differences in drought characteristics make it challenging to apply a single index consistently across diverse ecological zones. As highlighted by Wambua et al., the spatial variation in drought intensity within the same basin emphasizes methodological uncertainties and index sensitivity to local climatic nuances [15]. This lack of consensus on the most suitable drought indicators underscores the need for further research that transcends generic applications. Instead, there is a critical need for refined, ASAL-specific approaches that provide reliable forecasting and enable localized planning and intervention [29].

The Standardized Precipitation Index (SPI) is a widely used metric for characterizing meteorological droughts. It is calculated by dividing the difference between observed and mean precipitation over a specified period by the standard deviation. This statistical normalization facilitates the comparison of rainfall deficits across diverse temporal scales and geographic regions. To account for the typically non-normal distribution of precipitation over short durations, transformations can be applied to ensure accuracy, particularly for periods shorter than one year. Notably, computing SPI requires a continuous monthly precipitation dataset spanning at least 30 years [30].

The 2009 Lincoln Declaration on Drought Indices formally recommended the SPI for classifying meteorological droughts [1]. Leveraging historical precipitation records, the SPI can quantify drought probability across multiple timescales, from one month up to 48 months or longer [30, 31]. According to McKee et al., drought onset is defined as the point when the SPI drops below -1.0 and persists until it returns to a positive value. Prolonged negative SPI values indicate ongoing drought conditions, with the severity, duration, intensity, and magnitude determined by specific SPI thresholds [30].

To support improved drought monitoring in Kenya's arid and Semi-Arid Lands (ASALs), this study employed the Climate Hazards Group InfraRed Precipitation with Station (CHIRPS) dataset. CHIRPS provides high-resolution (0.05°) precipitation estimates from 1981 to the present [32]. It is particularly advantageous in data-sparse regions such as Kenya's ASALs, where ground-based observations are limited [33].

Numerous studies have validated the performance of CHIRPS against other open-source climate datasets across a variety of geographies. Duan et al. evaluated its reliability in Italy's Adige Basin [34], while Macharia et al. demonstrated its suitability in Kenya [35]. Further validation has been conducted in the Central Andes of Argentina [36], Pakistan [37], and Mainland China [38]. These studies consistently highlight the superiority of CHIRPS in terms of historical coverage, spatial accuracy, and application in climate monitoring [39]. Specifically, Macharia et al. concluded that CHIRPS 2.0 provides the most accurate rainfall estimates in tropical warm arid regions, reinforcing its appropriateness for drought prediction and related studies in Kenya's ASALs [35]. While CHIRPS addresses the data challenge, modeling future drought events requires an approach capable of capturing spatial patterns—thus motivating the use of CNNs.

To address class imbalance in the classification of drought events using Convolutional Neural Networks (CNNs), this study employed class-weight adjustment techniques. These adjustments improved the model's ability to correctly identify the minority class—drought conditions—thereby enhancing predictive performance. The exclusive focus on CHIRPS precipitation data enabled a comprehensive understanding of climatic variability, supporting more precise drought forecasting and evidence-based mitigation planning.

Convolutional Neural Networks (CNNs) have gained increasing traction in weather forecasting, demonstrating significant potential to address the limitations of traditional predictive models. In the domain of monthly rainfall forecasting, CNNs have outperformed both operational models and conventional machine learning approaches. For instance, Haidar and Verma [40] proposed a deep CNN architecture for predicting monthly rainfall in eastern Australia, reporting superior performance compared to a traditional multilayer perceptron (MLP) model and the Australian Community Climate and Earth-System Simulator–Seasonal Prediction System (ACCESS).

One of the key advantages of CNNs lies in their ability to effectively model the spatial heterogeneity and complex nonlinear patterns characteristic of climate data, areas where traditional models often underperform [41, 42]. By automatically learning spatiotemporal features from input data, CNNs enhance the predictive capacity of seasonal forecasting models and enable the modeling of long-range spatial dependencies across multiple timescales [42].

Despite these advances, the application of deep learning to drought forecasting continues to face notable challenges. As highlighted by Gyaneshwar et al. [18], these include the need for more resilient evaluation frameworks, the potential for enhanced generalization through transfer learning or domain adaptation, and the exploration of alternative data sources in regions with limited observational data. To address these challenges, [18] advocates for the development of advanced deep learning architectures, such as CNNs, that can improve forecast accuracy, particularly over extended timescales. The study also emphasizes the importance of integrating alternative datasets, including remotely sensed products, to overcome the limitations of sparse ground-based networks.

In this context, the use of CHIRPS precipitation data in the present study aligns with the recommendations for mitigating data scarcity challenges. The high spatial and temporal resolution of CHIRPS makes it a valuable input for CNN-based drought forecasting in data-scarce environments such as Kenya's ASAL regions. Moreover, the current body of research underscores an urgent need for a robust, localized drought forecasting framework tailored to the Kenyan context. Given the region-specific manifestations of climate change, this study seeks to bridge that gap by employing a CNN-driven model utilizing CHIRPS precipitation data, thereby offering a contextually relevant and data-efficient solution for drought prediction in Kenya's ASALs.

In summary, the integration of high-resolution CHIRPS precipitation data and advanced CNN-based modeling presents a promising pathway for improving drought forecasting in Kenya's ASALs. By addressing the limitations of traditional models and capturing the spatiotemporal nuances of meteorological droughts, this approach contributes to early warning systems, policymaking, and climate resilience for some of the most vulnerable communities in East Africa.

2 Methodology

2.1 Data

The study utilized Climate Hazards Group InfraRed Precipitation with Station data (CHIRPS), a global gridded precipitation dataset that integrates satellite data with ground station observations. It provides high-resolution (0.05°) monthly rainfall estimates from 1981 to the present [32], offering comprehensive coverage for drought monitoring in Kenya's ASALs region. The data is available in NetCDF format, with more than 16 million observations for the period of January 1981 to November 2024 [43].

2.2 Data Processing

A mixed-method approach was adopted, combining retrospective analysis of CHIRPS data with a deep learning-based forecasting model. Missing values, mainly in coastal areas, were addressed using Inverse Distance Weighting (IDW) interpolation, defined as:

$$\hat{V}_p = \frac{\sum_{i=1}^{n} \frac{V_i}{h_i^p}}{\sum_{i=1}^{n} \frac{1}{h_i^p}} \tag{1}$$

where $\hat{V}_p$ is the estimated value, V_i are known values, h_i are distances to known points, p is the power parameter, and n is the number of known values [44].

The interpolated data was exported in CSV format for further analysis.

The Standardized Precipitation Index (SPI) was calculated at a 3-month timescale to characterize droughts. For each grid point, a gamma distribution was fitted to historical precipitation records. Zero-precipitation probabilities (p_0) were included, and the cumulative distribution function (CDF) was computed as:

$$\text{CDF} = (p_0 + (1 - p_0)) \times \text{Gamma CDF} \tag{2}$$

To ensure numerical stability, CDF values were clipped within $(1 * 10^{-10})$ and $(1 - (1 * 10^{-10}))$ before converting to standard normal scores using the inverse normal distribution (Percent Point Function). Drought was flagged where SPI $\leq -$ 1.0, with binary labels: 1 for drought, 0 otherwise [30].

For tabular models, SPI, precipitation, and drought flag data were transformed into a flat tabular structure with spatial coordinates (time, latitude, longitude) stacked. A 3-month rolling window was applied to generate lagged features, capturing temporal dependencies. Additional temporal features such as month and year were extracted.

For the CNN-based approach, precipitation data was standardized using Z-score normalization:

$$Z = \frac{X - \mu}{\sigma} \tag{3}$$

where X is the precipitation value, μ is the mean, and σ is the standard deviation across the dataset. Data was then reshaped into 5D arrays with dimensions: time, latitude, longitude, sequence length, and channels.

To preserve temporal continuity, the dataset was chronologically split into training (60%), validation (20%), and test (20%) sets. This ensured no data leakage and realistic model evaluation. The binary drought flag exhibited class imbalance, so class weights were applied during training. Weights were calculated as:

$$w_i = \frac{N}{ni^* C} \tag{4}$$

where w_i is the weight for class i, N is the total number of samples, n_i is the number of samples in class iii, and C is the number of classes.

The final input dataset for the CNN was structured into a 5D array with dimensions: time, latitude, longitude, sequence length, and channels. Target labels were structured as a 4D array of the same spatial and temporal dimensions, with the binary drought flag as the output. This approach ensured that the model was trained on a temporally consistent, spatially representative dataset, while mitigating class imbalance and enabling effective forecasting of drought occurrences across Kenya's ASALs.

2.3 Model Development

Random Forest, Logistic Regression, and XGBoost classifiers were implemented to establish baseline drought prediction models. The Random Forest model employed an ensemble of decision trees with class weights set to "balanced" to address class imbalance, while ensuring reproducibility through a fixed random state. Logistic Regression was similarly trained with balanced class weights to enhance its sensitivity to the minority class. For the XGBoost classifier, the scale_pos_weight parameter was calculated based on the ratio of majority to minority classes in the training dataset. Its performance was further optimized through hyperparameter tuning, including the number of estimators, learning rate, and maximum tree depth.

To capture the complex spatiotemporal dynamics of drought, a 3D Convolutional Neural Network (CNN) was developed. The network architecture consisted of two convolutional blocks, each comprising a 3D convolutional layer with 16 and 32 filters respectively, followed by batch normalization to stabilize and accelerate training, ReLU activation, and 3D max-pooling layers to reduce spatial dimensionality. The output feature maps were flattened and passed through fully connected layers, with dropout regularization applied to mitigate overfitting. A final dense layer with sigmoid activation produced binary predictions for each grid cell, which were reshaped to align with the original spatial dimensions.

The model was compiled using the binary cross-entropy loss function and optimized using the Adam optimizer. A batch size of eight was selected to balance memory consumption and model convergence. Training was conducted for up to 50 epochs, with early stopping employed to halt training when validation loss plateaued. The best-performing model was saved via checkpointing, and training history was visualized to monitor learning dynamics. Hyperparameter tuning was performed using the Keras Tuner with the Hyperband search algorithm, optimizing dropout rates, dense layer units, filter sizes, and learning rates.

2.4 Spatial Post-Processing of CNN Predictions

To evaluate the spatial distribution of predicted drought events, a geospatial analysis was conducted. A shapefile containing Kenya's administrative boundaries was used to extract polygons representing the semi-arid counties of interest. Binary drought predictions from the CNN model for a selected time step were merged with corresponding latitude and longitude coordinates to form a GeoDataFrame, which was aligned with Kenya's coordinate reference system (CRS).

Each prediction point was spatially joined with its corresponding administrative region using containment logic. Regional-level summaries were then computed to quantify the proportion of drought-affected areas. The same spatial post-processing approach was applied to the test dataset to ensure methodological consistency.

3 Results

3.1 SPI Analysis

The Standardized Precipitation Index (SPI) was calculated to characterize the drought conditions across the entire Kenyan region during the study period. The statistical summary of the SPI values is presented in Table 1.

Table 1 Summary of SPI statistics

SPI statistic	Value
Min	− 6.36
Max	4.42
Mean	− 0.0049
Standard deviation	1.02

Source Author's own computation, 2025

3.2 Model Performance

The performance of the various machine learning models, including Random Forest, Logistic Regression, XGBoost, and CNN, was evaluated using key performance metrics: precision, recall, F1-score, and accuracy. These metrics were calculated on the validation set to assess the models' ability to predict drought events. The performance metrics for each model are summarized in Table 2.

The Convolutional Neural Network (CNN), identified as the best-performing model, was evaluated on the test set using precision, recall, F1-score, and accuracy. The performance metrics for the CNN model on the test set are summarized in Table 3.

Table 2 Performance metrics for the models

Model	Class	Precision	Recall	F1 score	Support	Accuracy
Logistic regression	0	0.97	0.82	0.89	2,709,479	0.84
	1	0.55	0.89	0.68	657,377	
Random Forest classifier	0	0.93	0.95	0.94	2,709,479	0.91
	1	0.78	0.72	0.75	657,377	
XGBoost	0	0.99	0.85	0.91	2,709,479	0.87
	1	0.65	0.91	0.74	657,377	
CNN	0	0.96	0.97	0.96	2,727,370	0.94
	1	0.85	0.83	0.84	645,974	

Source Author 2025

Table 3 Performance metrics for the best model (CNN)

Model	Class	Precision	Recall	F1-score	Support	Accuracy
CNN	0	0.96	0.96	0.96	2,802,104	0.93
	1	0.80	0.82	0.81	603,676	

Source Author 2025

3.3 Spatial Comparison of Predicted and Actual Drought Events

Spatial comparisons between predicted and actual drought events were visualized for eight months in 2023: January, February, March, May, June, September, October, and December. These months correspond to the transitions between Kenya's key climatic seasons: the long rains (March–May), short rains (October–December), and dry periods (January–February, June–September).

For each month, as shown in Fig. 1 that follows, three county-level maps were generated for the 14 semi-arid counties of Kenya: (i) predicted drought proportions, (ii) actual observed proportions, and (iii) the difference (Prediction − Actual). These three maps for each month, presented in Figs. 2, 3, 4 and 5, provide a detailed visual assessment of model accuracy across time and regions.

Semi-Arid Counties in Kenya

Spatial and Temporal Analysis of Predicted Drought Events

4 Discussion

The SPI analysis indicates a wide range of drought conditions across Kenya, with extreme drought at -6.36 and a mean value suggesting near-normal precipitation. Using SPI ≤ -1.0 to define drought, 3,268,905 drought events were identified, compared to 13,759,995 non-drought events, resulting in a drought-to-non-drought ratio of approximately 0.24. While droughts are less frequent, their intensity and periodic nature highlight the need for improved monitoring, particularly in the ASALs. The observed class imbalance underscores the region's vulnerability to recurring dry spells, emphasizing the importance of accurate drought prediction models.

Among the tabular models, the Random Forest classifier exhibited the highest accuracy (91%) and achieved a balanced performance in terms of both precision and recall for the two classes. XGBoost, while demonstrating a high recall (91%) for drought events (Class 1), showed a trade-off with increased false positives, making it more sensitive to detecting droughts. Logistic Regression, although simple and interpretable, resulted in the lowest precision for Class 1, indicating a higher rate of false positives. Despite the class imbalance in the dataset, Random Forest provided the best trade-off, effectively balancing precision and recall.

In comparison to the traditional tabular models, the CNN, leveraging 3D convolutional layers, significantly outperformed all other models. As shown in Table 2, it achieved an overall accuracy of 94%, demonstrating strong performance in classifying both drought and non-drought events. The CNN achieved 96% precision and 97% recall for non-drought events (Class 0), indicating high accuracy with minimal false positives and negatives. For drought events (Class 1), it recorded 85% precision and 83% recall, offering a good balance despite the class imbalance.

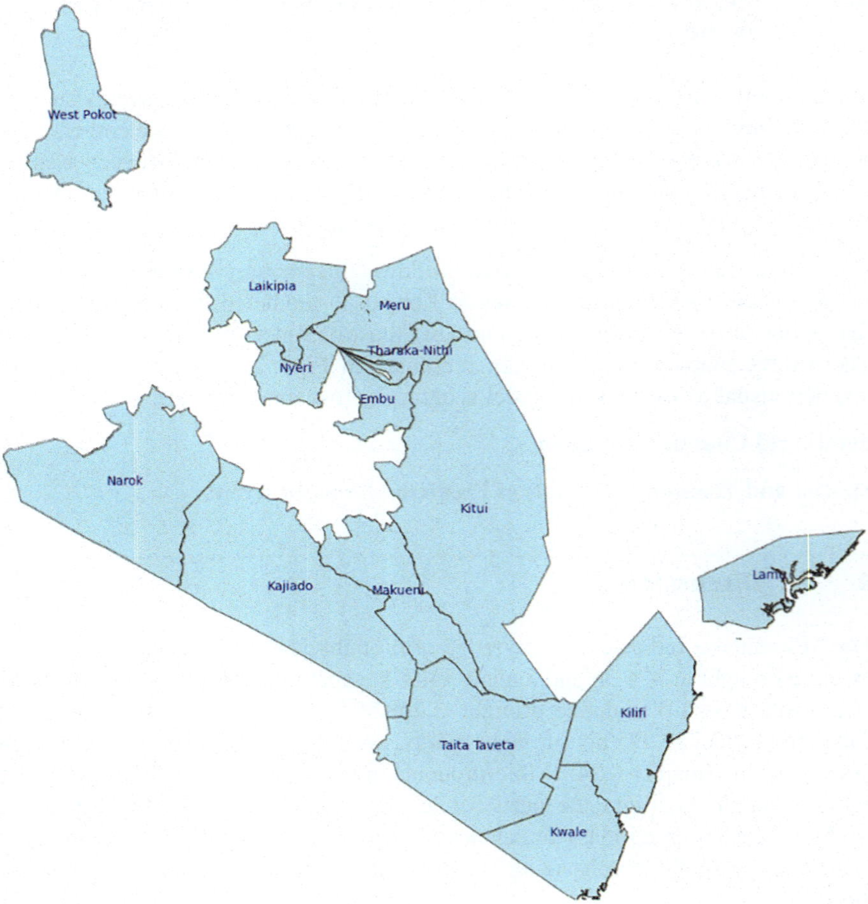

Fig. 1 Semi-arid counties in Kenya. *Source* Author's own compilation using GADM administrative boundary data [47]

The CNN's confusion matrix confirmed its strong classification ability, with minimal false positives and negatives. The model's performance on the final test set, shown in Table 3, demonstrated minimal degradation and supported its generalizability to unseen data. In contrast to the tabular models, which struggled with class imbalance—especially XGBoost's trade-off between high recall and false positives, and Logistic Regression's lower precision for drought events—the CNN model exhibited a more stable and balanced performance across both classes.

The CNN model's strong performance, especially in identifying the minority class (drought events), is attributed to its ability to capture spatiotemporal features, a crucial aspect of drought forecasting. The model demonstrated high precision and recall for both non-drought (Class 0) and drought (Class 1) events, with 96%

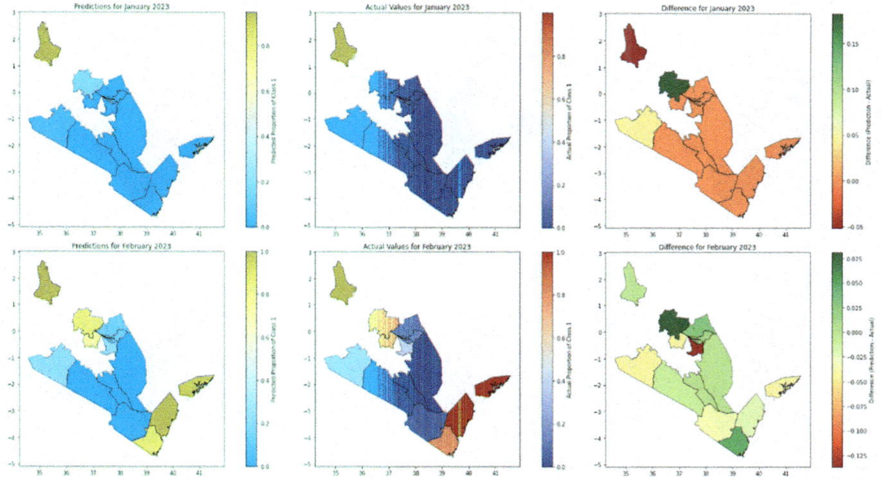

Fig. 2 Predicted versus actual drought events in the semi-arid areas for Jan and Feb 2023. *Source* Author 2025

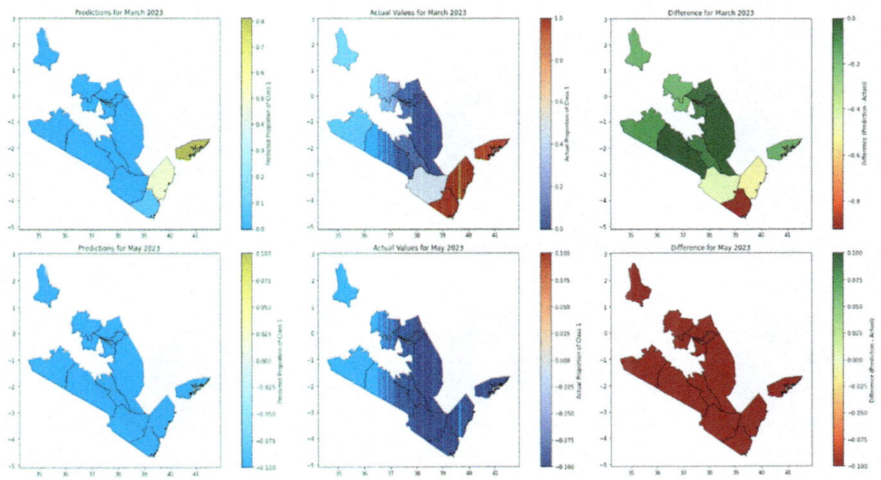

Fig. 3 Predicted versus actual drought events in the semi-arid areas for March and May 2023. *Source* Author 2025

precision and recall for non-drought events and 80% precision and 82% recall for drought events.

Additionally, the CNN showed strong generalization, with its performance on the test set confirming its resilience to overfitting. Given these advantages—high accuracy and robust detection across both classes—the CNN is well-suited for operational deployment in drought prediction.

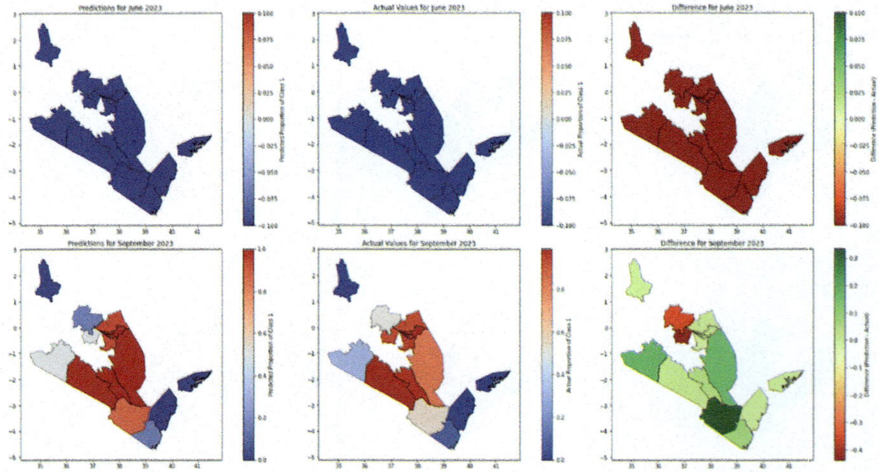

Fig. 4 Predicted vs actual drought events in the semi-arid areas for June and September 2023. *Source* Author 2025

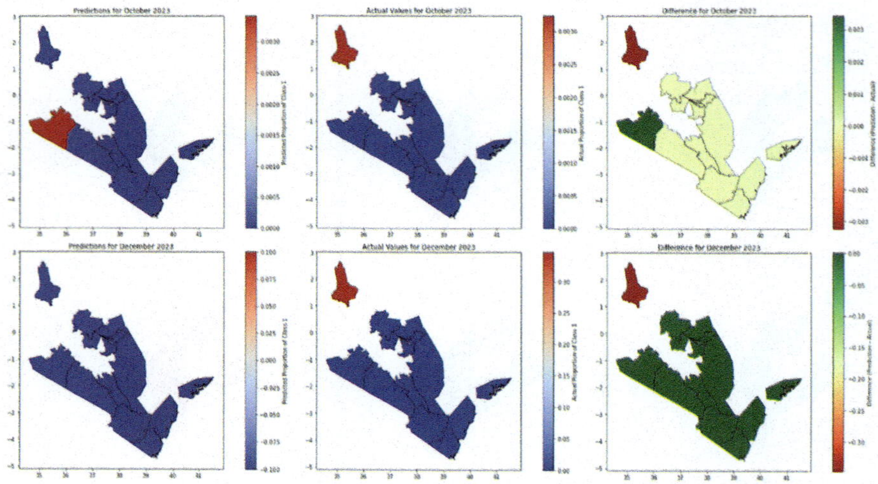

Fig. 5 Predicted versus actual drought events in the semi-arid areas for October and December 2023. *Source* Author 2025

The analysis began by examining the temporal patterns of drought predictions and observed occurrences, focusing on differences between predicted and actual drought episodes across selected months and counties in 2023. The model demonstrated varying levels of overestimation and underestimation across space and time.

In January (Fig. 2), the model slightly underestimated drought severity in West Pokot, while overestimating conditions in Laikipia. In February, overestimations were observed in Laikipia, Kwale, and Meru, whereas Nyeri experienced slight underestimation. By March (Fig. 3), mild underestimations were noted in Kwale, Taita Taveta, and Kilifi, with a general trend of underprediction persisting into June (Fig. 4) across most counties. In September (Fig. 4), underestimations were again observed in Taita Taveta, Kitui, and Narok, while Nyeri and Laikipia were slightly overestimated. October (Fig. 5) exhibited underestimation in West Pokot and overestimation in Narok, with a similar pattern in December (Fig. 5) where West Pokot was again underestimated.

These monthly patterns reflect Kenya's seasonal climate variability. Most drought events were concentrated during the dry months—January to February and September—though sporadic droughts were also observed during the short rainy season (October to December), consistent with the region's historical climatic variability.

From a spatial perspective, the results reveal persistent drought conditions in several vulnerable counties. West Pokot experienced droughts in four distinct months—January, February, October, and December—indicating recurring susceptibility. In September, widespread drought events were noted in counties such as Kajiado, Makueni, Embu, Nyeri, Tharaka Nithi, and Meru, reflecting significant dryness during the seasonal transition. Along the coastal region, Kwale, Lamu, and Kilifi experienced droughts in both February and March.

These spatial patterns highlight the need for localized drought forecasting systems and tailored early warning mechanisms, particularly for semi-arid regions facing recurrent rainfall deficits. The findings also underscore the importance of addressing overestimations and underestimations in model predictions, especially for vulnerable counties.

Building on the demonstrated effectiveness of CNN-based drought forecasting in Kenya's ASALs, several directions for future research are proposed. To enhance the performance and applicability of drought forecasting models, future research should consider the integration of additional environmental datasets such as soil moisture and evapotranspiration. These variables have the potential to enrich model inputs and improve prediction accuracy, particularly in water-scarce regions.

Incorporating socio-economic indicators, such as agricultural productivity, land use, and water resource availability, could further support a holistic assessment of drought impacts on livelihoods and ecosystems.

From a modeling perspective, the adoption of hybrid deep learning architectures that combine Convolutional Neural Networks (CNNs) with Recurrent Neural Networks (RNNs) could more effectively capture the long-term temporal dependencies characteristic of climatic time series. Adaptive regridding techniques should also be explored to better represent spatial heterogeneity, especially in regions characterized by complex topography or highly variable precipitation patterns.

To validate generalizability, future work should expand the study to neighboring countries or agro-climatically similar zones. Finally, the development of real-time drought monitoring and early warning systems—leveraging satellite-based observations and scalable cloud computing infrastructure—would significantly support timely decision-making and contribute to sustainable drought mitigation and resource management strategies across Kenya's ASALs.

While the study provides valuable insights into historical drought patterns and establishes a reliable CNN-based forecasting framework, it does not incorporate or analyze the broader socio-economic implications of drought events on affected communities. Understanding these impacts is essential for informing integrated drought mitigation and adaptation strategies. Additionally, although the model proved effective for forecasting droughts within Kenya's ASAL region, its applicability to areas with different climatic conditions remains uncertain and should be explored in future research.

This study offers important insights into historical drought patterns and spatial distributions while establishing a reliable CNN-based drought forecasting model. It focuses on the months of January, February, March, May, June, October, and December—critical transitional periods within Kenya's seasonal calendar that are essential for understanding drought dynamics and planning timely mitigation strategies.

The analysis centers on the year 2023, identified by UNOCHA and other sources [45, 46] as the peak of a prolonged three-year drought (2020–2023)—the most severe in East Africa since the 1980s, causing widespread agricultural losses, food insecurity, and livelihood disruptions.

By translating model outputs into spatial visualizations, the study provides actionable, location-specific insights that can guide resource prioritization and early warning efforts. These tools are especially valuable for policymakers at both national and county levels, supporting the design of proactive drought response strategies and adaptation plans tailored to vulnerable regions. The CNN model outperforms traditional approaches such as Random Forest, Logistic Regression, and XGBoost by effectively capturing complex spatiotemporal patterns and addressing data imbalances—key requirements for informing evidence-based policies.

Ultimately, the research contributes directly to Kenya's climate resilience agenda by offering a scalable, data-driven framework that can inform national development planning, strengthen drought contingency financing, and enhance institutional capacities for disaster risk reduction. Its relevance extends to international donors, agricultural policy advisors, and climate adaptation initiatives seeking to align interventions with timely, scientifically grounded forecasts.

Acknowledgements Not Applicable.

Author's Contributions BM conceptualized the study, analyzed the data, and drafted the initial manuscript. JB verified the data analysis and reviewed the final draft. CM ensured data accessibility, integrity, and accurate interpretation of results. All authors contributed to manuscript revisions.

Availability of Data and Materials The CHIRPS dataset is publicly available at https://data.chc.ucsb.edu/products/CHIRPS/v3.0/. This research utilized the African Centre of Excellence in Data Science's HPC cluster and Jupyter Lab, equipped with NVIDIA GPUs and Intel Xeon processors, enabling efficient data processing, SPI calculation, and CNN training. Access to these resources is available by request at https://aceds.ur.ac.rw/facilities.

Funding Not Applicable.

Declarations

Ethics Approval and Consent to Participate Not applicable.

Consent for Publication Not applicable.

Competing Interests The authors declare that they have no competing interests. All authors approved the final manuscript.

Author Details 1. African Center of Excellence in Data Science, University of Rwanda, Kigali Rwanda.

2. University of Rwanda, Kigali, Rwanda, j.batamuliza@ur.ac.rw, bajelo2000@yahoo.co.uk.

3. Department of Mathematics, Physics and Computing, Moi University, P.O. Box 3900, Eldoret, Kenya.

Abbreviations

CHIRPS	Climate Hazards Center InfraRed Precipitation with Station data
ASALs	Arid and Semi-Arid Lands
UNOCHA	United Nations Office for the Coordination of Humanitarian Affairs
RNN	Recurrent Neural Network
XGBoost	extreme Gradient Boosting
SPI	Standardized Precipitation Index
SPEI	Standardized Precipitation Evapotranspiration Index
CRS	Coordinate Reference System
ReLU	Rectified Linear Unit
3D	Three-Dimensional
4D	Four-Dimensional
5D	Five-Dimensional
Z-Score	Standard Score
CDF	Cumulative Distribution Function
CSV	Comma-Separated Values
IDW	Inverse Distance Weighting
NetCDF	Network Common Data Form
MLP	Multilayer Perceptron
ACCESS	Australian Community Climate and Earth-System Simulator–Seasonal Prediction System
EDI	Effective Drought Index
NDVI	Normalized Difference Vegetation Index

ARC2	African Rainfall Climatology Version 2
TAMSAT2	Tropical Applications of Meteorology using Satellite Data and Ground-based Observations Version 2
HoA	Horn of Africa

References

1. Hayes MJ, Svoboda M, Wall NA, Widhalm M (2011) The Lincoln declaration on drought indices: universal meteorological drought index recommended. Bull Am Meteorol Soc 92:4
2. Promping T, Tingsanchali T (2020) Meteorological drought hazard assessment under future climate change projection for agriculture area in Songkhram River Basin, Thailand. In: 020 International Conference and Utility Exhibition on Energy, Environment and Climate Change (ICUE). Pattaya, Thailand
3. Brázdil R, Kiss A, Luterbacher J, Nash DJ, Řezníčková L (2018) Documentary data and the study of past droughts: a global state of the art. Clim Past 14:1915–1960
4. D'Alessandro SP, Caballero J, Lichte J, Simpkin S. Kenya Agricultural Risk Assessment
5. Indeje M, Semazzi FH, Ogallo LJ (2000) ENSO signals in east African rainfall seasons. Int J Climatol 20(1):19–46
6. Orindi V, Nyong A, Herrero M (2007) Pastoral livelihood adaptation to drought and institutional interventions in Kenya. UNDP, Nairobi(Kenya)
7. PDNA (2012) Kenya Post-Disaster Needs Assessment (PDNA) 2008–2011 Drought
8. Ouma JO, Olang LO, Ouma GO, Oludhe C, Ogallo L, Artan G (2018) Magnitudes of climate variability and changes over the arid and semi-arid lands of Kenya between 1961 and 2013 period. Am J Clim Chang 7:27–39
9. KMD (2021) State of the climate in Kenya 2020. Kenya Meteorological Department, Nairobi
10. I. R. Committee. What is drought? Causes, impact & countries most affected, 13 July 2023. [Online]. Available: https://www.rescue.org/article/what-drought-causes-impact-countries-most-affected#:~:text=Over%2040%20million%20people%20have,to%20the%20brink%20of%20famine. Accessed on 5 June 2024
11. Ayugi B, Tan G, Niu R, Dong Z, Ojara M, Mumo L, Babaousmail H, Ongoma V (2020) Evaluation of meteorological drought and flood scenarios over Kenya, East Africa. Atmosphere 11(3):307
12. Gebremeskel Haile G, Tang Q, Sun S, Huang Z, Zhang X, Liu X (2019) Droughts in East Africa: causes, impacts and resilience. Earth Sci Rev 193:146–161
13. Guha-Sapir D, Hargitt D, Hoyois P (2004) Thirty years of natural disasters 1974–2003: the numbers, Louvain. Presses universitaires de Louvain, Belgium
14. Balint Z, Mutua F, Muchiri P, Omuto CT (2013) Monitoring drought with the combined drought index in Kenya. Develop Earth Surface Process 16:341–356
15. Wambua RM, Mutua BM, Raude JM (2018) Detection of spatial, temporal and trend of meteorological drought using standardized precipitation index(SPI) and effective drought index(EDI) in the upper Tana River basin, Kenya. Open J Mod Hydrol 8:83–100
16. Paredes-Trejo F, Barbosa HA, Venkata Lakshmi Kumar T, Kumar Thakur M, de Oliveira Buriti C (2020) Assessment of the CHIRPS-based satellite precipitation estimates. In: Devlin A, Pan J, Shah MM (eds) Inland waters-dynamics and ecology. IntechOpen
17. Dinku T, Funk C, Peterson P, Maidment R, Tadesse T, Gadain H, Ceccato P (2018) Validation of the CHIRPS satellite rainfall estimates over eastern Africa. Q J R Meteorol Soc 144
18. Gyaneshwar A, Mishra A, Chadha U, Raj Vincent PMD, Rajinikanth V, Pattukandan Ganapathy G, Srinivasan K (2023) A contemporary review on deep learning models for drought prediction. Sustainability 15(7):6160

19. Nabipour N, Dehghani M, Mosavi A, Shamshirband S (2020) Short-term hydrological drought forecasting based on different nature-inspired optimization algorithms hybridized with artificial neural networks. IEEE Access 8:15210–15222
20. Campbell EP (2005) Statistical modeling in nonlinear systems. J Climate 18(16):3388–3399
21. W. U. a. Research (2018) Climate change in East Africa; towards a methodological framework on adaptation and mitigation strategies of natural resources. Alterra
22. Kisaka MO, Mucheru-Muna M, Ngetich FK, Mugwe JN, Mugendi DN, Mairura F (2015) Rainfall variability, drought characterization, and efficacy of rainfall data reconstruction: case of eastern Kenya. Adv Meteorol
23. Djikeng A, Rao IM, Njarui D, Mutimura M, Caradus J, Ghimire SR, Johnson L, Cardoso JA, Ahonsi M, Kelemu S (2014) Climate-smart Brachiaria grasses for improving livestock production in East Africa. Tropical Grasslands- Forrajes Tropicales:38–39
24. Assan N (2015) Focusing on livestock improvement strategies that enhance adaptive and coping mechanisms in the context of climate change in Southern Africa. Scientif J Anim Sci, pp 72–80
25. Karanja A, Ondimu K, Recha CW (2017) Analysis of temporal drought characteristic using SPI drought index based on rainfall data in Laikipia west Sub-County, Kenya. OALib:1–11
26. Polong F, Chen H, Sun S, Ongoma V (2019) Temporal and spatial evolution of the standard precipitation evapotranspiration index (SPEI) in the Tana River Basin, Kenya, pp 777–792
27. Mutsotso RB, Sichangi AW, Makokha GO (2018) Spatio-temporal drought characterization in Kenya from 1987 to 2016. Adv Remote Sens:125–143
28. Frank A, Armenski T, Gocic M, Popov S, Popovic L, Trajkovic S (2017) Influence of mathematical and physical backgroud of drought indices on their complementarity and drought recognition ability, pp 268–280
29. Zargar A, Sadiq R, Naser B, Khan FI (2011) A review of drought indices, pp 333–349
30. McKee TB, Doesken NJ, Kleist JR (1993) The relationship of drought frequency and duration to time scales. In: Proceedings of 8th Conference on Applied Climatology, Anaheim
31. I. D. M. Programme (n.d.) Standardized Precipitation Index (SPI). [Online]. Available: https://www.droughtmanagement.info/standardized-precipitation-index-spi/. Accessed 14 July 2024
32. Funk C, Peterson P, Landsfeld M, Pedreros D, Verdin J, Shukla S, Husak G, Rowland J, Harrison L, Hoell A, Michaelsen J (2015) The climate hazards infrared precipitation with stations—a new environmental record for monitoring extremes. Sci Data 2:150066
33. Muigai Njoroge E, Muthama NJ, Ouma GO, Lukorito CB (2010) Validation of satellite derived rainfall estimates over Kenya. In: Proceedings of the second RUFORUM biennial meeting, Entebbe
34. Duan Z, Liu J, Tuo Y, Chiogna G, Disse M (2016) Evaluation of eight high spatial resolution gridded precipitation products in Adige Basin (Italy) at multiple temporal and spatial scales. Sci Total Environ 573:1536–1553
35. Macharia JM, Ngetich FK, Shisanya CA (2020) Comparison of satellite remote sensing derived precipitation estimates and observed data in Kenya. Agric For Meteorol 284
36. Rivera JA, Marianetti G, Hinrichs S (2018) Validation of CHIRPS precipitation dataset along the Central Andes of Argentina. Atmos Res 213:437–449
37. Ullah W, Wang G, Ali G, Hagan DFT, Bhatti AS, Lou D (2019) Comparing multiple precipitation products against in-situ observations over different climate regions of Pakistan. Remote Sens 11(6)
38. Bai L, Shi C, Li L, Yang Y, Wu J (2018) Accuracy of CHIRPS satellite-rainfall products over mainland China. Remote Sens 10(3)
39. Guo J, Su X (2019) Parameter sensitivity analysis of SWAT model for streamflow simulation with multisource precipitation datasets. Hydrol Res 50(3):861–877
40. Haidar A, Verma B (2018) Monthly rainfall forecasting using one-dimensional deep convolutional neural network. IEEE Access:69053–69063
41. LeCun Y, Bengio Y, Hinton G (2015) Deep learning. Nature, pp 436–444

42. Reichstein M, Camps-Valls G, Stevens B, Jung M, Denzler J, Carvalhais N, Prabhat (2019) Deep learning and process understanding for data-driven Earth system science. Nature, pp 195–204
43. C. H. Center. CHIRPS: climate hazards group infrared precipitation with station data. UC Santa Barbara. [Online]. Available: http://www.chc.ucsb.edu/data/chirps/. Accessed on 8 July 2024
44. G. Geography. Inverse Distance Weighting (IDW) interpolation. [Online]. Available: https://gisgeography.com/inverse-distance-weighting-idw-interpolation/. Accessed on 03 July 2024
45. Jernberg OL. Data: a critical tool for farmers to fight drought in Kenya. United Nations Office for Disaster Risk Reduction (UNDRR), 28 Aug 2023. [Online]. Available: https://www.preventionweb.net/news/data-critical-tool-farmers-fight-drought-kenya. Accessed on 22 Jan 2025
46. U. N. O. f. t. C. o. H. A. (UNOCHA) (2024) Kenya drought response dashboard (January–December 2023). United Nations Office for the Coordination of Humanitarian Affairs (UNOCHA)
47. Hijmans RJ (2022) GADM: database of global administrative areas. [Online]. Available: https://gadm.org. Accessed on 2024

Assessment of Water Quality in Selected Fish Farms in Uasin Gishu County, Kenya

Patricia Jepchirchir Kosgei, David Mbeva Liti, Edwin Kipyegon Mutai, and Emily Jepyegon Chemoiwa

Abstract

Most studies on aquaculture focus on production techniques or market development, often overlooking the critical roles of water quality management in sustaining fish growth. This study was conducted between May 2023 and April 2024 and evaluated key water quality parameters across five farms to determine their suitability for fish farming. Physico-chemical parameters, including pH, biological oxygen demand (BOD), Chemical Oxygen Demand (COD), salinity, conductivity, and temperature, were measured in situ using a Multi-probe HQ40D meter. Data analysis involved ANOVA and Kruskal–Wallis tests, with Tukey's and Dunn's tests used to identify significant variations, and Fisher's exact test applied to assess associations between water quality parameters and farm, aquaculture facility, aquaculture facility type, or sampling date ($P \leq 0.05$). Results showed that most farms maintained water temperatures within the optimal range (25–30 °C), with salinity, pH, and TDS also within acceptable limits. However, dissolved oxygen levels often fell below the recommended threshold (60–100% or 0.1–0.2 mmol/L), posing risks to aquatic life. While elevated COD (> 50 mg/L) and BOD (> 20 mg/L) values indicated organic pollution. Most water quality parameters were suitable for fish farming, low dissolved oxygen,

P. J. Kosgei (✉) · D. M. Liti · E. K. Mutai · E. J. Chemoiwa
Department of Biological Sciences, School of Science, University of Eldoret, Eldoret, Kenya
e-mail: Chemoiwaej@uoeld.ac.ke

D. M. Liti
e-mail: davidmbevaliti@gmail.com

E. K. Mutai
e-mail: edwinmutaikp@gmail.com

E. J. Chemoiwa
e-mail: chemoiwaej@gmail.com

R. Ramkat et al. (eds.), *Towards the Sustainability of our Ecosystems*, Springer Proceedings in Geography and Sustainability,
https://doi.org/10.1007/978-3-032-30076-8_2

high BOD, and COD, and microbial contamination in some farms could negatively affect fish health and productivity. To improve water quality management, regular aquaculture facility cleaning, controlled feeding practices, and continuous monitoring are recommended to reduce organic pollution, stabilize BOD and COD levels, and limit microbial risks thereby supporting sustainable fish development.

Keywords

Water quality • Aquaculture • Coliforms • Nitrates • Biodiversity • Water pollutants

1 Introduction

Aquaculture has emerged as a fundamental sector contributing to global food security, economic development, and employment creation. According to Hisano et al. [1], aquaculture production reached a record 130.9 million tons in 2022, surpassing capture fisheries in aquatic animal production. This rapid expansion highlights the sector's significance in national development strategies and poverty alleviation initiatives worldwide. Aquaculture encompasses the breeding, rearing, and harvesting of aquatic organisms in controlled environments for purposes such as food production, species conservation, restocking wild populations, the aquarium trade, and habitat restoration [2]. Given its exponential growth, aquaculture is increasingly viewed as a sustainable solution to global food insecurity and a viable alternative to wild fisheries.

Water quality is a critical factor influencing aquaculture productivity, affecting fish health and overall yield. The interaction between water chemistry and environmental factors significantly impacts fish survival and growth [3]. Poor water quality, often caused by organic pollutants, has led to mass fish mortality events, such as the February 2010 incident in Lake Naivasha, Kenya, where critically low dissolved oxygen levels were recorded. According to Misman et al. [4], variations in aquaculture facility management practices, including stocking density, fertilization, and feeding regimes, influence water quality and fish health. Addressing these issues through improved aquaculture facility management and water quality monitoring is crucial for sustainable aquaculture development. This study therefore aims to assess water quality in selected fish farms in Uasin Gishu County by examining key physico-chemical parameters such as temperature, dissolved oxygen, pH, Biological Oxygen Demand (BOD), and Chemical Oxygen Demand (COD).

2 Materials and Methods

2.1 Study Sites

The study was conducted across five selected fish farms in Uasin Gishu County, Kenya, from May 2023 to April 2024. Within the Great Rift Valley region, Uasin Gishu County lies approximately 330 km Northwest of Nairobi along the Kenya-Uganda highway (Fig. 1). It shares borders with Trans-Nzoia County to the North, Elgeyo-Marakwet and Baringo counties to the East, Kericho County to the South, Nandi County to the southwest, and Kakamega County to the West. Covering an area of 2955 km^2, the county's geographical coordinates range between Longitude 34°50″E and 35°37″W, and Latitude 0°03″S and 0°55″ N (KNBS, 2021). Positioned on a plateau, Uasin Gishu experiences a cool and temperate climate, with annual temperatures ranging from a minimum of 8.4 °C to a maximum of 27 °C. The county benefits from high and reliable rainfall, averaging between 900 mm and 1200 mm per annum, with the wettest months typically occurring between April and May and the driest months between January and February.

The geographical coordinates and elevation of selected fish farms Chebororwa, Talget, Cheplaskei A, Cheplaskei B, and Turbo are shown in Table 1. The sites had tanks, earthen, and liner facilities as the main types of aquaculture facilities used in the region.

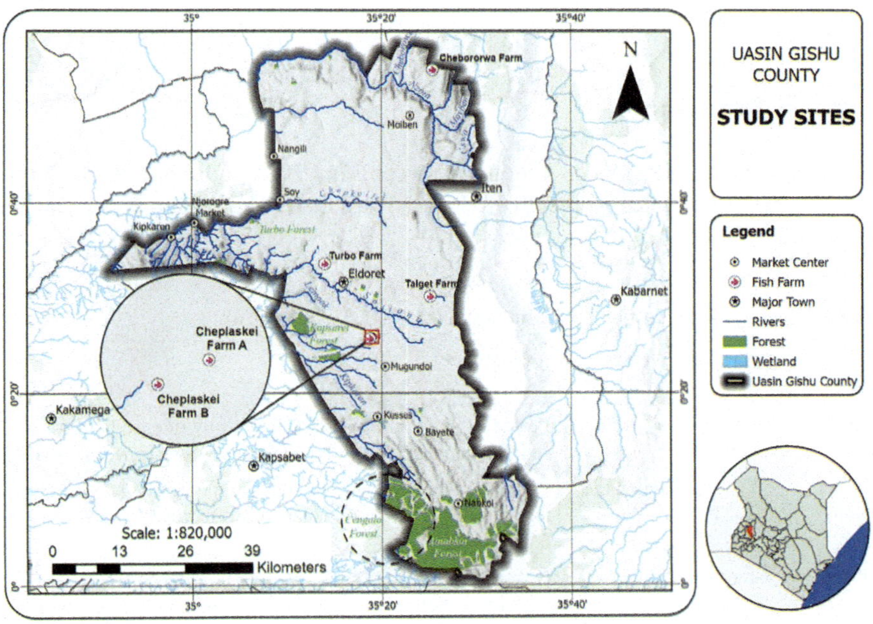

Fig. 1 Study area showing the spatial distribution. Geographic coordinates and elevations of the Five [5] sampled Fish farms in Uasin Gishu County

Table 1 Geographic coordinates and elevation of selected fish farms

FARM	Latitude (N)	Longitude (E)	Altitude (m ASL)
Chebororwa	N 0°54′0.504″	E 35°25′25.0392″	2108
Talget	N 0°30′9.576″	E 35°25′9.516″	2278
Cheplaskei A	N 0°25′56.0784″	E 35°19′11.5608″	2181
Cheplaskei B	N 0°25′44.1336″	E 35°18′46.2348″	2169
Turbo	N 0°33′37.0404″	E 35°14′3.1308″	2029

2.2 Selection of the Study Sites

The locations were selected for the study because they were good representatives of the area based on the interest of the study, the information to be collected, and accessibility.

2.3 Research Design

A longitudinal research design was adopted for this study in which measurements were done monthly for twelve months (1 yr). A total of five operational aquaculture fish farms were selected, through stratified random sampling. The selection was based on the type, size and the location of the farm, and number of year(s) of operation (minimum 1 year).

2.4 Sampling

Samples were collected once a month between May 2023 and April 2024. Purposive sampling was used to identify selected fish farms. The season was classified as shown in Table 2.

Table 2 Season classification for Uasin Gishu (May 2023–April 2024)

Season label	Months covered	Rationale
Long rains	March–May 2024 May 2023	Period of peak rainfall characterized by intense nutrient runoff, increased turbidity, and fluctuations in water quality
Short rains	October–December 2023	Period of moderate and variable rainfall, often resulting in less runoff than long rains but still affecting aquaculture facility ecosystems
Low rains/dry season	June–September 2023 January–February 2024	Period of minimal rainfall, leading to reduced water exchange, thermal stratification, and possible accumulation of metabolites in aquaculture facilities

2.5 Measurement of Physico-Chemical Parameters

Physico-chemical water quality parameters were measured in situ using the Hach HQ40D multi-parameter portable meter fitted with Intellical digital probes to establish the environmental conditions within each aquaculture facility. The parameters monitored included dissolved oxygen (DO), Salinity, temperature, pH, Biological Oxygen Demand (BOD), and Chemical Oxygen Demand (COD), as these are important indicators of water quality and ecological balance in aquaculture systems. Sampling was conducted monthly between May 2023 and April 2024, with measurements taken once per month at three-hour intervals between 600 and 1800 h.

2.6 Statistical Analysis

Descriptive statistics (means, standard deviations, and ranges) were first generated for water quality. Normality of distributions was assessed using the Shapiro–Wilk test, and homogeneity of variances with Levene's test, to confirm the suitability of parametric tests. Differences in physico-chemical parameters across farms, facility types, and seasons were tested using one-way ANOVA. Where significant effects were observed, Tukey's HSD was applied for post-hoc comparisons. All statistical analyses were done using R software 4.4.2 with $p \leq 0.05$ being statistically significant.

3 Results

3.1 Means and Ranges of Values for Physico-Chemical Parameters in Different Farms

The study examined key physico-chemical parameters across five fish farms: Chebororwa, Talget, Cheplaskei A, Cheplaskei B, and Turbo. The findings revealed marked variations across farms, in temperature, dissolved oxygen (DO), salinity, pH, Chemical Oxygen Demand (COD) and biological oxygen demand (BOD). The means and the ranges of Physico-chemical parameters across farms are shown in Table 3.The temperature for the entire study period was low for the growth of *Oreochromis niloticus*, ranging between 21.0 and 23 °C, compared to the temperature required for optimum growth of Tilapia, which ranges between 25 and 30 °C. The means of DO for most of the farms were above the limiting incipient level (3 mg/L) for *O. niloticus* except for Cheplaskei Farm B (1.8 mg/L), which was below the limiting requirement level for *O. niloticus*. The salinity levels among aquaculture facilities were low with all the values being less than 1%. A level of 0.07% was recorded in Chebororwa farm aquaculture facilities, which was tenfold higher than the values recorded in all other aquaculture facilities during the

Table 3 Means and ranges of values for physico-chemical parameters in different farms

Farm	Temp (°C)	DO (mg/L)	Salinity (%)	pH	COD (mg/L)	BOD (mg/L)
Chebororwa	21.0 ± 1.31 (18.1–23.2)	3.9 ± 1.64 (1.48–7.23)	0.072 ± 0.13 (0.40–0.99)	8.77 ± 0.6 (6.62–10.1)	70.3 ± 56.6 (2.77–234)	49.6 ± 39.9 (1.54–172)
Talget	22.7 ± 2.00 (20.0–30.6*)	4.2 ± 1.45 (1.25–7.60)	0.005 ± 0.02 (0.020–0.13)	9.04 ± 0.482 (8.01–10.2)	53.7 ± 31.1 (10.0–115)	58.4 ± 33.5 (3.00–150)
Cheplaskei A	22.8 ± 1.22 (20.2–25.5)	3.5 ± 0.98 (1.84–5.30)	0.04 ± 0.01 (0.02–0.08)	8.49 ± 0.833 (6.94–10.1)	95.2 ± 35.0 (45.0–172)	103.0 ± 55.2 (34.0–212)
Cheplaskei B	21.6 ± 2.50 (18.6–25.1)	1.8 ± 0.66 (0.83–3.24)	0.04 ± 0.01 (0.02–0.06)	8.02 ± 0.718 (6.60–9.14)	32.1 ± 26.4 (3.00–112)	33.8 ± 27.8 (1.00–103)
Turbo	22.4 ± 1.29 (20.2–24.6)	4.5 ± 1.63 (1.43–6.83)	0.04 ± 0.01 (0.02–0.06)	8.08 ± 0.641 (6.77–9.34)	89.7 ± 39.0 (30.0–168)	78.9 ± 31.1 (16.0–123)
Overall	22.0 ± 1.85 (18.1–30.6)	3.7 ± 1.61 (0.83–7.60)	0.25 ± 0.33 (0.02–0.99)	8.57 ± 0.767 (6.60–10.2)	67.1 ± 46.2 (2.77–234)	61.7 ± 44.0 (1.00–212)

study period. The mean pH ranged between 8 and 9, which was within the optimum range (6.5–9) for the growth of *O. niloticus*. COD is a measure of oxidizable organic matter. During the study period, the mean COD levels for all the aquaculture facilities ranged between 32 to 95 mg/L. BOD is a measure of the Biological activity of organisms, especially bacteria. The mean BOD levels were similar to those of COD, which ranged between 33 and 103 mg/L.

The means and the ranges of the physico-chemical water quality parameters for the three different aquaculture facility types: earthen aquaculture facilities, liner aquaculture facilities, and tank systems are shown in Table 4. Temperature values exhibited relatively narrow ranges across all facilities. Tank systems and liner aquaculture facilities recorded slightly higher mean temperatures (22.56 ± 1.10 °C and 22.57 ± 1.89 °C, respectively) compared to earthen aquaculture facilities (21.3 ± 1.84 °C). The observed temperature range was widest in liner systems (20.02–30.55 °C), suggesting higher exposure to external temperature fluctuations, possibly due to limited insulation.

Dissolved oxygen (DO) concentrations showed considerable variability, particularly in tank systems, which recorded moderately high mean DO level (3.67±

Table 4 Means and ranges of values for physico-chemical parameters in different aquaculture facility types

Facility type	Temperature (°C)	DO (mg/L)	pH	COD (mg/L)	BOD (mg/L)
Earthen	21.3 ± 1.8 (18.07–25.1)	3.18 ± 1.55 (0.83–7.23)	8.64 ± 0.84 (6.6–10.18)	66.16 ± 48.91 (3.0–234.0)	49.68 ± 34.38 (1.0–172.0)
Liner	22.57 ± 1.89 (20.02–30.55)	4.30 ± 1.58 (1.25–7.60)	8.4 ± 0.79 (6.77–10.18)	68.39 ± 39.96 (10.0–168.0)	73.62 ± 33.98 (3.0–150.4)
Tank	22.56 ± 1.1 (20.34–25.53)	3.67 ± 1.61 (0.83–7.60)	11.83 ± 19.57 (7.23–125.92)	67.82 ± 48.78 (2.8–172.0)	72.21 ± 63.44 (1.5–212.0)

1.61mg/L) and an extensive range (0.83–7.60 mg/L). Liner aquaculture facilities had highest DO levels (4.30 ± 1.58 mg/L), while earthen aquaculture facilities recorded the lowest average (3.18 ± 1.55mg/L), reflecting the typically lower oxygen availability in natural or semi-natural aquaculture facility environments. pH levels were generally alkaline in all systems, but tank systems showed an abnormally high mean value (11.83 ± 19.57) with an extensive range (7.23–125.92). Earthen and liner aquaculture facilities exhibited more stable and comparable pH values, with means of 8.64 ± 0.84 and 8.4 ± 0.79, respectively, within the optimal range for tilapia culture.

COD values were relatively consistent across all facility types, with mean values ranging from 66.16 mg/L in earthen aquaculture facilities to 68.39 mg/L in liner systems. These levels, along with their wide ranges (3.0–234.0 mg/L in earthen aquaculture facilities). BOD levels followed a similar trend to COD, with liner and tank systems showing higher average BOD values (73.62 ± 33.98 mg/L and 72.21 ± 63.44 mg/L, respectively) than earthen aquaculture facilities (49.68 ± 34.38 mg/L).

3.2 One-Way ANOVA Results for Water Quality Parameters Across Farms

The one-way ANOVA results for key water quality parameters across farms are shown in Table 5. Temperature showed a highly significant difference across farms ($F = 8.84$, $p < 0.001$). Dissolved Oxygen (DO) also displayed a significant variation ($F = 2.48$, $p = 0.046$), though with a smaller F-statistic and marginal significance. pH did not differ significantly across farms ($F = 1.43$, $p = 0.227$)**,** indicating relatively stable acidity-alkalinity levels. Chemical Oxygen Demand (COD) shows strong statistical significance ($F = 9.58$, $p < 0.000$)**,** highlighting variability in the presence of oxidizable organic matter. Biological oxygen demand (BOD) presents the most pronounced variation ($F = 12.77$, $p < 0.0000$).

Table 5 One-way ANOVA results for water quality parameters across farms

Parameter	*F*-statistic	*p*-value
Temp	8.84	0.000
DO	2.48	0.046
pH	1.429	0.227
COD (Mg/L)	9.58	0.000
BOD (Mg/L)	12.77	0.000

Table 6 One-way ANOVA results for water quality parameters across aquaculture facilities

Parameter	*F*-statistic	*p*-value
Temp	2.23	0.09
DO	0.74	0.53
pH	0.42	0.74
COD (Mg/L)	1.40	0.25
BOD (Mg/L)	2.75	0.05

3.3 One-Way ANOVA Results for Water Quality Parameters Across Aquaculture Facilities

The one-way ANOVA results for water quality parameters across aquaculture facilities are shown in Table 6. The results revealed that there were no statistically significant differences in temperature, dissolved oxygen (DO), or pH among the sampled aquaculture facilities. Temperature shows the highest F-statistic ($F = 2.23$), but the corresponding *p*-value ($p = 0.087$) is above the conventional 0.05 threshold, indicating that any observed temperature differences between aquaculture facilities are not statistically significant. Dissolved oxygen (DO) exhibited less variability across aquaculture facilities ($F = 0.74$, $p = 0.529$). The pH shows no significant differences ($F = 0.42$, $p = 0.738$), implying that acidity-alkalinity levels are stable across aquaculture facilities.

3.4 One-Way ANOVA Results for Water Quality Parameters Across Facility Types

The one-way ANOVA results for water quality parameters across different aquaculture facility types (earthen, liner, tanks) are shown in Table 7. The results revealed significant variation for some parameters while others remained consistent across facilities. Temperature shows a highly significant difference ($F = 10.51$, $p < 0.001$), indicating that facility type has a strong influence on thermal conditions. Biological oxygen demand (BOD) also showed statistically significant differences across facility types ($F = 5.80$, $p = 0.004$). Dissolved oxygen (DO) did not show statistically significant differences across facility types ($F = 1.35$, $p = 0.262$) and pH ($F = 1.71$, $p = 0.185$), indicating that these parameters are relatively stable regardless of the facility type. Chemical Oxygen Demand (COD) showed no significant variation across facility types ($F = 0.037$, $p = 0.963$).

The one-way ANOVA results for water quality parameters across seasons are shown in Table 8. BOD exhibits a significant difference across seasons ($F = 2.52$, $p = 0.0061$), indicating that Biological oxygen demand (BOD) is the only parameter that shows statistically significant variation, while all other parameters remained relatively stable throughout the year. The other parameters, Temperature ($F = 1.00$, $p = 0.447$), Dissolved Oxygen (DO) ($F = 1.02$, $p = 0.429$), pH ($F = 1.02$, $p =$

Table 7 One-way ANOVA results for water quality parameters across facility types

Parameter	F-statistic	*p*-value
Temp	10.506	0.000
DO	1.352	0.262
pH	1.705	0.185
COD (Mg/L)	0.037	0.963
BOD (Mg/L)	5.798	0.004

Table 8 One-way ANOVA results for water quality parameters across seasons

Parameter	*F*-statistic	*p*-value
Temperature	1.0031	0.4466
Dissolved oxygen (DO)	1.0237	0.4286
pH	1.0151	0.4361
COD (Mg/L)	0.6226	0.8073
BOD (Mg/L)	2.5234	0.0061

0.436), and Chemical Oxygen Demand (COD) ($F = 0.62$, $p = 0.807$) all exhibited non-significant variation across seasons.

4 Discussion

4.1 Water Quality Parameters Across Farms

The study assessed key physico-chemical parameters in five fish farms (Chebororwa, Talget, Cheplaskei A, Cheplaskei B, and Turbo) showing notable differences in temperature, dissolved oxygen (DO), salinity, pH, Chemical Oxygen Demand (COD), and biological oxygen demand (BOD). Such variations are critical in shaping water quality, which directly affects fish growth, health, and productivity. According to Odende et al. [5] effective water quality management is vital in aquaculture systems. Temperature varied by location, pond depth, and shading. Chebororwa farm recorded the lowest mean values, while Talget farm had the highest, reflecting differences in solar exposure and water retention. Temperature influences fish metabolism, feeding, and DO availability [6]. Most farms maintained temperatures within the recommended 25–30 °C, consistent with [6], who noted stable temperatures promote growth and immunity [7]. Further observed that tilapia thrive above 24 °C, while trout and salmon require below 16 °C, showing the suitability of these conditions for warm-water species.

DO levels varied, with Cheplaskei farm B being highest and Turbo lowest. Low DO, often due to high stocking, poor exchange, or organic buildup, can induce hypoxia, stunting growth and raising disease risk [8]. Several farms recorded values below 3.0 mg/L, indicating stress conditions. This aligns with [9], who emphasize adequate aeration for fish health. Similar cases, such as fish kills linked to low DO in Hinchinbrook Shire highlight the importance of proper water circulation. Salinity remained generally low, with Chebororwa farm being highest. Freshwater farms typically range between 0 and 15 ppt, suitable for osmotic balance [10]. Fluctuations may result from rainfall, evaporation, or source water differences [11]. Observed values were within safe limits and unlikely to cause stress.

pH levels ranged from slightly to highly alkaline, with Talget recording the highest. Elevated pH can result from nutrient enrichment and photosynthesis [12]. Most farms were within the recommended 6.0–9.0 range, supporting normal physiological and microbial processes. However, extreme pH can disrupt metabolism and waste breakdown [5, 13], hence routine monitoring remains important. COD and BOD peaked at Cheplaskei A and were lowest at Cheplaskei B, reflecting organic load from uneaten feed, feces, and farm runoff [5, 14]. In several farms, values exceeded the recommended thresholds of 0–50 mg/L (COD) and 0–20 mg/L (BOD), potentially reducing oxygen availability. Excess organic matter may also release harmful compounds such as ammonia and hydrogen sulfide [15, 16]. Improved waste management, including better feeding and water exchange, is essential to reduce these risks [10].

4.2 Water Quality Across Aquaculture Facility Types

The comparison of physico-chemical parameters across earthen, liner, and tank aquaculture systems revealed distinct water quality dynamics. Temperature varied narrowly across systems, with earthen ponds recording the lowest means, while liner and tank systems showed slightly higher values due to reduced buffering capacity and greater exposure to climatic fluctuations [17]. Dissolved oxygen (DO) levels were highest in tank and liner systems but lower in earthen ponds, reflecting differences in water exchange, aeration efficiency, and organic matter decomposition [18]. pH values were generally within optimal limits in earthen and liner systems, whereas tanks exhibited higher mean values.

COD and BOD levels were broadly comparable across systems, though higher averages in tank and liner facilities suggested increased microbial activity from feed inputs or limited water exchange [12, 14]. Earthen systems displayed wider COD variation, consistent with organic accumulation and sediment interactions typical of semi-natural facilities. Earthen ponds provided more stable conditions but with reduced oxygen availability, liner systems exhibited higher yet more fluctuating values, and tank systems showed the greatest variability, particularly in DO and pH. These findings emphasize the need for targeted management practices including aeration, feeding control, and routine water quality monitoring to maintain conditions within optimal thresholds and minimize stress or mortality risks [8, 15].

4.3 Water Quality Parameters Across Seasons

Biological Oxygen Demand (BOD) varied significantly across seasons, suggesting that organic matter decomposition and microbial activity were influenced by seasonal dynamics. These fluctuations may be attributed to changes in runoff, nutrient loading, and temperature, which alter the concentration of organic pollutants in aquaculture systems [12, 14]. In contrast, temperature, dissolved oxygen (DO), pH, and Chemical Oxygen Demand (COD) showed no significant seasonal variation ($p > 0.05$), indicating relative stability throughout the year. This consistency implies that the ponds maintained balanced physico-chemical conditions, likely due to effective management and the buffering capacity of the aquatic environment [17, 19]. While most parameters remained steady, the seasonal shifts in BOD highlight the importance of continuous monitoring, as elevated organic load and oxygen demand can negatively affect fish health, growth, and overall productivity [20, 21].

5 Conclusion

The study highlights that key physico-chemical parameters (temperature, DO, salinity, pH, COD and BOD) varied significantly across the five fish farms. These variations directly impact fish health, growth, and productivity. Temperature differences were influenced by environmental factors, while DO variations pointed to management practices like aeration and stocking density. Although salinity levels remained low, slight fluctuations were observed. High pH levels, particularly at Talget farm, suggest excessive photosynthetic activity. Elevated COD and BOD levels in certain farms, especially Cheplaskei farm A, indicate high organic loads, requiring better waste management. The findings emphasize that careful monitoring and management of these parameters are critical for maintaining optimal water quality and supporting sustainable aquaculture.

6 Recommendation

Capacity building for farmers through training on water quality monitoring, will help improve their technical skills, contributing to increased productivity, sustainability, and food security. Access to quality resources, such as water testing kits and aerators should be made more affordable through subsidies or credit facilities.

References

1. Hisano H, Pinheiro VR, Losekann ME, Moura e Silva MSG (2021) Effect of feeding frequency on water quality, growth, and hematological parameters of Nile tilapia *Oreochromis niloticus* reared using biofloc technology. J Appl Aquac 33(2):96–110

2. Fonda JA, Mary AO, Kevin OO, Jonathan MM, Jacob A, Betty MN et al (2021) Aquaculture extension service in Kenya: farmers and extension officers perspectives. J Agric Ext Rural Dev 13(1):14–22
3. Nwuba LA, Ude EF, Ogbonnaya HF (2022) Current trends in fisheries and aquaculture. Int J Agric Food Biodivers 1(1):64–69
4. Misman NA, Sharif MF, Chowdhury AJK, Azizan NH (2023) Water pollution and the assessment of water quality parameters: a review. Desalin Water Treat 294:79–88
5. Odende T, Ogello EO, Iteba JO, Owori H, Outa N, Obiero KO et al (2022) Promoting sustainable smallholder aquaculture productivity through landscape and seascape aquapark models: a case study of Busia County, Kenya. Front Sustain Food Syst 6:898044
6. Yusoff F, Umi WAD, Ramli NM, Harun R (2024) Water quality management in aquaculture. Cambridge Prisms Water 2:e6. https://doi.org/10.1017/wat.2024.6
7. Kibenge FSB (2020) Introduction to aquaculture. In: Cabello FC, Godfrey HP (eds) Aquaculture Toxicology. Academic Press, Cambridge, pp 1–16
8. Hildah J, Josephine M, Emily C, Bett J (2022) Influence of water quality on distribution patterns and diversity of *Enteromius* fish species in small water bodies of Uasin Gishu County, Kenya. Afr J Aquat Sci 47(3):255–264
9. Welker TL, Overturf K, Abernathy J (2019) Effect of aeration and oxygenation on growth and survival of rainbow trout in a commercial serial-pass, flow-through raceway system. Aquac Rep 14:100194
10. Khanjani MH, Sharifinia M (2020) Biofloc technology as a promising tool to improve aquaculture production. Rev Aquac 12(3):1836–1850
11. Boyd CE, D'Abramo LR, Glencross BD, Huyben DC, Juarez LM, Lockwood GS et al (2020) Achieving sustainable aquaculture: historical and current perspectives and future needs and challenges. J World Aquac Soc 51(3):578–633
12. Araujo AN, San Andres CF, Nguyen KQ, Corby TL, Rhodes MA, García J et al (2025) Effects of minimum dissolved oxygen setpoints for aeration in semi-intensive pond production of Pacific white shrimp (*Litopenaeus vannamei*). Aquaculture 594:741376
13. Zhang K, Ye Z, Qi M, Cai W, Saraiva JL, Wen Y et al (2025) Water quality impact on fish behavior: a review from an aquaculture perspective. Rev Aquac 17(2):e12985
14. Lindholm-Lehto P (2023) Water quality monitoring in recirculating aquaculture systems. Aquac Fish Fish 3(2):113–131
15. Lubembe SI, Walumona JR, Hyangya BL, Kondowe BN, Kulimushi JDM, Shamamba GA et al (2024) Environmental impacts of tilapia fish cage aquaculture on water physico-chemical parameters of Lake Kivu, Democratic Republic of the Congo. Front Water 6:1325967
16. Alam MR, Sharmin S, Islam SMM, Alam MA, Ehiguese FO, Pattadar SN et al (2020) Salinity intrusion affects early development of freshwater aquaculture species pabda. Ompok pabda Aquac Rep 18:100460
17. Mramba RP, Kahindi EJ (2023) Pond water quality and its relation to fish yield and disease occurrence in small-scale aquaculture in arid areas. Heliyon 9(6):e16753
18. Onyango K, Sulo T, Mose J (2023) Determinants of aquaculture productivity among smallholder farmers in Bungoma County, Kenya. J Bus Econ Manag 6(2):45–55
19. Gautam A, Dubey S, Mahajan SK, Khare A, Chatterjee D, Shakya P, Katara S (2025) Impact of unsustainable aquaculture practices on water quality parameters in fish pond ecosystem in Balaghat, Madhya Pradesh, India. Int J Environ Climate Chang 15(3):10–9734
20. Mutiah M, MSA M, Seftarita C, Yahya Y (2022) Exploring aquaculture fish production: the case of South Aceh District. J Aquac Fish Health 11(1):13–23
21. Charo-Karisa H, Gichuri M (2010) Overview of fish farming enterprise program: end of year report fish farming enterprise productivity program phase 1. Ministry of Fisheries Development, Nairobi

Spatial and Temporal Variation in the Diversity of Parasite Species Infecting Fish Cultured in Selected Fish Farms in Uasin Gishu County, Kenya

Edwin Kipyegon Mutai, Moses Mwajar Ngeiywa,
David Mbeva Liti, and Patricia Jepchirchir Kosgei

Abstract

Aquaculture plays a critical role in improving food security livelihoods, but parasitic infections remain a major constraint to fish health and productivity. This study investigated the diversity, distribution, and seasonal variation of parasites infecting *Oreochromis niloticus* cultured in different aquaculture farms in Uasin Gishu County, Kenya. A total of 468 fish samples were collected from earthen ponds, liner ponds, and tanks between May 2023 and April 2024. Parasites were identified through Physical and microscopic examination, and their prevalence and diversity analyzed across farms, organs, and seasons. A wide range of parasites were identified, including protozoans (*Trichodina nigra, Ichthyophthirius multifiliis*), monogeneans (*Dactylogyrus vastator, Cichlidogyrus tilapiae*), digenean trematodes (*Clinostomum complanatum, Haplorchis spp.*), and nematodes (*Contracaecum spp.*). Gills and intestines were the most frequently parasitized organs, while the kidneys showed lowest infestation. Significant differences ($p \leq 0.05$) were observed in parasite prevalence across fish organs, farms, pond types, and sampling seasons. Earthen and liner ponds supported higher parasite diversity compared to tanks, and infestations peaked during warmer and wetter months. The frequent co-occurrence of multiple parasite species, particularly in the gills and intestines, suggests complex infection dynamics that may exacerbate disease severity. These findings emphasize the strong influence of aquaculture system type, farm management, and seasonal changes on parasite infections in cultured fish. To minimize production losses and safeguard public health, integrated parasite management strategies, including routine monitoring,

E. K. Mutai (✉) · M. M. Ngeiywa · D. M. Liti · P. J. Kosgei
Department of Biological Sciences, School of Science, University of Eldoret, Eldoret, Kenya
e-mail: edwinmutaikp@gmail.com; emutai@tonp.ac.ke

R. Ramkat et al. (eds.), *Towards the Sustainability of our Ecosystems*, Springer Proceedings in Geography and Sustainability,
https://doi.org/10.1007/978-3-032-30076-8_3

biosecurity, sanitation, and farmer training, are recommended. This study provides important baseline data to guide sustainable fish health management in Uasin Gishu County and beyond.

Keywords

Fish parasites • *Oreochromis niloticus* • Uasin Gishu County • Aquaculture systems • Fish health management

1 Introduction

Aquaculture is one of the fastest-growing food production sectors worldwide and plays a central role in improving nutrition, food security, and household incomes, especially in developing nations [1, 2]. With the rising demand for affordable sources of animal protein, farmed fish have become a key component of the human diet and a major livelihood option for rural and peri-urban communities [3, 4]. In Kenya, aquaculture has been promoted through national development initiatives, resulting in the expansion of fish farming enterprises across several counties, including Uasin Gishu [5]. However, the sustainability of this growth is increasingly constrained by fish health challenges, particularly parasitic infestations that undermine productivity.

Parasitic infections are among the most widespread and damaging health problems in cultured fish [6]. They affect growth performance, feed conversion, and survival by causing extensive tissue damage, disrupting metabolic processes, and inducing anemia. In addition, parasites can weaken the immune system, making fish more susceptible to opportunistic pathogens and secondary infections, ultimately leading to significant mortalities [7]. Common examples include protozoans such as *Ichthyophthirius multifiliis* and helminths like *Dactylogyrus* spp. and *Gyrodactylus* spp., which are notorious for infesting gills, skin, and internal organs, thereby impairing normal physiological functions and reducing aquaculture profitability [8].

Despite their impact, information on the diversity, organ-specific distribution, and prevalence of fish parasites in aquaculture systems in Uasin Gishu County remains limited. Most available studies emphasize general production bottlenecks or water quality, with little focus on parasitological diagnostics. This knowledge gap restricts farmers' ability to adopt targeted parasite control strategies. The study was therefore designed to determine the diversity and prevalence of parasites in farmed fish from selected aquaculture facilities in Uasin Gishu County. The findings are expected to provide baseline data for developing effective health management measures and contribute to a more resilient and sustainable aquaculture industry in the region.

2 Materials and Methods

2.1 Study Area

The study was conducted from May 2023 to April 2024 in selected fish farms across five sub-counties in Uasin Gishu County (UG) as shown in Fig. 1. Uasin Gishu County, located in the northwest of the Great Rift Valley region, lies approximately 330 km northwest of Nairobi. It covers an area of 2955 km^2 and is characterized by a highland plateau with several perennial rivers and swampy wetlands [9, 10]. The region experiences a temperate climate, with annual temperatures ranging between 7 °C and 29 °C and rainfall between 624.9 mm to 1560.4 mm per year [11]. Aquaculture has been expanding in Uasin Gishu in the last decade, with 1728 operational fish ponds covering 486,000 m^2, producing approximately 593,000 kg of fish annually [12]. Most farmers practicing fish farming use earthen ponds, liner ponds, and other water reservoirs. Nile tilapia (*Oreochromis niloticus*) is the predominant species farmed, followed by African catfish (*Clarias gariepinus*) [13].

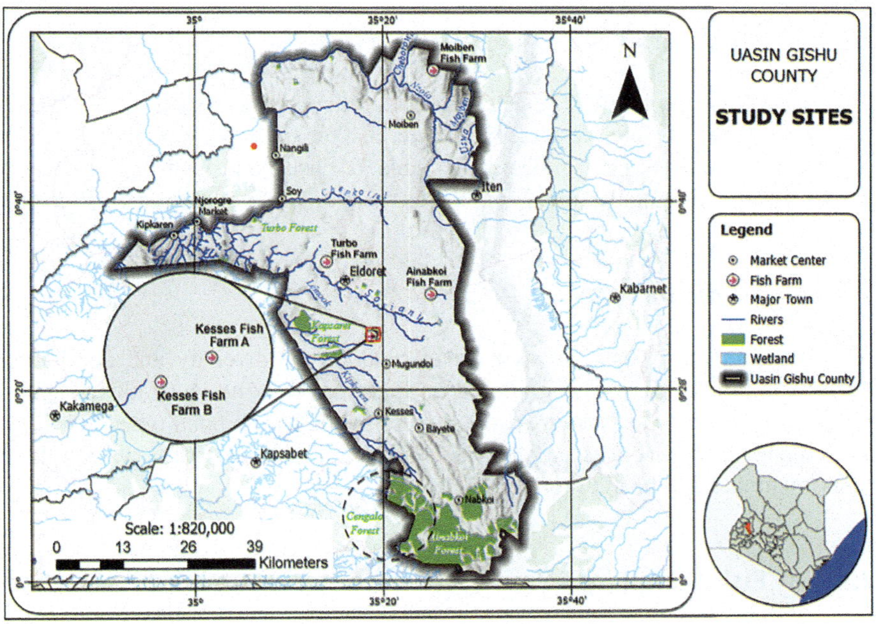

Fig. 1 Map of Uasin Gishu showing its location in western Kenya with perennial rivers and swampy wetlands

Table 1 Location and features of the selected fish farms in Uasin Gishu County where the study was carried out

Sub-County	Farm site name	Pond	Type of ponds
Moiben	Moiben Farm (Farm A)	P1	Raised liner pond
		P2	Earthen pond
		P3	Earthen pond
		P4	Earthen pond
Ainapkoi	Ainapkoi Farm (Farm C)	P1	Ground liner pond
		P2	Ground liner pond
		P3	Earthen pond
Kesses	Kesses Farm 1 (Farm K)	P1	Raised liner pond
		P2	Raised liner pond
Kesses	Kesses Farm 2 (Farm S)	P1	Earthen pond
		P2	Earthen pond
Turbo	Turbo Farm (Farm T)	P1	Ground liner pond
		P2	Ground liner pond

2.2 Study Sites

Five aquaculture farms were selected based on their features such as: pond type, size, location, inputs (e.g., fertilizers, manure, fish feeds), and duration of operation (minimum of one year) as indicated in Table 1. Each pond in the selected farm site was sampled monthly for 12 months to monitor water quality and fish health.

2.3 Study Design

A longitudinal study design was used to evaluate the diversity and distribution of parasite infestations in fish cultured in the selected farms. A stratified random sampling approach was used to ensure representation across pond types, locations, and management practices. Each pond was sampled monthly for water quality assessment and fish health monitoring.

2.4 Fish Sampling

The sample size for fish was determined following the Thrusfield formula [14], with a 95% confidence interval and 5% absolute precision. Fish were sampled using seine nets with a mesh size of 1 mm, a length of 15 m, and a width of 2 m, allowing for the collection of individuals across different size classes. Three (3) fish were selected randomly per pond from the sampled fish and transported in cool boxes to the laboratory for further analysis. The sampled fish from the 13 ponds from all the fish farms in a month was 39 and a total of 468 for the entire sampling period of 12 months. Fish were collected using seine nets and transported in cool boxes to the laboratory. The total length (cm) was measured using a measuring board, and weighed (g) using a digital balance (Model GX-200, A&D Company Limited, Tokyo, Japan). The condition factor (K) was determined using Froese's formula [15].

2.5 Parasite Examination and Identification

The outer layer of the skin was scraped from the right and left sides of the back and posterior of the fish body, transferred on to a microscope slide, diluted with a drop of sterilized water, cover-slipped and examined under a microscope (Olympus CX40). Large parasites were expressed by their absolute numbers, while the microscopic parasites were expressed by decanting serially and determining their minimum, maximum and average numbers in each field of view of microscope at a definite magnification. The intestines of fish samples were removed and separated into stomach and intestine sections. Parasite cysts located on their surfaces were located and examined microscopically and recorded by their location. The parasites were then transferred using a dissection needle to a slide containing sterilized water after getting rid of their slime and examined at high magnification (400X). Smears were made from samples collected from the skin, gills, stomach and intestines. These specimens were maintained in petri dishes with few drops of 9% saline solution to keep the parasites alive. Smears were prepared on sterilized slides and viewed microscopically under low and high power (400X) magnification. The body cavity, pericardial cavity and muscle tissues of the fish were also examined for endoparasites. The collected parasites were counted and preserved in either 4% formalin or 96% Ethanol for further identification and analysis later. The parasites were identified morphologically using standard identification keys and pictorial guides as previously described [16–18]. The parameters of infection such as prevalence, mean intensity of infection and abundance of parasites were calculated according to Bush et al. methods [19].

2.6 Determination of the Bacterial Diversity

Shannon–Wiener diversity index for the species community was calculated using biodiversity calculator [20]. The diversity of the parasitic organisms infecting fish was determined using Shannon–Wiener diversity index as shown in Equation 1.

$$H = \sum_{I=1}^{R} \text{Pi In Pi} \tag{1}$$

where H' = Shannon's diversity index, R = Total number of species, i = the ith species, In = natural logarithm and Pi = the proportion of individuals belonging to the ith species and n = Total number of individuals of all species of parasites.

2.7 Statistical Analysis

Data were tested for normality (Shapiro–Wilk test) and homogeneity (Levene's test). Non-parametric tests were used where necessary. Descriptive statistics (prevalence, mean abundance, and mean intensity) were computed to quantify parasite distribution across fish organs, farms, pond types, and seasons [19]. Shannon diversity indices (H') were calculated to assess community diversity. Differences in bacterial prevalence were tested using the Kruskal–Wallis test, followed by Dunn's post hoc test with Bonferroni adjustment. The effects of farm, pond/tank, facility type, and season on parasite occurrence were further examined with ANOVA on presence/absence data ($p < 0.05$). Pairwise comparisons identified significant parasite variations across farms and organs. All statistical analyses were conducted using R software 4.4.2 (Pile of Leaves) with $p \leq 0.05$ considered statistically significant.

3 Results

3.1 Distribution and Occurrence of Parasite Species

Identification of the Parasite Species

Examination of the fish samples revealed the occurrence of parasitic species affecting various tissues, including the gills, muscles, and internal organs as shown in Table 2. The parasites include monogeneans, digenean trematodes, and ciliate protozoans, with organ-specific distributions. Monogeneans were predominantly found on the gills and skin, digenean trematodes in the intestines, and ciliate protozoans in both gills and skin.

Table 2 Classification of parasite species identified in fish organs

Organ	Parasite group	Parasite genera	Parasite species
Gills	Ciliate protozoan	*Trichodina*	*Trichodina nigra*
	Monogenean	*Cichlidogyrus*	*Cichlidogyrus tilapiae*
	Monogenean	*Dactylogyrus*	*Dactylogyrus vastator*
Intestines	Digenean trematode	*Clinostomum*	*Clinostomum complanatum*
	Digenean trematode	*Haplorchis*	*Haplorchis spp.*
Skin	Monogenean	*Gyrodactylus*	*Gyrodactylus elegans*
	Ciliate protozoan	*Ichthyophthirius*	*Ichthyophthirius multifiliis*

Nematode larvae, primarily belonging to the *Contracaecum* spp., were frequently observed in the visceral cavity, intestines, and gills (Fig. 2A). The parasites were particularly concentrated in the peritoneal cavity, causing visible inflammation and tissue damage (Fig. 2B). *Harpochilus* species, also found within the gastrointestinal tract of fish (Fig. 2C).

Encysted trematode metacercariae, identified as belonging to *C. complanatum*, were observed within the gills and muscle tissues (Fig. 3A and B).

Monogenean parasites, identified as *D. vastator*, were found predominantly in the gill filaments (Fig. 4A and B). Affected fish exhibited increased mucus secretion and hyperplasia of gill tissues.

Protozoan parasites, including *I. multifiliis* ("Ich") and *T. nigra*, were identified on the skin and gills of fish (Fig. 5A and B). The affected fish showed

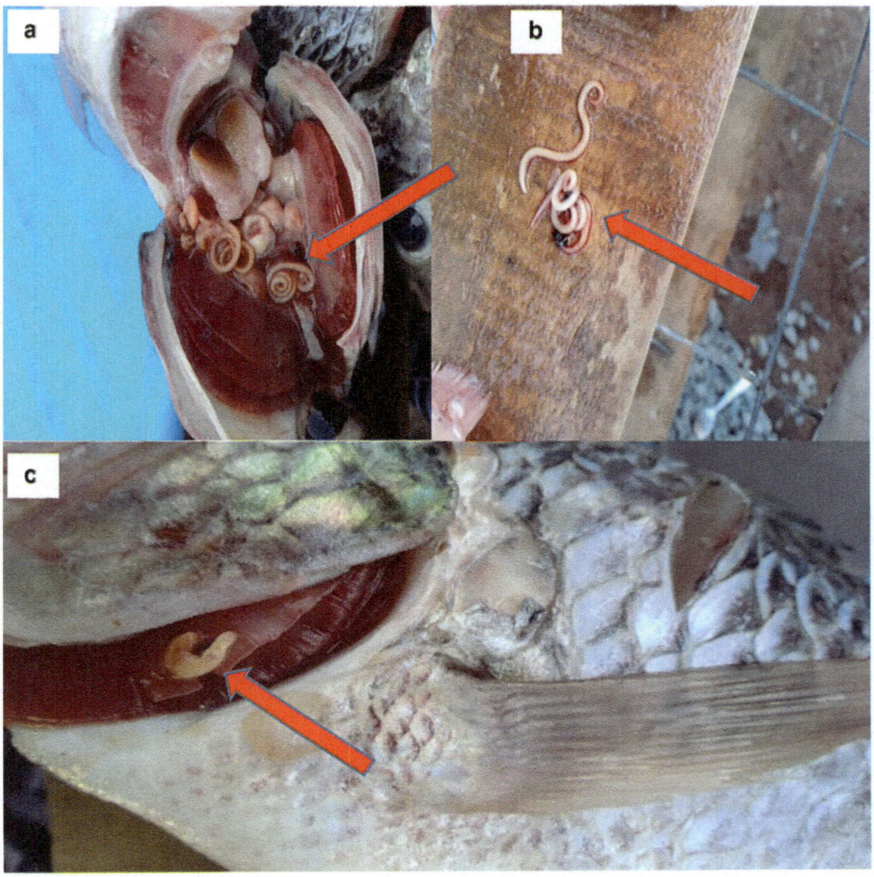

Fig. 2 **A** Larval and adult stage *Contracaecum spp.* In the Gills, **B** *Contracaecum sp.* Extracted from the fish's intestines and **C** larval-stage (L3 or L4) *Contracaecum sp.* embedded in the gills of fish

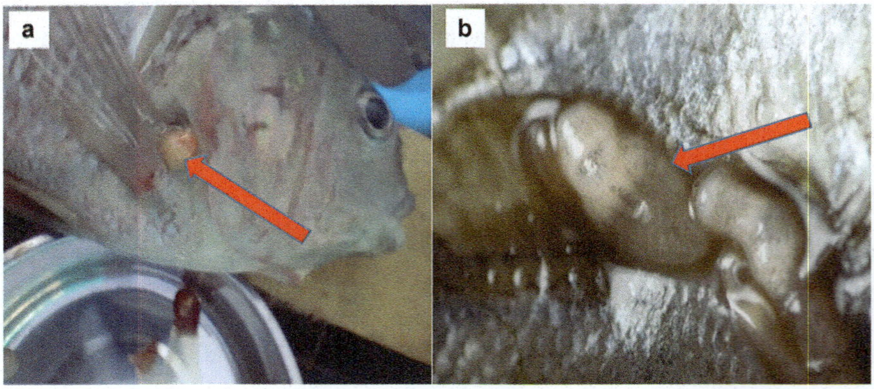

Fig. 3 **A** Parasitic cyst on the skin Caused by *Clinostomum sp* and **B** *Clinostomum* spp. metacercariae within fish tissue

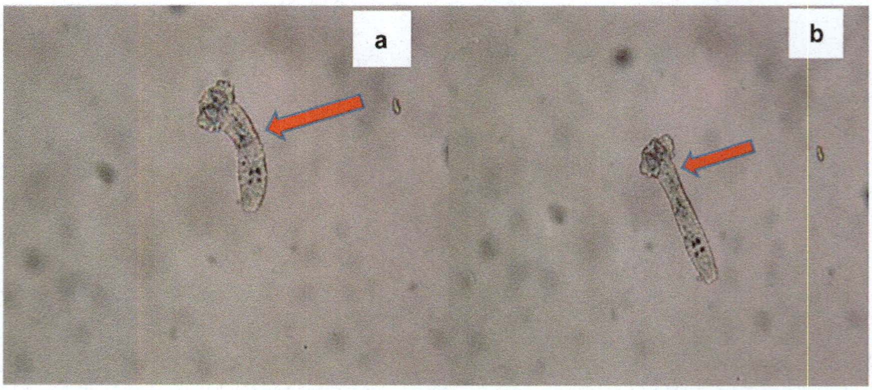

Fig. 4 Microscopic view of *Dactylogyrus spp.* with **A** Elongated body and attachment showing the internal structures and **B** Haptor, a specialized organ with hooks used for attachment to fish gills

signs of irritation, erratic swimming, and excessive mucus production. *T. nigra* was found adhering to the gill filaments and skin surface of infected fish (Fig. 5C and D). Their presence was associated with epithelial erosion and increased mucus secretion.

Distribution and Occurrence of Parasite Species

Shannon diversity results revealed that **parasite diversity** varied considerably across farms and Ponds/Tanks, with certain locations consistently exhibiting higher levels of parasitic complexity than others, as shown in Table 3. **Turbo Farm** recorded the **highest overall parasite diversity**, with a Shannon index of **0.88**,

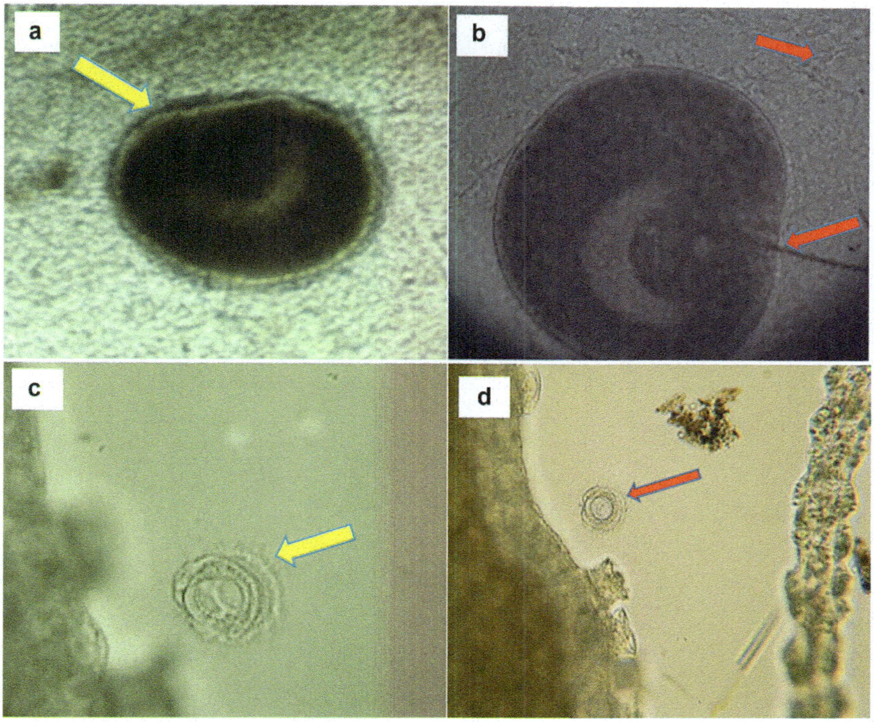

Fig. 5 **A** Trophonts of *I. multifiliis* in the skin epidermis, **B** *Ichtyoptirius spp.* (Top Left) With *T. nigra,* **C** Microscopic image of *T. nigra.* circular body with radial cilia and attachment disc visible and **D**. *T. nigra* with surrounding organic debris and fish epithelial tissue

followed by **Moiben** and **Ainapkoi Farm**, each with an index around **0.60**, reflecting moderate richness and evenness of parasite species. In contrast, **Kesses Farm A** demonstrated the **lowest diversity (0.06)**, suggesting limited parasite species richness or dominance by one or two species. At the **pond level**, noticeable differences in parasite diversity were observed even within the same farm. For instance, in **Moiben Farm, Pond 2** had a **Shannon index of 1.38**, significantly higher than **Pond 1 (0.50)** and **Pond 4 (0).** Similarly, **Kesses B Farm Pond 2** showed elevated diversity (**1.19**), while **Pond 1** had none recorded.

Facility type comparison revealed that parasite diversity varied notably across aquaculture systems as shown in Table 4. Liner ponds exhibited the highest parasite diversity, with a Shannon index of 0.75, indicating a more diverse and evenly distributed parasite population. This was followed by earthen ponds, which showed moderate diversity at 0.61. In contrast, tank systems recorded the lowest parasite diversity, with a Shannon index of just 0.20, suggesting a more limited parasite community or dominance by a few species.

The ANOVA results indicate that for parasites, both season and farm had a statistically significant influence on their presence ($p < 0.05$) as shown in Table 5.

Table 3 Shannon diversity index of parasites by farm and pond

Farm	Pond	Parasite Shannon	Parasite richness (S)	Rank
Moiben Farm		**0.596**	1.00	1
	1	0.496	3.42	2
	2	1.379	1.75	3
	3	0.548	0.50	1
	4	0.000	0.50	2
Ainapkoi Farm		**0.595**	0.83	1
	1	0.126	4.25	2
	2	1.101	1.42	1
	3	0.558	5.17	2
Kesses Farm A		**0.055**	1.83	3
	1	0.058	0.50	4
	2	0.053	2.67	1
Kesses Farm B		**0.596**	3.75	2
	1	0.000	1.00	1
	2	1.191	3.42	2
Turbo Farm		**0.877**	1.75	3
	1	0.777	0.50	1
	2	0.977	0.50	2

Table 4 Shannon diversity index of parasites by facility type

Farm	Parasite Shannon index	Parasite richness (S)	Rank
Earthen ponds	0.612727	2.15	2
Liner ponds	0.745431	2.88	1
Tanks	0.202231	0.50	3

Table 5 ANOVA results for parasite occurrence by farm, pond, facility type, and season

Source	sum_sq	df	*F*	PR(>*F*)
Across Farm	2.336051	4	2.726663	0.027837
Across Pond	0.011256	3	0.017517	0.996845
Across Facility Type	0.005956	2	0.013904	0.986192
Across Seasons (date)	45.33965	11	19.24397	9.29×10^{-38}

In contrast, pond and facility type did not show significant effects when considered alongside parasite occurrence, suggesting that their influence may be more context-specific or limited under seasonal factors.

Significant variations in parasite infestation were observed across the five fish farms, with distinct patterns evident in different organs and among parasite species as shown in Table 6. In the gills, *T. nigra.* showed markedly higher infestation levels in fish from Turbo farm compared to Ainapkoi, Kesses A, Kesses B, and Moiben. The difference between Ainapkoi and Turbo was especially large (difference = 2.528, $p < 0.001$), while Kesses A also showed significantly higher infestation than Kesses B, Moiben, and Turbo. On the skin, *G. elegans* infestation was significantly higher in Turbo compared to Ainapkoi ($p < 0.001$) and showed notable differences in other farm pairs, including between Kesses A and both Kesses B and Turbo. Moiben also had higher infestation levels compared to Turbo. Another skin parasite, *I. multifilis*, showed a significant difference only between Kesses A and Kesses B ($p = 0.009$). In the intestines, *C. complanatum* was significantly more prevalent in fish from Turbo compared to Kesses A ($p = 0.008$). Overall, Turbo emerged as the farm with the highest infestation levels across multiple parasite species and organs, especially in the gills and skin. In contrast, Ainapkoi often showed lower parasite occurence.

The analysis of mean parasite presence across different facility types and seasons revealed clear and consistent trends as shown in Fig. 6. Liner and earthen systems exhibited markedly higher levels of parasite presence compared to tank

Table 6 Significant pairwise differences in parasite species infestation across fish organs and farms

Organ	Parasite Spp	Farm		Difference	*p*-value
		Group 1	Group 2		
Gills	*T. nigra*	Ainapkoi	Turbo	2.528	< 0.001
		Kesses A	Kesses B	2	0.002
		Kesses A	Moiben	1.479	0.013
		Kesses A	Turbo	3.75	< 0.001
		Kesses B	Turbo	1.75	0.01
		Moiben	Turbo	2.271	< 0.001
	D. vastator	Ainapkoi	Moiben	0.771	< 0.001
		Kesses A	Moiben	1.104	< 0.001
		Kesses B	Moiben	0.812	< 0.001
		Moiben	Turbo	− 0.771	< 0.001
Skin	*G. elegans*	Ainapkoi	Turbo	0.875	< 0.001
		Kesses A	Kesses B	0.583	0.048
		Kesses A	Turbo	1.167	< 0.001
		Kesses B	Turbo	0.583	0.048
		Moiben	Turbo	0.958	< 0.001
	I. multifilis	Kesses A	Kesses B	0.417	0.009
Intestines	*C. complanatum*	Kesses A	Turbo	0.5	0.008

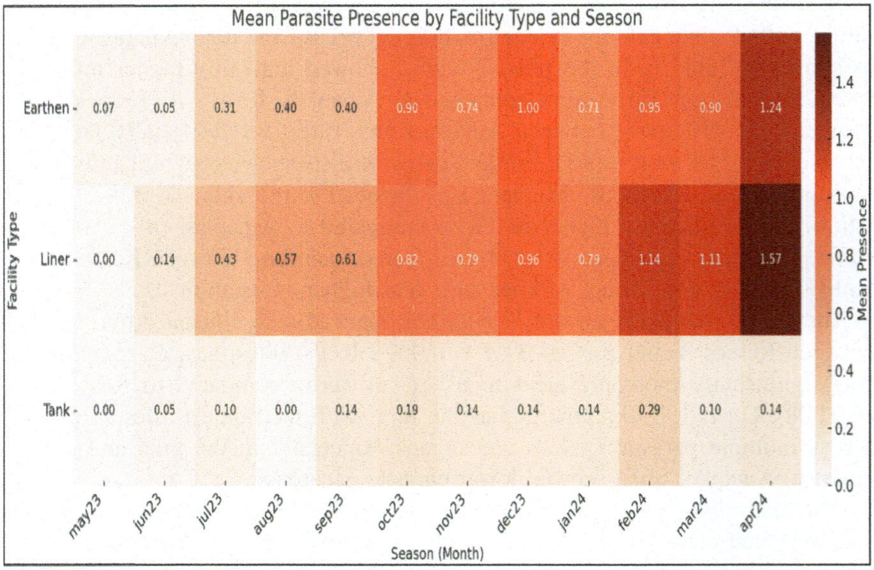

Fig. 6 Heatmap of bacterial species occurrence by facility type and season

systems, which maintained relatively low parasite loads throughout the year. Seasonal dynamics were also prominent. Parasite levels increased steadily from July onward, with significant peaks observed in October, December, February, and April. The highest mean parasite presence was recorded in April 2024, particularly in liner systems (1.57) and earthen ponds (1.24), highlighting a clear seasonal buildup that may be tied to temperature, rainfall, and fish stress conditions.

The analysis of mean parasite presence across farms, fish organs, and seasons, as shown in Fig. 7, reveals substantial spatial and temporal variation. **The gill tissues** consistently experienced the highest parasite loads, particularly in **Turbo Farm**, where presence peaked at **3.17 in April 2024**, and the highest observed value across all combinations. **Moiben** and **Kesses B Farms** also exhibited elevated Gill infections across multiple months. In contrast, **intestinal infections** remained relatively low and stable across farms and time. These trends highlight distinct seasonal and anatomical patterns of parasitic infestation, with **Turbo** and **Moiben farms** emerging as hotspots, especially during late rainy and early dry seasons.

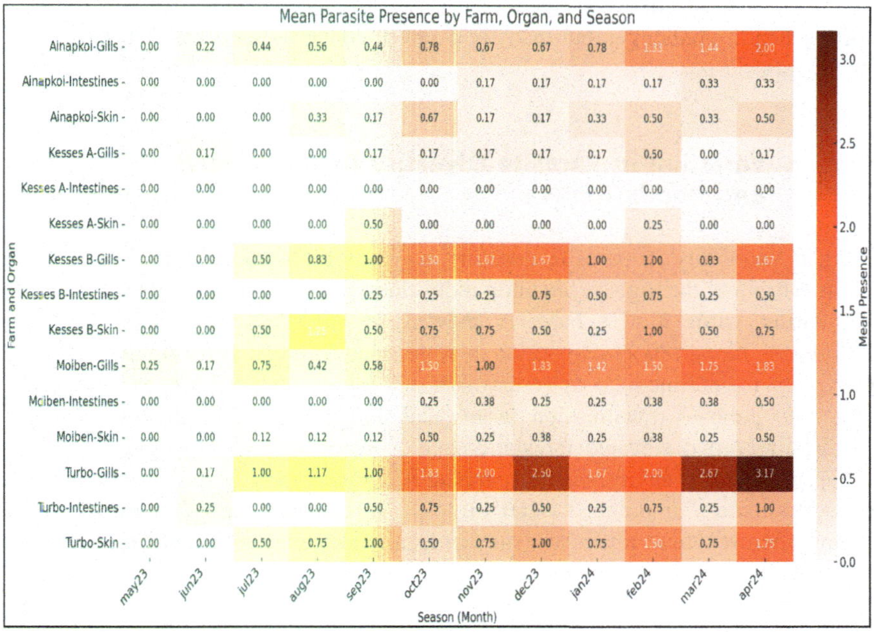

Fig. 7 Mean parasite presence across fish farms, organs, and sampling seasons (May 2023–April 2024)

4 Discussion

4.1 Parasite Diversity and Tissue-Specific Occurrence

The present study revealed a broad spectrum of parasitic taxa in cultured tilapia, including ciliate protozoans (*Trichodina nigra, Ichthyophthirius multifiliis*), monogeneans (*Cichlidogyrus tilapiae, Dactylogyrus vastator, Gyrodactylus elegans*), digenean trematodes (*Clinostomum complanatum, Haplorchis spp.*), and nematode larvae (*Contracaecum spp.*). These taxa exhibited clear tissue preferences, reinforcing the concept of organ-specific parasite ecology in aquaculture systems. Gills were the most heavily colonized, harboring protozoans and monogeneans that caused epithelial erosion, mucus hypersecretion, and reduced respiratory efficiency. Such pathology is consistent with earlier findings that highlight the gills as the primary target for parasites due to their vascularity, constant water exposure, and relatively lower immune protection [21]. Skin infestations were dominated by *Gyrodactylus spp.* and *Ichthyophthirius spp.*, while intestinal infections were primarily associated with digenean trematodes. The visceral cavity carried Contracaecum larvae, whose presence indicates complex transmission cycles involving avian hosts. Importantly, encysted metacercariae of *C. complanatum* and *Haplorchis spp.* were detected in gills and muscle tissues, raising zoonotic risks when fish are consumed undercooked [22]. These findings underscore not only aquaculture health challenges but also food-safety concerns with direct public health implications.

4.2 Variation in Parasite Diversity Across Farms and Ponds/Tanks

Marked variability in parasite diversity was observed among farms, with Turbo Farm exhibiting the highest Shannon index values, followed by Moiben and Ainapkoi Farms, while Kesses Farm A showed very low diversity. Such farm-level differences likely reflect disparities in water management, stocking densities, and proximity to intermediate or reservoir hosts, factors that have been consistently linked to parasite transmission in tropical aquaculture [23]. Turbo Farm consistently recorded elevated infestations of *T. nigra*, *D. vastator*, and *G. elegans*, suggesting higher environmental parasite pressure, possibly due to organic matter accumulation or reduced water turnover. In contrast, Ainapkoi and Kesses A Farms often recorded lower parasite burdens, which may reflect stronger biosecurity practices or environmental constraints limiting parasite propagation. Intestinal infestations were relatively low overall, though *C. complanatum* was significantly more prevalent in Turbo compared to Kesses Farm A, a finding that underscores the role of local snail populations and bird vectors in sustaining digenean transmission cycles [22]. Intra-farm variability was also evident. For instance, Moiben Farm Pond 2 had significantly higher parasite diversity than other ponds, suggesting that pond-specific ecological features, such as shade, snail abundance, organic load, or water exchange, play a critical role in shaping parasite communities. This aligns with observations that microhabitat heterogeneity within farms can influence both parasite survival and host exposure [24, 25].

4.3 Influence of Facility Type on Parasite Occurrence

Facility design strongly influenced parasite occurrence. Liner ponds harbored the most diverse parasite communities, followed by earthen ponds, while tanks showed minimal infestations. The elevated diversity in liner and earthen systems likely reflects their semi-open nature, exposure to runoff and soil, and higher likelihood of harboring intermediate hosts such as mollusks and crustaceans [26]. Conversely, tank systems, with their greater biosecurity and controlled water quality, presented less favorable environments for sustaining parasite lifecycles. This facility-type gradient highlights an important management trade-off: while semi-open systems enhance productivity through natural food inputs, they also increase vulnerability to parasitic infections compared to closed, controlled tank systems.

4.4 Seasonal and Spatial Trends in Parasite Occurrence

Seasonal dynamics were pronounced, with parasite presence peaking during warmer and wetter months (October, December, February, and April). These trends are consistent with the influence of temperature and rainfall on parasite reproduction, intermediate host proliferation, and fish stress responses [21]. Cooler months (May–June) corresponded with suppressed parasite loads, supporting evidence that thermal conditions are major regulators of parasite transmission in aquaculture systems. Gill tissues consistently carried the heaviest parasite burdens across farms, with Turbo Farm showing particularly high infestation peaks. This pattern reflects both the ecological vulnerability of gills and farm-specific factors, such as stocking density and water management that enhance pathogen proliferation [27].

5 Conclusion

This study revealed a diverse community of parasites infecting cultured Oreochromis niloticus in Uasin Gishu County, with protozoans, monogeneans, digenean trematodes, and nematodes identified across various fish organs. The gills emerged as the most heavily parasitized organ, supporting protozoans such as *Trichodina nigra* and *Ichthyophthirius multifiliis* as well as monogeneans like *Dactylogyrus vastator* and *Cichlidogyrus tilapiae.* Intestinal tissues were dominated by digenean trematodes including *Clinostomum complanatum* and *Haplorchis spp.*, while nematode larvae (*Contracaecum* spp.) were frequently detected in the visceral cavity. These findings highlight the organ-specific ecology of parasites in aquaculture systems and confirm their significant impact on fish health and survival. Marked differences in parasite prevalence and diversity were observed across farms, pond types, and seasons, with liner and earthen ponds supporting higher parasite burdens compared to tanks. Seasonal trends showed peaks in parasite infestations during warmer and wetter months, underscoring the role of environmental factors such as temperature and rainfall in parasite transmission. Farms with poor water turnover and higher organic loads, such as Turbo Farm, consistently exhibited greater parasite diversity and intensity, suggesting that farm management practices directly shape infection dynamics. The parasites may significantly compromise fish health by damaging vital tissues, impairing feeding and respiration, and predisposing fish to secondary infections and posing zoonotic risks when fish are consumed raw or undercooked, raising public health concerns. To address these threats, integrated parasite management strategies are essential, including routine parasitological screening, farm-level biosecurity enhancement, regular pond sanitation, and strategic treatments targeting high-risk periods. Adoption of controlled aquaculture systems such as tanks, alongside farmer education on parasite lifecycles and transmission pathways, could further reduce infection risks.

Acknowledgements We sincerely thank the University of Eldoret for providing the necessary resources and facilities that enabled us to carry out this study effectively. We are also grateful to the Eldowas Laboratory, Ollessos National Polytechnic, and the Microbiology Technicians for their support and cooperation during the study. Our appreciation extends to the Uasin Gishu County Fisheries Department for their assistance and valuable contributions. We further thank the farmers whose farms were used in this study for their participation and willingness to support our research efforts.

References

1. FAO (2022) The state of world fisheries and aquaculture 2022: towards blue transformation. Food and Agriculture Organization of the United Nations, Rome
2. Pradeepkiran JA (2019) Aquaculture role in global food security with nutritional value: a review. Transl Anim Sci 3(2):903–910
3. Edwards P, Zhang W, Belton B, Little DC (2019) Misunderstandings, myths and mantras in aquaculture: its contribution to world food supplies has been systematically over-reported. Mar Policy 106:103547
4. Makori AJ, Abuom PO, Kapiyo R, Anyona DN, Dida GO (2017) Effects of water physico-chemical parameters on tilapia (*Oreochromis niloticus*) growth in earthen ponds in Teso North Sub-County Busia County. Fish Aquat Sci 20(1):1–10
5. Munguti JM, Mboya JB, Iteba JO, Kirimi JG, Obiero KO, Kyule DN et al (2024) Status and prospects of the ornamental fish industry in Kenya. Aquac Fish Fish 4(3):147–158
6. Wanja DW, Mbuthia PG, Waruiru RM, Mwadime JM, Bebora LC, Nyaga PN et al (2020) Fish husbandry practices and water quality in central Kenya: potential risk factors for fish mortality and infectious diseases. Vet Med Int 2020:6839354
7. Jacobs L, Chenia HY (2007) Characterization of integrons and tetracycline resistance determinants in *Aeromonas* spp. isolated from South African aquaculture systems. Int J Food Microbiol 114(3):295–306
8. Hussein MA, Eissa AE, El-Tarabili RM, Attia ASA, Zaki MM, Ibrahim TB et al (2024) Impact of climate change on some seasonal bacterial eruptions among cultured marine fishes from Egyptian coastal provinces. J Appl Vet Sci 9(2):18–30
9. Basweti EA, Turk PJ, Rayburn EB, Bryan WB (2009) No-till sequential cropping of summer and fall annual forage species compared with grassland. Agron J 101(6):1497–1502
10. Petrikova I, Otieno MA, Were G, Eltholth M, Mateos AR, Harding S et al (2024) Peri-urban agriculture and household food and nutrition security around Eldoret Kenya. Food Secur 16:1235–1263
11. Njonge T (2023) Influence of psychological well-being and school factors on delinquency during the COVID-19 period among secondary school students in selected schools in Nakuru County, Kenya. Int J Res Innov Soc Sci 7(7):1175–1189
12. Uasin Gishu County Government (2015) Livestock and fisheries. County Government of Uasin Gishu, Eldoret. Available from: https://uasingishu.go.ke/index.php/tag/livestock-and-fisheries
13. Mwangi J, Maina J, Gachuiri C, Wahome R, Waruiru R (2024) Management practices, farmers' perceptions and coping strategies to climate change in integrated fish farms in Kenya. East Afr J Sci Technol Innov 5(4):1–12
14. Thrusfield M (2013) Veterinary epidemiology, 3rd edn. Wiley-Blackwell, Oxford
15. Froese R (2006) Cube law, condition factor and weight–length relationships: history, meta-analysis and recommendations. J Appl Ichthyol 22(4):241–253
16. Kuchta R, Burianová A, Jirků M, De Chambrier A, Oros M, Brabec J, et al (2012) Bothriocephalidean tapeworms (Cestoda) of freshwater fish in Africa, including erection of *Kirstenella* n. gen. and description of *Tetracampos martinae* n. sp. Folia Parasitol 59(1):1–35

17. Scholz T, Hanzelová V, Škeříková A, Shimazu T, Rolbiecki L (2007) An annotated list of species of the *Proteocephalus* Weinland, 1858 aggregate sensu de Chambrier et al. (2004) (Cestoda: Proteocephalidea), parasites of fishes. Syst Parasitol 67(2):139–56
18. Paperna I (1996) Parasites, infections and diseases of fishes in Africa: an update. CIFA Technical Paper. Food and Agriculture Organization of the United Nations, Rome
19. Bush AO, Lafferty KD, Lotz JM, Shostak AW (1997) Parasitology meets ecology on its own terms: Margolis et al. revisited. J Parasitol 83(4):575–583
20. Nolan K, Callan J (2006) Beachcomber Biology: the Shannon-Weiner species diversity index. ABLE Proc 27:334–338
21. Rahaman R, Asikur M, Roy N, Khan Manon MR, Ullah MR, Islam MM et al (2022) Temporal distribution of fishery resources in Payra River: relationship with climatological changes, ecological assessment, and threat assessment. Heliyon 8(9):e10584
22. Menconi V, Lazzaro E, Bertola M, Gustinelli A (2023) The occurrence of freshwater fish-borne zoonotic helminths in Italy and neighbouring countries: a systematic review. Animals 13(7):1178
23. Yadav M, Ranjan R, Sah TS, Upadhyay U (2023) Fish farming practices and disease occurrence in the fish farms of Dhanusha district Nepal. J Appl Aquac 35(2):189–202
24. Abbas F, Hafeez-Ur-Rehman M, Mubeen N, Bhatti A, Daniel I (2023) Parasites of fish and aquaculture and their control. In: Aquaculture parasitology, pp 248–266. CABI, Wallingford
25. Mzula A, Wambura PN, Mdegela RH, Shirima GM (2021) Present status of aquaculture and the challenge of bacterial diseases in freshwater farmed fish in Tanzania: a call for sustainable strategies. Aquac Fish 6(4):391–399
26. Onyango J, Kipkorir N, van Bruggen J, Schmitz J (2024) Agricultural intensification in Lake Naivasha catchment in Kenya and associated nutrient and pesticide pollution. Sci Rep 14:18539
27. Consuegra S, Webster TU, Anka I (2023) Microbiome, epigenetics and fish health interactions in aquaculture. In: Epigenetics in aquaculture, pp 245–262. Wiley, Hoboken

Variations in the Spatial and Temporal Diversity of Bacteria Isolated from Farm Cultured Fish in Uasin Gishu County, Kenya

Edwin Kipyegon Mutai, Moses Mwajar Ngeiywa, David Mbeva Liti, and Patricia Jepchirchir Kosgei

Abstract

Fish farming plays a crucial role in food security and economic development, yet bacterial infections threaten productivity and the fish health. This study assessed the spatial and temporal diversity of bacteria isolated from *Oreochromis niloticus* and *Clarias gariepinus* in Uasin Gishu County, Kenya. A total of 468 fish samples were analyzed through cultural, biochemical, and microscopic characterization. A total of 154 bacterial isolates belonging to 12 species were identified. *Escherichia coli* was the most prevalent, followed by *Pseudomonas aeruginosa*, *Vibrio harveyi*, and *Vibrio alginolyticus*. Organ-specific analysis showed that gills and intestines harbored the highest bacterial diversity, while kidneys recorded the lowest. Significant differences ($p \leq 0.05$) were observed in bacterial prevalence across fish organs, farms, pond types, and sampling seasons. Earthen and liner ponds supported richer bacterial communities than tanks, and higher bacterial loads were recorded during warmer months. Co-occurrence of multiple species, particularly in gills and intestines, suggested complex infection dynamics that may exacerbate disease severity. These findings underscore the influence of aquaculture system type, farm management practices, and environmental conditions on bacterial diversity and distribution. To safeguard fish health and enhance productivity, regular water quality monitoring, improved biosecurity, and routine health assessments are recommended. This study provides essential baseline data to guide sustainable aquaculture practices in the region. Moreover, the presence of opportunistic and zoonotic bacteria such as *E. coli* highlights potential risks, reinforcing the need for integrated fish health management and food safety interventions. Addressing these risks is crucial for sustaining aquaculture yields and protecting public health.

E. K. Mutai (✉) · M. M. Ngeiywa · D. M. Liti · P. J. Kosgei
Department of Biological Sciences, School of Science, University of Eldoret, Eldoret, Kenya
e-mail: emutai@tonp.ac.ke; edwinmutaikp@gmail.com

R. Ramkat et al. (eds.), *Towards the Sustainability of our Ecosystems*, Springer Proceedings in Geography and Sustainability,
https://doi.org/10.1007/978-3-032-30076-8_4

Keywords

Bacterial diversity • Fish health • Aquaculture systems • Oreochromis niloticus • Uasin Gishu County

1 Introduction

Aquaculture has become one of the most rapidly expanding sectors in global food production, contributing significantly to food security, income generation, and nutrition, particularly in developing countries [1, 2]. With the growing demand for affordable animal protein, farmed fish now form a substantial component of the human diet and serve as a vital source of livelihood for rural and peri-urban communities [3, 4].

In Kenya, the government has promoted aquaculture as a key economic activity under national development programs, leading to increased investment and expansion of fish farming across various counties, including Uasin Gishu [5]. While the sector shows strong potential, it is increasingly threatened by fish health-related challenges, particularly infections caused by bacterial pathogens.

Farmed fish, like other animals, are susceptible to a range of infectious agents, among which bacteria and parasites are the most prevalent and economically significant [6]. These pathogens contribute to reduced growth, poor feed conversion, high morbidity, and mortality in aquaculture systems. Bacterial species such as *Aeromonas* spp., *Pseudomonas* spp., and *Vibrio* spp. are commonly associated with fish infections and are known to cause septicemia, skin ulcerations, fin rot, and systemic infections [7]. On the other hand, parasites such as protozoans (e.g., *Ichthyophthirius multifiliis*) and helminths (e.g., *Dactylogyrus* spp., *Gyrodactylus* spp.) can cause tissue damage, anemia, and immunosuppression, making fish more vulnerable to secondary infections [8].

Despite the known risks, there is limited data on the diversity and prevalence distribution of bacterial pathogens in aquaculture systems in Uasin Gishu County. Most local studies have focused broadly on production constraints or water quality, with very few studies specifically addressing pathogen identification and distribution. As such, farmers often lack access to accurate diagnostic information, which hinders the implementation of timely and targeted disease management strategies.

Despite the increasing importance of aquaculture in Uasin Gishu, little is known about the organ-specific distribution and temporal diversity of bacterial pathogens, particularly in relation to pond type and farm management practices. This study, therefore, aimed to assess the diversity of bacteria infecting fish in selected aquaculture farms in Uasin Gishu County. The isolation and identification of these pathogens will provide baseline data necessary for developing effective fish health management practices. Ultimately, the findings are expected to support the growth of a healthier, more sustainable aquaculture industry in the region.

2 Materials and Methods

2.1 Study Area

The study was conducted from May 2023 to April 2024 in selected fish farms across five sub-counties in Uasin Gishu County (UG) as shown in Fig. 1. Uasin Gishu County, located in the north west of the Great Rift Valley region, lies approximately 330 km northwest of Nairobi. It covers an area of 2955 km^2 and characterized by a highland plateau with several perennial rivers and swampy wetlands [9, 10]. The region experiences a temperate climate, with annual temperatures ranging between 7 °C to 29 °C and rainfall between 624.9 mm to and 1560.4 mm per year [11].

Aquaculture has been expanding in Uasin Gishu in the last decade, with 1728 operational fish ponds covering 486,000 m^2, producing approximately 593,000 kg of fish annually [12]. Most farmers practicing fish farming use earthen ponds, liner ponds, and other water reservoirs. Nile tilapia (*Oreochromis niloticus*) is the predominant species farmed, followed by African catfish (*Clarias gariepinus*) [13].

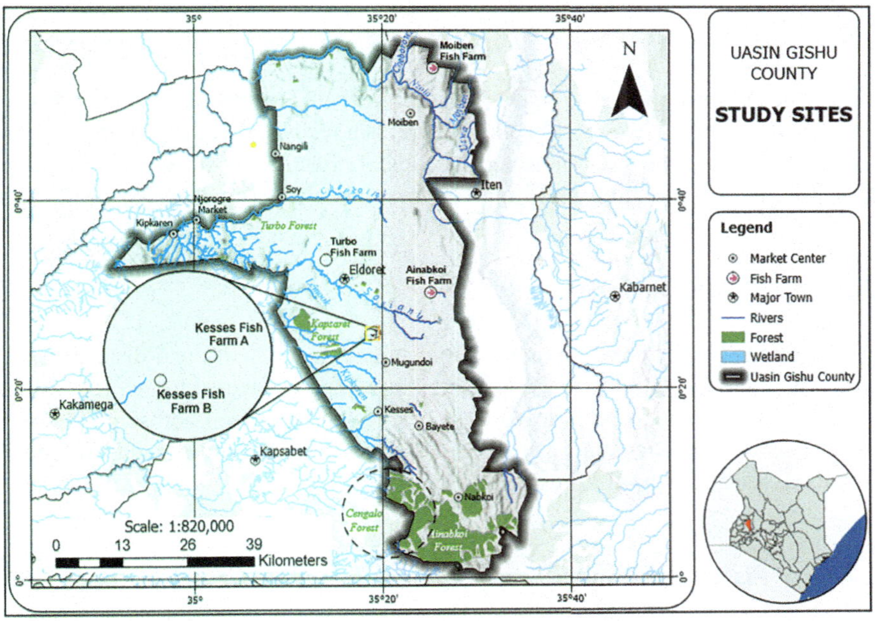

Fig. 1 Map of Uasin Gishu showing its location in western Kenya with perennial rivers and swampy wetlands

Table 1 Location and features of the selected fish farms in Uasin Gishu County where the study was carried out

Sub-county	Farm site name	Pond	Type of ponds
Moiben	Moiben Farm (Farm A)	P1	Raised liner pond
		P2	Earthen pond
		P3	Earthen pond
		P4	Earthen pond
Ainapkoi	Ainapkoi Farm (Farm C)	P1	Ground liner pond
		P2	Ground liner pond
		P3	Earthen pond
Kesses	Kesses Farm 1 (Farm K)	P1	Raised liner pond
		P2	Raised liner pond
Kesses	Kesses Farm 2 (Farm S)	P1	Earthen pond
		P2	Earthen pond
Turbo	Turbo Farm (Farm T)	P1	Ground liner pond
		P2	Ground liner pond

2.2 Study Sites

Five aquaculture farms were selected based on their features such as; pond type, size, location, inputs (e.g., fertilizers, manure, fish feeds), and duration of operation (minimum of one year) as indicated in Table 1. Each pond in the selected farm site was sampled monthly for 12 months to monitor water quality and fish health.

2.3 Study Design

A longitudinal study design was used to evaluate the diversity and distribution of bacterial and parasite infestations in fish cultured in the selected farms. A stratified random sampling approach was used to ensure representation across pond types, locations, and management practices. Each pond was sampled monthly for water quality assessment and fish health monitoring.

2.4 Fish Sampling

The sample size for fish was determined following the Thrusfield formular (Thrusfield, 2013), with a 95% confidence interval and 5% absolute precision. Fish were sampled using seine nets with a mesh size of 1 mm, a length of 15 m, and a width of 2 m, allowing for the collection of individuals across different size classes. Three (3) fish were selected randomly per pond from the sampled fish and transported in cool boxes to the laboratory for further analysis. The sampled fish from the 13 ponds from all the fish farms in a month was 39 and a total of 468 for the entire sampling period of 12 months. Fish were collected using seine nets and transported in cool boxes to the laboratory. The total length (cm) was measured using a measuring board; and weighed (g) using a digital balance (Model GX-200, A&D Company Limited, Tokyo, Japan). The condition factor (K) was determined using Froese's formula [14].

2.5 Bacterial Isolation and Identification

Sample Processing and Enrichment of Bacteria

Aseptic measures were maintained during the sampling procedure for bacteria to prevent contamination of the samples. Three types of specimens were collected from different organs and locations of infected fish such as intestine, skin and gill, and examined using microbiological test. These specimens were placed on sterile chopping board and then properly minced and ground together. Ten (10) gm of samples were homogenized with 90 ml(ml) of freshly prepared 0.1% peptone water and 0.1 ml of the homogenized sample were inoculated according to standard methods on to selective media such as: Rimler Shotts Medium Base agar (*Aeromonas spp.),* Pseudomonas Base agar (for Pseudomonas spp), Thiosulfate citrate bile salt sucrose (TCBS) agar (*Vibrio* spp.), Tryptic Soy Agar (TSA) for enrichment of bacterial isolates, Brain Heart Infusion (BHI) Agar (for fastidious organisms) and incubated at 37 °C for 24 h. Pure isolates were characterized using morphological and biochemical tests, including Gram staining, oxidase, catalase, motility, indole, urease and citrate tests. The identification followed Bergey's manual of bacteriology [15].

Identification of Bacterial Pathogens

Suspected bacterial colonies obtained from different culture plates were isolated and then streaked on TSA slants, MIU medium, Simon citrate agar slant and incubated overnight at 37 °C. The pure isolates were characterized by bacterial cell morphology, alkaline and acidic reaction, H_2S (hydrogen sulfide production) and gas production, motility test, indole production, urease test, oxidase test, catalase test, Methyl Red (MR) test, and Voges Praskaure (VP) test. The Gram staining techniques were performed to identify Gram positive and Gram negative bacteria, while biochemical tests were carried out to identify the pathogens. The growth and identification of bacterial pathogens was done following Bergey's manual of Bacteriological classification [15] and bacterial isolates were subsequently stored for antibiotic susceptibility testing.

Determination of the Bacterial diversity

Shannon–Wiener diversity index for the species community was calculated using biodiversity calculator [16]. The diversity of the parasitic and bacterial organisms infecting fish was determined using Shannon–Wiener diversity index as shown in

Eq. 1.

$$H = \sum_{I=1}^{R} \text{Pi In Pi} \quad (1)$$

where H' = Shannon's diversity index, R = Total number of species, i = the ith species, In = natural logarithm and Pi = the proportion of individuals belonging to the ith species and n = Total number of individuals of all species of parasites or bacterial colonies.

2.6 Statistical Analysis

Data were tested for normality (Shapiro–Wilk test) and homogeneity (Levene's test). Non-parametric tests were used where necessary. Descriptive statistics (prevalence, mean abundance, and mean intensity) were computed to quantify bacterial distribution across fish organs, farms, pond types, and seasons [17]. Shannon diversity indices (H') were calculated to assess community diversity. Differences in bacterial prevalence were tested using the Kruskal–Wallis test, followed by Dunn's post hoc test with Bonferroni adjustment. The effects of farm, pond/tank, facility type, and season on bacterial occurrence were further examined with ANOVA on presence/absence data ($p < 0.05$). Co-occurrence patterns were analyzed using cross-tabulations and percentage distributions to distinguish single from multi-species infections.

3 Results

3.1 Bacterial Organisms Isolated from Selected Fish Farms

A total of 468 fish samples collected from selected fish farms in Uasin Gishu County were investigated for bacterial characterization over a one-year period, resulting in 154 bacterial colony isolates. The 154 bacterial colony isolates displayed a variety of distinct cultural and biochemical characteristics that enabled for their identification. These isolates were distributed across 12 bacterial species, including both Gram-positive and Gram-negative organisms, each exhibiting unique colony morphology and biochemical properties as shown in Table 2.

3.2 Distribution and Prevalence of Bacterial Species

Organ-Specific Distribution

The distribution of bacterial isolates across different organs (gills, intestines, skin, and kidney), highlighting both shared and organ-specific bacterial occurrence as shown in Table 3. *E coli* and *P. aeruginosa* were detected in all four tissues. *P.*

Table 2 Classification of Isolated Bacteria from Fish Samples

Gram positive bacteria		Gram negative bacteria	
Genera	Species	Genera	Species
Bacillus	*Bacillus* spp.	*Escherichia*	*Escherichia coli*
Streptociccus	*Streptociccus* spp.	*Pseudomonas*	*Pseudomonas aeruginosa*
Micrococcus	*Micrococcus* spp.	*Aeromonas*	*Aeromonas hydrophila*
			Aeromonas caviae
			Aeromonas sobria
		Vibrio	*Vibrio harveyi*
			Vibrio alginolyticus
		Flavobacterium	*Flavobacterium columnare*
		Plesiomonas	*Plesiomonas shigelloides*

Table 3 Summary of isolated species by Organ

Gills	Intestines	Skin	Kidney
E. coli	*E. coli*	*E. coli*	*E. coli*
P. aeruginosa	*P. aeruginosa*	*P. aeruginosa*	*P. aeruginosa*
V. harveyi	*V. harveyi*	*P. fluorescens*	
V. alginoliticus	*A. hydrophila*	*A. hydrophila*	
	Flav. columnare	*Flav. columnare*	
	V. alginoliticus		

fluorescens was uniquely found in the skin while *Flav. columnare* was limited to the skin and intestines. Among all tissues, the intestines exhibited the highest bacterial diversity.

Farm Level and Pond Level Variability

Shannon diversity by farm and pond/tank, revealed that bacterial diversity varied notably across farms, with some consistently exhibiting greater microbiological complexity than others, as summarized in Table 4. Turbo Farm recorded the highest bacterial diversity with a Shannon index of 1.94, indicating a rich and relatively even bacterial community. This was followed closely by Kesses Farm B (1.75) and Moiben Farm (1.75). In contrast, Kesses Farm A showed the lowest bacterial diversity at 1.67. At the pond/tank level, diversity differences were also apparent within farms as shown in Table 4. For instance, Pond 1 in Kesses Farm B had a higher Shannon index (1.76) than Pond 2 (1.75), while in Ainapkoi Farm, bacterial diversity ranged from 1.69 to 1.72 across ponds.

Table 4 Shannon Diversity index of bacterial communities by farm and pond/tank

Farm	Pond/tank	Bacterial shannon index	Bacterial richness (S)	Rank
Moiben Farm		**1.753949**	6.25	7
	1	1.723553	6.00	16
	2	1.712425	6.17	12
	3	1.723553	6.14	15
	4	1.712425	5.58	18
Ainapkoi Farm		**1.709018**	6.25	7
	1	1.723553	5.92	17
	2	1.689764	6.50	4
	3	1.713737	6.42	6
Kesses Farm A		**1.67201**	6.46	5
	1	1.619156	6.25	7
	2	1.724865	6.17	12
Kesses Farm B		**1.753949**	6.25	7
	1	1.760372	6.17	12
	2	1.747526	6.21	11
Turbo Farm		**1.93536**	8.00	1
	1	1.941783	7.92	3
	2	1.928937	7.96	2

Facility Type Distribution and Diversity

Facility type comparison revealed that bacterial diversity varied across different aquaculture systems, as shown in Table 5. Liner ponds exhibited the highest bacterial diversity with a Shannon index of **1.82**, suggesting a richer and more even microbial community. This was followed by **earthen ponds**, which recorded a slightly lower but still substantial diversity index of **1.73. Tank systems** demonstrated the lowest bacterial diversity at **1.69**, indicating a more limited bacterial population structure.

The ANOVA results, presented in Table 6, revealed that both season and farm exerted a statistically significant influence on bacterial presence ($p < 0.001$). While pond and facility type also demonstrated statistically significant effects ($p < 0.001$), their influence was comparatively smaller.

Table 5 Shannon diversity index of bacterial communities by facility type

Farm	Bacterial shannon index	Rank	Estimated species richness (S)
Earthen Pond	1.728	1	6
Liner Pond	1.821	2	6
Tank	1.689	3	5

Table 6 Analysis of variance (ANOVA) of bacterial presence by farm, pond, facility type, and season

Source	sum_sq	df	F	PR(>F)
Across Farm	47.59864	4	11.15822	0.0000
Across Ponds/Tanks	93.74603	3	29.30164	0.0000
Across Facility Type	18.30575	2	8.582582	0.0002
Across Seasons (date)	103.739	11	8.843206	0.0000

Significant differences in bacterial infestation were observed across the sampled fish farms, with patterns varying by organ and bacterial species as shown in Table 7. In the kidney, *P. aeruginosa* was found to be significantly more prevalent in fish from Turbo farm than in all other farms ($p < 0.001$). Similarly, in the gills, *V. alginolyticus* exhibited significantly higher infestation levels in Turbo compared to Ainapkoi, Kesses A, Kesses B, and Moiben Farms ($p < 0.001$). In the intestines, multiple bacterial species displayed significant variation between farms. *P. aeruginosa* levels were notably higher in fish from Turbo Farm compared to Ainapkoi and Moiben Farms. For *Flavobacterium columnare*, significant differences were found between Kesses Farm B and Moiben Farm, as well as between Moiben and Turbo Farms. Additionally, *V. alginolyticus* was significantly more prevalent in intestines of fish from Turbo and Kesses B Farms than in those from Ainapkoi and Moiben Farms. The skin showed the greatest diversity in bacterial infestation patterns. All five bacteria analyzed (*P. fluorescens*, *A. hydrophila*, *P. aeruginosa*, *V. alginolyticus*, and *F. columnare*) exhibited significant differences across farms. Notably, *P. fluorescens* and *A. hydrophila* infestations were significantly lower in Ainapkoi Farm, while Moiben and Turbo Farms displayed elevated infection levels. *P. aeruginosa* and *V. alginolyticus* followed similar patterns, with Turbo Farm showing consistently higher infestations, especially when compared with Ainapkoi and Kesses A Farms.

Seasonal and Temporal Variation

The mean bacterial presence by facility type and season reveals important trends in microbial dynamics within aquaculture systems as illustrated in Fig. 2. Across all facility types, Earthen, Liner, and Tank, the highest bacterial loads were observed during the months of August (Earthen: 0.52, Liner: 0.56, Tank: 0.52) and April (Earthen: 0.47, Liner: 0.51, Tank: 0.46), with additional peaks in September (Earthen: 0.40, Liner: 0.44, Tank: 0.39) and November (Earthen: 0.43, Liner: 0.50, Tank: 0.41). Among the facility types, liner systems consistently recorded the highest bacterial presence, followed closely by earthen ponds. Tank systems exhibited relatively lower bacterial levels. Notably, bacterial presence remained elevated in liner and earthen systems even during cooler months, indicating that open systems may retain residual microbial load longer than closed ones.

The spatial and temporal distribution of bacterial occurrence across farms, fish organs, and seasons, as illustrated in Fig. 3, revealed consistent trends of **higher**

Table 7 Significant pairwise differences in bacterial species infestation across fish organs and farms

Organ	Bacterial Spp	Farm		Difference	*p*-value
		Group 1	Group 2		
Kidney	*P. aeruginosa*	Ainapkoi	Turbo	0.25	< 0.001
		Kesses A	Turbo	0.25	< 0.001
		Kesses B	Turbo	0.25	< 0.001
		Moiben	Turbo	0.25	< 0.001
Gills	*V. alginolyticus*	Ainapkoi	Turbo	0.5	< 0.001
		Kesses A	Turbo	0.5	< 0.001
		Kesses B	Turbo	0.5	< 0.001
		Moiben	Turbo	0.5	< 0.001
Intestines	*P. aeruginosa*	Ainapkoi	Turbo	0.417	0.008
		Kesses B	Moiben	– 0.333	0.038
		Moiben	Turbo	0.417	0.004
	Flav. columnare	Kesses B	Moiben	0.25	0.038
		Moiben	Turbo	– 0.25	0.038
	V. alginolyticus	Ainapkoi	Kesses B	0.333	< 0.001
		Ainapkoi	Turbo	0.333	< 0.001
		Kesses B	Moiben	– 0.333	< 0.001
		Moiben	Turbo	0.333	< 0.001
Skin	*P. fluorescens*	Ainapkoi	Kesses A	– 0.444	< 0.001
		Ainapkoi	Kesses B	– 0.444	< 0.001
		Ainapkoi	Turbo	– 0.444	< 0.001
		Kesses A	Moiben	0.5	< 0.001
		Kesses B	Moiben	0.5	< 0.001
		Moiben	Turbo	– 0.5	< 0.001
	A. hydrophila	Ainapkoi	Kesses A	– 0.25	0.031
		Ainapkoi	Kesses B	– 0.25	0.031
		Ainapkoi	Turbo	– 0.25	0.031
		Kesses A	Moiben	0.25	0.019
		Kesses B	Moiben	0.25	0.019
		Moiben	Turbo	– 0.25	0.019
	P. aeruginosa	Ainapkoi	Kesses A	0.333	< 0.001
		Ainapkoi	Kesses B	0.583	< 0.001
		Ainapkoi	Turbo	0.833	< 0.001
		Kesses A	Kesses B	0.25	0.046
		Kesses A	Moiben	– 0.333	< 0.001

(continued)

Table 7 (continued)

Organ	Bacterial Spp	Farm		Difference	*p*-value
		Group 1	Group 2		
		Kesses A	Turbo	0.5	< 0.001
		Kesses B	Moiben	− 0.583	< 0.001
		Kesses B	Turbo	0.25	0.046
		Moiben	Turbo	0.833	< 0.001
	V. alginolyticus	Ainapkoi	Kesses A	0.333	0.001
		Ainapkoi	Kesses B	0.333	0.001
		Ainapkoi	Turbo	0.75	< 0.001
		Kesses A	Moiben	− 0.333	< 0.001
		Kesses A	Turbo	0.417	< 0.001
		Kesses B	Moiben	− 0.333	< 0.001
		Kesses B	Turbo	0.417	< 0.001
		Moiben	Turbo	0.75	< 0.001

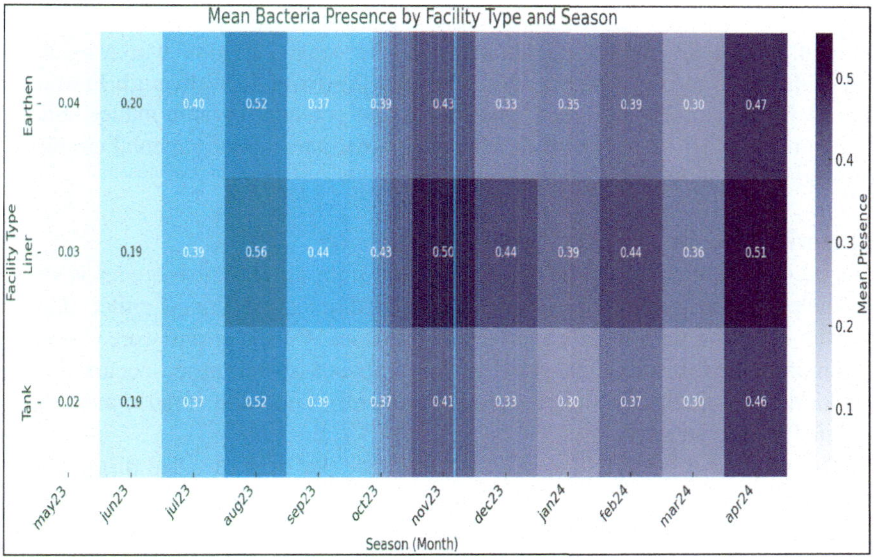

Fig. 2 Mean bacterial presence by facility type and season

bacterial loads in gills, followed by intestines and skin. Among the farms, **Turbo Farm** stood out with the **highest mean bacterial occurrence**, particularly in gill samples, which maintained a **mean value of 1.00** from **August 2023 through April 2024**. Other farms, including **Moiben, Kesses B, and Ainapkoi Farms**, also showed moderately elevated bacterial levels in the gills, with peak values ranging

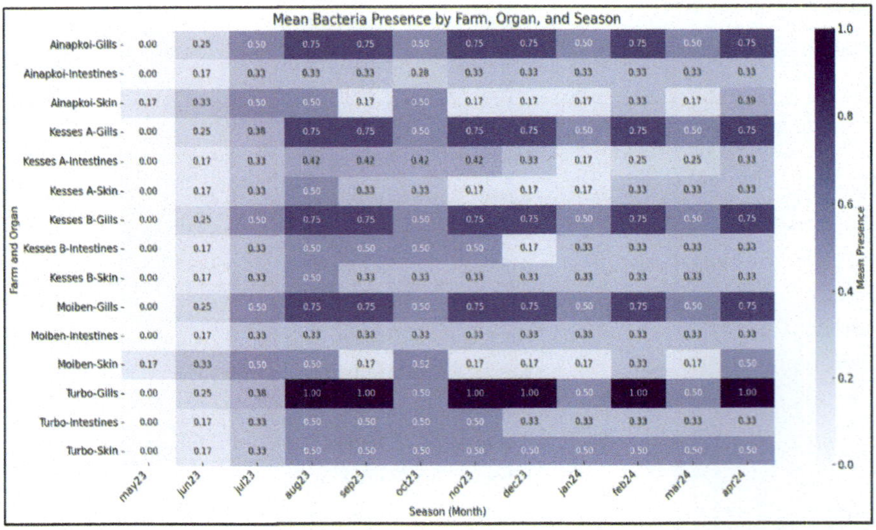

Fig. 3 Mean bacterial presence by farm, tissue type, and season

from **0.75 to 0.83** during the same period. Seasonal variation was also apparent with bacterial occurence generally **increasing from July through February**, aligning with periods of higher temperature and potential organic matter buildup. In contrast, values were **lowest in May and June**, when environmental conditions may be less favorable for bacterial growth.

Bacterial Species Co-occurrence

The co-occurrence of bacterial species isolated from different fish organs is shown in Table 8. *Escherichia coli* consistently isolated across all organs either alone or as a co infection with most bacterial combinations. Co-infections were especially common in the **intestines** and **skin**, where complex associations co-occurred while the kidney did not exhibit any co-occurrence with E. coli and *P. aeruginosa* being isolated independently.

The results for bacterial species co-occurrence per farm revealed distinct farm-specific trends as shown in Table 9. **Moiben Farm** consistently showed the highest overall bacterial diversity and frequency of co-infections. **Turbo Farm** stood out for its dominance in complex co-occurrence patterns. It recorded the highest presence of *P. aeruginosa* in kidney samples (27.3%), a bacterium that was absent from kidneys in all other farms. The skin samples from Turbo Farm also revealed a high rate (72.7%) of complex co-infections involving *E. coli*, *P. aeruginosa*, and *V. alginolyticus*. **Ainapkoi Farm** showed simpler co-occurrence patterns while **Kesses Farms A and B** displayed moderate co-occurrence levels.

The co-occurrence of bacterial species per facility type varied **notably** as presented in Table 10. **Earthen ponds** supported the most complex bacterial interactions, especially in gill tissues where the combination of *E. coli*, *P. aeruginosa*,

Table 8 Distribution of bacterial species co-occurrence across fish organs

Bacterial Co-occurrence	Organ involved			
	Gills	Intestines	Skin	Kidney
E. coli	✓	✓	✓	✓
E. coli/P. aeruginosa	✓	✓	✓	
E. coli/P. aeruginosa/V. harveyi	✓			
E. coli/P. aeruginosa/V. alginolyticus	✓	✓	✓	
E. coli/A. hydrophila		✓	✓	
E. coli/Flav. Columnare		✓		
E. coli/V. harveyi		✓		
E. coli/V. alginolyticus		✓	✓	
P. aeruginosa				✓
E. coli/P. fluorescens			✓	
E. coli/P. fluorescens/A. hydrophila			✓	
E. coli/P. fluorescens/Flav. columnare			✓	

and *V. harveyi* was most prevalent (**50.0%**)**.** In the intestines, co-occurrence involving *E. coli* with *Flav. columnare* or *P. aeruginosa* was also relatively high (**16.7%** and **30.6%,** respectively). **Tanks**, in contrast, showed simpler bacterial profiles. While *E. coli* remained commonly detected in all pond types, especially in the **skin** (**38.9%** in Tanks and **36.1%** in Earthen ponds). Complex co-infections were less frequent For example, the trio of *E. coli*, *P. aeruginosa*, and *V. alginolyticus* appeared predominantly in **liner ponds** (**34.8%**) but was rare in the others.

4 Discussion

4.1 Isolation and Identification of Bacterial Species

The identification of 154 bacterial isolates from 468 fish samples across aquaculture farms in Uasin Gishu County revealed a broad spectrum of bacterial taxa, spanning both Gram-positive and Gram-negative groups. This microbial diversity aligns with the established understanding that fish and aquaculture systems harbor a dynamic microbiota influenced by water quality, stocking density, feeding practices, and environmental hygiene [18, 19].

Among the Gram-positive bacteria, *Bacillus spp.* and *Streptococcus spp.* were the most prevalent. *Bacillus* species are known for their dual role in aquaculture: while generally regarded as beneficial due to their probiotic properties and ability to degrade organic waste [20], certain strains have been implicated in opportunistic infections under stressed conditions. *Streptococcus spp.* have been reported in streptococcal outbreaks in East African aquaculture systems, particularly affecting tilapia species [21]. The identification of *Micrococcus spp.*

Table 9 Percentage distribution of bacterial species co-occurrence across farms

Organ/species	Farm				
	Moiben	Ainapkoi	Kesses A	Kesses B	Turbo
Bacterial species in Kidney					
E. coli	37.5	38.9	37.5	37.5	45.5
P. aeruginosa	0	0	0	0	27.3
Bacterial species in Gills					
E. coli	8.3	8.3	8.3	8.3	9.1
E. coli, P. aeruginosa	33.3	33.3	33.3	33.3	36.4
E. coli, P. aeruginosa, V. harveyi	50	50	50	50	0
E. coli, P. aeruginosa, V. alginolyticus	0	0	0	0	45.5
Bacterial species in Intestines					
E. coli	8.3	8.3	16.7	16.7	9.1
E. coli, A. hydrophila	8.3	8.3	0	0	0
E. coli, Flav. Columnare	25	25	12.5	0	0
E. coli, P. aeruginosa	25	25	33.3	41.7	45.5
E. coli, V. harveyi	25	25	8.3	0	0
E. coli, P. aeruginosa, V. alginolyticus	0	0	8.3	16.7	18.2
E. coli, V. alginolyticus	0	0	8.3	16.7	18.2
Bacterial species in Skin					
E. coli	50	50	33.3	8.3	9.1
E. coli, A. hydrophila	8.3	8.3	0	0	0
E. coli, P. fluorescens	16.7	16.7	0	0	0
E. coli, P. fluorescens, A. hydrophila	16.7	16.7	0	0	0
E. coli, P. fluorescens, F. columnare	8.3	8.3	0	0	0
E. coli, P. aeruginosa	0	0	25	50	9.1
E. coli, P. aeruginosa, V. alginolyticus	0	0	8.3	8.3	72.7
E. coli, V. alginolyticus	0	0	25	25	0

and *Staphylococcus spp.*, although less commonly associated with pathogenicity, suggests environmental contamination, given their ubiquity in air and water surfaces.

Gram-negative isolates dominated the microbial profile, with *Escherichia coli*, *Pseudomonas aeruginosa*, and *Aeromonas spp.* being the most frequently encountered. The presence of *E. coli*, a well-established fecal contamination indicator, raises public health concerns and reflects potential runoff from agricultural or domestic sources [22]. *P. aeruginosa* was distinguishable by its β-hemolytic activity and green pigmentation due to pyocyanin production. This species is a known opportunistic pathogen capable of causing ulcerative lesions in fish and has been increasingly reported in Kenyan aquaculture systems [18, 23]. Its detection in

Table 10 Percentage co-occurrence of bacterial species by pond type and organ

Organ/species	Facility type		
	Earthen pond	Liner pond	Tank
Bacterial_species_kidney			
E. coli	37.5	41.3	38.9
P. aeruginosa	0	13	0
Bacterial_species_gills			
E. coli	8.3	8.7	8.3
E. coli, P. aeruginosa	33.3	34.8	33.3
E. coli, P. aeruginosa, V. harveyi	50	26.1	50
E. coli, P. aeruginosa, V. alginolyticus	0	21.7	0
Bacterial_species_intestines			
E. coli	11.1	8.7	13.9
E. coli, A. hydrophila	5.6	4.3	2.8
E. coli, Flav. columnare	16.7	13	16.7
E. coli, P. aeruginosa	30.6	34.8	30.6
E. coli, P. aeruginosa, V. alginolyticus	5.6	8.7	5.6
E. coli, V. alginolyticus	5.6	8.7	5.6
E. coli, V. harveyi	16.7	13	13.9
Bacterial_species_skin			
E. coli	36.1	30.4	38.9
E. coli, A. hydrophila	5.6	4.3	2.8
E. coli, P. aeruginosa	16.7	4.3	16.7
E. coli, P. aeruginosa, V. alginolyticus	2.8	34.8	5.6
E. coli, P. fluorescens	11.1	8.7	5.6
E. coli, P. fluorescens, A. hydrophila	11.1	8.7	5.6
E. coli, P. fluorescens, F. columnare	5.6	4.3	2.8
E. coli, V. alginolyticus	8.3	0	16.7

all organs underlines its adaptive versatility and persistence in aquatic environments. Several *Aeromonas* species were also isolated, including *A. hydrophila, A. caviae*, and *A. sobria*. These bacteria are frequently implicated in hemorrhagic septicemia, especially in warm, nutrient-rich waters [24]. Their identification through hemolysis, oxidase positivity, and fermentative behavior confirms their pathogenic potential in both fish and humans.

Other Gram-negative bacteria included *Vibrio alginolyticus, Vibrio harveyi, Flavobacterium columnare*, and *Plesiomonas shigelloides*. The *Vibrio* species, identifiable via yellow colonies on TCBS agar, are halophilic organisms commonly associated with vibriosis and skin ulceration in marine and estuarine fish but have increasingly been isolated from freshwater farms [25]. Their presence in Uasin Gishu freshwater systems may reflect introduction from contaminated water sources or their known ability to adapt and persist in low-salinity environments under favorable nutrient and temperature conditions. *Flav. columnare,* the causative agent of columnaris disease and a major concern in tilapia production, was among the pathogens detected in this study, highlighting its continued presence in regional aquaculture systems. Its considerable genomic variability and environmental resilience, as documented in earlier research [26, 27], likely contribute to its persistence under the diverse water quality and management conditions observed in Uasin Gishu County. The identification of *Plesiomonas shigelloides*, a less frequently reported aquatic bacterium, is noteworthy. Though typically associated with gastrointestinal infections in humans, its occurrence in aquaculture systems has been linked to high organic loads and poor hygiene [28].

4.2 Tissue-Specific Distribution and Diversity of Bacterial Species Across Farms, Facility Types, and Seasons

Organ-Specific Distribution of Bacteria

The study demonstrated distinct patterns of bacterial distribution across fish tissues. *E. coli* and *P. aeruginosa* were consistently isolated from gills, intestines, skin, and kidneys, confirming their status as dominant opportunistic pathogens in aquaculture systems. This widespread colonization has been attributed to their ability to form biofilms, exhibit high motility, and tolerate environmental stressors, which collectively enable persistence across multiple tissue types [29]. The detection of these species in all examined organs aligns with their frequent association with fecal contamination and environmental exposure. Similarly, *Flav. columnare* was restricted to the intestines and skin, reinforcing its tropism for mucosal tissues where it exploits epithelial and digestive tract surfaces [27]. Notably, the intestines exhibited the greatest bacterial richness, reflecting their role as a primary microbial reservoir shaped by host diet, water intake, and metabolic processes. These findings emphasize the significance of organ-specific niches in structuring microbial communities and highlight the importance of targeted monitoring and intervention strategies.

Farm Level and Pond Level Variability

Clear inter-farm differences in bacterial diversity were observed. Turbo Farm recorded the highest Shannon diversity index (1.94), indicative of a rich and evenly distributed bacterial community, while Kesses Farm A exhibited the lowest diversity (1.67). These patterns likely reflect divergent management practices, such as differences in feed composition, stocking densities, and water exchange regimes, which are known to influence microbial complexity [30]. Elevated diversity at Turbo Farm may be attributable to increased organic loads or less effective disinfection practices, both of which favor microbial proliferation. Within farms, intra-pond differences were also detected. For example, in Kesses Farm B, Pond 1 exhibited slightly higher bacterial diversity (1.76) than Pond 2 (1.75), while in

Ainapkoi, diversity ranged from 1.69 to 1.72 across ponds. Such fine-scale variability is consistent with the findings of Horn et al. [31], who reported that even subtle environmental differences, including shading, aeration efficiency, and sedimentation dynamics, can alter microbial community structure within aquaculture ponds.

Pairwise comparisons further revealed significant differences in bacterial infestation patterns across farms and organs. For example, *P. aeruginosa* was significantly more prevalent in kidneys at Turbo Farm compared to all other farms ($p < 0.001$). Similarly, *V. alginolyticus* showed higher gill infestation in Turbo Farm relative to Ainapkoi, Kesses A, Kesses B, and Moiben Farms ($p < 0.001$). In the intestines, multiple species displayed significant variation, including higher *P. aeruginosa* levels at Turbo and distinct differences in *F. columnare* and *V. alginolyticus* distributions among farms. Skin tissues showed the greatest variability, with all five analyzed species exhibiting significant inter-farm differences. These findings suggest that specific farm environments strongly mediate bacterial colonization pressures, likely through combined effects of water quality, management practices, and local ecological conditions.

Facility Type Distribution and Diversity

Facility type also influenced bacterial diversity. Liner ponds recorded the highest Shannon index (1.82), followed by earthen ponds (1.73), with tanks showing the lowest diversity (1.69). The richer bacterial assemblages in liner and earthen systems may reflect greater exposure to environmental inputs such as soil, runoff, and rainwater, as well as more dynamic ecological interactions [28, 32]. By contrast, tanks, with their more controlled environments and limited organic inputs, supported comparatively simplified microbial populations. These findings align with previous research demonstrating that recirculating aquaculture systems often harbor lower alpha diversity compared to semi-open or open systems [33].

Seasonal and Temporal Variation

Seasonal variation exerted a significant effect on bacterial dynamics. Bacterial loads peaked during August and April, with additional increases observed in September and November across all facility types. These peaks correspond to warmer months, when elevated water temperatures accelerate microbial metabolism, enzymatic activity, and organic matter decomposition [34]. Conversely, bacterial presence was lowest during May and June, suggesting that cooler conditions suppress microbial growth. Notably, gill tissues consistently exhibited the highest bacterial loads during warmer periods (Fig. 3). This trend was most pronounced in Turbo Farm, where gill colonization remained elevated from August through April. These findings support the work of Consuegra et al. [35], who highlighted the gills' large surface area and continuous water contact as key factors predisposing them to bacterial adherence and colonization.

4.3 Bacterial Species Co-occurrence in Organs Across Farms, Facility Types, and Seasonal Patterns

Organ-Level Co-occurrence

Bacterial species co-occurrence displayed strong organ-specific patterns. The gills supported the most complex associations, frequently involving *E. coli, P. aeruginosa,* and *Vibrio* spp. This complexity reflects the gills' continuous interaction with the external environment, making them highly susceptible to multi-species colonization. Comparable interactions among these bacteria have been identified in aquaculture systems with limited water management and poor biosecurity [36]. The intestines exhibited the highest overall diversity, with recurring co-infections involving *E. coli, P. aeruginosa, V. alginolyticus,* and *Flav. columnare.* This highlights the gut as a major microbial reservoir where nutrient availability and organic matter accumulation facilitate species interactions. Previous studies similarly identified intestinal microbial complexity as a marker of suboptimal water quality and nutrient loading [31, 37]. The kidneys exhibited the least complex bacterial associations, being dominated by *E. coli* across most farms, with *P. aeruginosa* largely confined to Turbo Farm. The site-specific presence of *P. aeruginosa* in kidneys likely reflects environmental and management conditions, such as reduced water exchange and organic matter accumulation, which can create favorable niches for this opportunistic pathogen [33].

Farm-Level, Facility-Type and Seasonal Variation Effects on Bacterial Co-occurrence

Marked differences were observed between farms. Some farms exhibited relatively simple microbial associations, whereas others displayed complex multi-species co-infections, particularly in skin and intestinal tissues. Sites characterized by greater organic loading or weaker biosecurity, such as those with limited water treatment, were associated with more complex co-occurrence networks. These patterns support evidence that farm-level practices strongly influence microbial diversity and the likelihood of pathogen interaction [38]. The type of facility significantly influenced co-occurrence dynamics. Earthen ponds supported complex microbial interactions, particularly in gill and intestinal tissues, reflecting the contribution of soil contact and organic matter accumulation [28, 39]. Liner ponds displayed higher frequencies of multi-species associations, especially combinations involving *E. coli* and *Pseudomonas spp.*, likely due to reduced water circulation and debris retention. In contrast, tank systems exhibited relatively simple bacterial profiles, usually dominated by *E. coli,* consistent with findings that closed, controlled systems suppress diversity but allow persistence of hardy pathogens [40].

Seasonal changes introduced an additional layer of variation in bacterial interactions. Opportunistic pathogens such as *P. aeruginosa* and *V. harveyi* increased in prevalence during warmer months, when higher temperatures and nutrient enrichment favor microbial proliferation. In contrast, *E. coli* maintained consistent presence throughout the year, reflecting its broad tolerance across environmental gradients. Seasonal dynamics of this kind have been linked to fluctuations in temperature, rainfall, and dissolved oxygen in aquaculture ecosystems [28, 41].

5 Conclusion

This study provides comprehensive evidence of the bacterial diversity present in farmed fish species, with a focus on *Oreochromis niloticus* and *Clarias gariepinus* in Uasin Gishu County, Kenya. The findings demonstrate that *Escherichia coli* and *Pseudomonas aeruginosa* were the most dominant bacterial isolates. Importantly, prevalence varied significantly across fish organs, with the gills and intestines consistently serving as key sites of infection. Temporal differences were also evident, with the highest bacterial loads occurring during warmer months (January and March), suggesting that seasonal environmental factors play a major role in disease dynamics. The observation that earthen ponds recorded higher microbial and parasitic loads compared to liner ponds highlights the influence of pond type and management practices on fish health. Additionally, the frequent co-occurrence of multiple bacterial and parasitic pathogens points to synergistic interactions that may amplify disease risks, compromise fish welfare, and ultimately reduce aquaculture productivity. From a practical perspective, these findings emphasize the urgent need for enhanced biosecurity measures, improved water quality management, and regular fish health monitoring in aquaculture systems. Targeted interventions, such as pond disinfection, disease control programs, and farmer training on hygienic practices, could significantly reduce the burden of infections. Furthermore, the identification of zoonotic pathogens such as *E. coli* underscores the potential public health risks associated with consumption of improperly handled or undercooked fish, highlighting the importance of food safety interventions.

Acknowledgements We sincerely thank the University of Eldoret for providing the necessary resources and facilities that enabled us to carry out this study effectively. We are also grateful to the Eldowas Laboratory, Ol'lessos National Polytechnic, and the Microbiology Technicians for their support and cooperation during the study. Our appreciation extends to the Uasin Gishu County Fisheries Department for their assistance and valuable contributions. We further thank the farmers whose farms were used in this study for their participation and willingness to support our research efforts.

References

1. FAO (2022) The state of world fisheries and aquaculture 2022: towards blue transformation. Food and Agriculture Organization of the United Nations, Rome
2. Pradeepkiran JA (2019) Aquaculture role in global food security with nutritional value: a review. Transl Anim Sci 3(2):903–910

3. Edwards P, Zhang W, Belton B, Little DC (2019) Misunderstandings, myths and mantras in aquaculture: its contribution to world food supplies has been systematically over-reported. Mar Policy 106:103547
4. Makori AJ, Abuom PO, Kapiyo R, Anyona DN, Dida GO (2017) Effects of water physico-chemical parameters on tilapia (*Oreochromis niloticus*) growth in earthen ponds in Teso North Sub-County Busia County. Fish Aquat Sci 20(1):1–10
5. Munguti JM, Mboya JB, Iteba JO, Kirimi JG, Obiero KO, Kyule DN et al (2024) Status and prospects of the ornamental fish industry in Kenya. Aquac Fish Fish 4(3):147–158
6. Wanja DW, Mbuthia PG, Waruiru RM, Mwadime JM, Bebora LC, Nyaga PN et al (2020) Fish husbandry practices and water quality in central Kenya: potential risk factors for fish mortality and infectious diseases. Vet Med Int 2020:6839354
7. Jacobs L, Chenia HY (2007) Characterization of integrons and tetracycline resistance determinants in *Aeromonas* spp. isolated from South African aquaculture systems. Int J Food Microbiol 114(3):295–306
8. Hussein MA, Eissa AE, El-Tarabili RM, Attia ASA, Zaki MM, Ibrahim TB et al (2024) Impact of climate change on some seasonal bacterial eruptions among cultured marine fishes from Egyptian coastal provinces. J Appl Vet Sci 9(2):18–30
9. Basweti EA, Turk PJ, Rayburn EB, Bryan WB (2009) No-till sequential cropping of summer and fall annual forage species compared with grassland. Agron J 101(6):1497–1502
10. Petrikova I, Otieno MA, Were G, Eltholth M, Mateos AR, Harding S et al (2024) Peri-urban agriculture and household food and nutrition security around Eldoret Kenya. Food Secur 16:1235–1263
11. Njonge T (2023) Influence of psychological well-being and school factors on delinquency during the COVID-19 period among secondary school students in selected schools in Nakuru County, Kenya. Int J Res Innov Soc Sci 7(7):1175–1189
12. Uasin Gishu County Government (2015) Livestock and fisheries. County Government of Uasin Gishu, Eldoret. Available from: https://uasingishu.go.ke/index.php/tag/livestock-and-fisheries
13. Mwangi J, Maina J, Gachuiri C, Wahome R, Waruiru R (2024) Management practices, farmers' perceptions and coping strategies to climate change in integrated fish farms in Kenya. East Afr J Sci Technol Innov 5(4):1–12
14. Froese R (2006) Cube law, condition factor and weight–length relationships: history, meta-analysis and recommendations. J Appl Ichthyol 22(4):241–253
15. Holt JG (ed) (1994) Bergey's manual of determinative bacteriology, 9th edn. Williams & Wilkins, Baltimore
16. Nolan K, Callan J (2006) Beachcomber Biology: the Shannon-Weiner species diversity index. ABLE Proc 27:334–338
17. Bush AO, Lafferty KD, Lotz JM, Shostak AW (1997) Parasitology meets ecology on its own terms: Margolis et al. revisited. J Parasitol 83(4):575–583
18. Ferreira T, Akter R, Hasan N, Reza F, Asaduzzaman M, Begum K et al (2023) Hydrobiology of saline agriculture ecosystem: a review of scenario change in South-West region of Bangladesh. Hydrobiology 1(1):1–15
19. Onyango J, Kipkorir N, van Bruggen J, Schmitz J (2024) Agricultural intensification in Lake Naivasha catchment in Kenya and associated nutrient and pesticide pollution. Sci Rep 14:18539
20. Naylor RL, Hardy RW, Buschmann AH, Bush SR, Cao L, Klinger DH et al (2021) A 20-year retrospective review of global aquaculture. Nature 591(7848):551–563
21. Akoll P, Mwanja W (2012) Fish health status, research and management in East Africa: past and present. J Appl Aquac 24(2):117–129
22. Xu W, Li W, Liu WC, Ma Y, Liao CY, Peng JJ et al (2021) Antimicrobial resistance of *Escherichia coli* from aquaculture farms and their environment in Zhanjiang, China. Front Vet Sci 8:806653
23. Igbinosa IH, Beshiru A, Igbinosa EO (2017) Antibiotic resistance profile of *Pseudomonas aeruginosa* isolated from aquaculture and abattoir environments in urban communities. Asian Pac J Trop Dis 7(1):47–52

24. Magouz FI, Moustafa EM, Abo-Remela EM, Halawa MR, Barakaat PM, Omar AA (2024) Summer mortality syndrome bacterial pathogens in farmed Nile tilapia (*Oreochromis niloticus*). Open Vet J 14(1):53–69
25. Shafiee NS, Mazlan N, Rahman NNA, Shapawi R, Shah MD (2024) Vibriosis in aquaculture: Pathogenic *Vibrio* species, affected animals, diagnosis and adverse effects. In: Advances in aquaculture health management, pp 113–138. Springer, Singapore
26. Duchaud E, Rochat T, Habib C, Barbier P, Loux V, Guérin C et al (2018) Genomic diversity and evolution of the fish pathogen *Flavobacterium psychrophilum*. Front Microbiol 9:138
27. Conrad RA, Evenhuis JP, Lipscomb RS, Birkett C, McBride MJ (2022) Siderophores produced by the fish pathogen *Flavobacterium columnare* strain MS-FC-4 are not essential for its virulence. Appl Environ Microbiol 88(17):e00854-e922
28. Kariuki S, Wairimu C, Mbae C (2021) Antimicrobial resistance in endemic enteric infections in Kenya and the region, and efforts toward addressing the challenges. J Infect Dis 224(S7):S883–S892
29. Zhang X, You Y, Peng F, Tang X, Zhou Y, Liu J et al (2022) Interaction of microbiota between fish and the environment of an in-pond raceway system in a lake. Microorganisms 10(6):1130
30. Debnath SC, McMurtrie J, Temperton B, Delamare-Deboutteville J, Mohan CV, Tyler CR (2023) Tilapia aquaculture, emerging diseases, and the roles of the skin microbiomes in health and disease. Rev Aquac 31(5):2945–2976
31. Horn M, Segner H, Buchtíková Š, Fuchs B, Schirmer K (2023) Chemically induced trout model of acute intestinal inflammation using TNBS. Fish Shellfish Immunol 136:108750
32. Wolińska A, Kruczyńska A, Grządziel J, Gałązka A, Marzec-Grządziel A, Szałaj K et al (2022) Functional and seasonal changes in the structure of microbiome inhabiting bottom sediments of a pond intended for ecological king carp farming. Biology (Basel) 11(6):900
33. Nsabimana A, Banzi W, Habimana V, Uwamariya C, Mutangana D (2021) Morphometric characteristics of four important fish species from the Rwandan side of Lake Kivu. Ethiop J Sci Technol 14(3):219–228
34. Jiang X, Wang J, Tang W, Sun Z, Pan B (2021) Non-native freshwater fish species in the Yellow River Basin: origin, distribution and potential risk. Environ Biol Fishes 104(3):253–264
35. Consuegra S, Webster TU, Anka I (2023) Microbiome, epigenetics and fish health interactions in aquaculture. In: Epigenetics in aquaculture, pp 245–262. Wiley, Hoboken
36. Lian Y, Zheng X, Xie S, Dan A, Wang J, Tang J et al (2023) Microbiota composition and correlations with environmental factors in grass carp (*Ctenopharyngodon idella*) culture ponds in South China. PeerJ 11:e15758
37. Clols-Fuentes J, Nguinkal JA, Unger P, Kreikemeyer B, Palm HW (2023) Bacterial community in African catfish (*Clarias gariepinus*) recirculating aquaculture systems under different stocking densities. Front Mar Sci 10:1122334
38. Opiyo MA, Marijani E, Muendo P, Odede R, Leschen W, Charo-Karisa H (2018) A review of aquaculture production and health management practices of farmed fish in Kenya. Int J Vet Sci Med 6(2):141–148
39. Kasozi N, Iwe GD, Langi S, Namulawa VT, Walakira J (2024) Histological features of the gastrointestinal tract of elongate tigerfish (*Hydrocynus forskahlii*) from Lake Albert. J Basic Appl Zool 85(1):11
40. Mia MJ, Mahmud R, Sadad MS, Asad HA, Hossain R (2022) An in-depth automated approach for fish disease recognition. J King Saud Univ Comput Inf Sci 34(9):7174–7183
41. Abd-ELrahman SM, Gareh A, Mohamed HI, Alrashdi BM, Dyab AK, El-Khadragy MF et al (2023) Prevalence and morphological investigation of parasitic infection in freshwater fish (*Nile tilapia*) from Upper Egypt. Animals 13(6):987

Restoration of Ecosystems and Mitigation Strategies

Status of Forest Landscape Restoration in Kenya

Benjamin Mutuku Kinyili

Abstract

Introduction Kenya's National Landscape and Ecosystem Restoration Strategy (2023–2032) targets 15 billion trees to restore ~ 10.6 million ha and lift tree cover to 30% by 2032, operationalized through the Forest Landscape Restoration (FLR) Implementation Action Plan (2022–2026) prioritizing 2.55 million ha. This review addresses a clear need to synthesize Kenya-specific evidence on restoration objectives, current initiatives and trends, and indicative economics of FLR options to inform implementation quality and progress reporting. *Methods* We synthesized peer-reviewed and gray literature, national policy and agency documents, and project practice notes related to FLR in Kenya. Inclusion emphasized Kenya-relevant programs with verifiable descriptions of objectives, actors, geographies, and (where available) progress/status. We mapped flagship initiatives (e.g., coastal mangroves, Mau, Mt. Kenya, Kakamega, Mukogodo, Makuli–Nzaui) and summarized economic signals drawing on published Kenyan analyses of benefits and costs of restoration interventions. *Results* Four objective clusters dominate Kenya's FLR: ecological (biodiversity, genetic resources, tree-cover recovery), socio-ecological (water security, soil stabilization), livelihood/economic (forest products, agroforestry), and climate (mitigation/carbon). Notable initiatives include a ~ 154,000 ha coastal mangrove/inland coastal forest program and community- or partner-led actions across Mau, Mt. Kenya, Kakamega, Mukogodo, and Makuli–Nzaui. However, status reporting is uneven (targets commonly stated; realized progress

Theme: Restoration of Ecosystems
Sub-Theme: Restoration of natural landscapes, to include mining areas, water catchment areas

B. M. Kinyili (✉)
Kenya Forest Service, Nairobi, Kenya

R. Ramkat et al. (eds.), *Towards the Sustainability of our Ecosystems*, Springer Proceedings in Geography and Sustainability,
https://doi.org/10.1007/978-3-032-30076-8_5

variably documented). Indicative economics consistently highlight comparatively strong performance for intensive agroforestry (e.g., Melia systems), Eucalyptus woodlots, commercial bamboo, and Grevillea/fruit-tree agroforestry, with riparian bamboo/grass strips attractive where erosion control and water quality co-benefits are valued. *Conclusion* Kenya's restoration drive is broad and accelerating, but its evidentiary base would benefit from standardized "status-to-date versus target" metrics, consistent sourcing/citation of program data, and use/permissioning of adapted figures. Priorities include: (i) tightening Methods transparency for stakeholder/gray sources, (ii) reframing tables to distinguish targets from verified progress, and (iii) embedding more critical analysis of barriers/enablers in Findings/Discussion to support policy and operational decision-making.

Keywords

Forest degradation • Forest restoration • Forest conservation • Kenya

1 Introduction

Forests perform vital ecological and socioeconomic functions by providing goods and services essential to human well-being and planetary health. Globally, they support livelihoods through the provision of timber, non-timber forest products, food, fuelwood, medicine, and construction materials [11, 32]. Beyond these provisioning services, forests regulate the Earth's climate through carbon sequestration and storage, stabilize soils, control erosion, and moderate hydrological cycles, thereby contributing critical regulating services [10, 33]. They also sustain supporting services, such as nutrient cycling, pollination, and soil formation, which maintain ecosystem productivity and biodiversity [23, 29]. Moreover, forests provide cultural services, including recreation, esthetic enjoyment, traditional knowledge, and spiritual values, especially among indigenous and local communities who regard forest landscapes as sacred or culturally significant [1, 12].

Despite these multiple benefits, global forest degradation remains widespread. According to the FAO's Global Forest Resources Assessment [10], the world continues to lose forests through unsustainable logging, wildfires, overgrazing, agricultural expansion, urbanization, and infrastructure development [27, 29]. It is estimated that approximately 420 million ha of forest have been lost globally since 1990 due to human-induced land-use changes [30]. Such degradation undermines climate stability, biodiversity, and human livelihoods, particularly in tropical regions where deforestation rates are accelerating [14].

In response, several international frameworks have been launched to promote forest landscape restoration (FLR). The Bonn Challenge, initiated in 2011, aims to restore 350 million ha of degraded and deforested landscapes by 2030 (http://www.bonnchallenge.org/). The United Nations Decade on Ecosystem Restoration (2021–2030) reinforces this commitment by fostering large-scale restoration efforts to achieve the Sustainable Development Goals (SDGs) (http://www.decadeonrestoration.org/). Similarly, the World Economic Forum's Trillion Trees Initiative (1t.org) consolidates global action to mobilize public and private partnerships for massive tree planting and restoration. The overarching objectives of these initiatives include climate change mitigation, biodiversity conservation, ecosystem service recovery, and enhanced rural livelihoods [33].

In Kenya, forest degradation remains a pressing environmental challenge. An estimated 22% of forested land is degraded, driven by agricultural encroachment, overgrazing, unsustainable land-use practices, and climate-related pressures [14, 31]. The economic cost of forest degradation is substantial, with annual losses estimated at USD 1.3 billion [4]. Degradation threatens the livelihoods of communities dependent on forests for food, fodder, and fuel, while also diminishing tourism and other nature-based economic sectors (Cheboiwo et al., [3, 22]). To counter these losses, the Government of Kenya has introduced ambitious restoration programs. The National Landscape and Ecosystem Restoration Strategy (2023–2032) targets the rehabilitation of 10.6 million ha of degraded land and aims to achieve 30% national tree cover through the planting of 15 billion trees by 2032 [5]. Implementation is guided by the Forest Landscape Restoration Implementation Action Plan (2022–2026), which prioritizes 2.55 million ha for restoration through multi-stakeholder partnerships led by the Kenya Forest Service (KFS), the Council of Governors, FAO, and World Agroforestry (ICRAF) [28]. However, available information on restoration progress in Kenya remains scanty and fragmented, limiting the ability to assess effectiveness and economic efficiency.

This paper therefore aims to consolidate existing evidence on forest landscape restoration in Kenya by addressing three focal areas: (i) the objectives of forest restoration, (ii) current trends and ongoing initiatives, and (iii) economic analysis of FLR options. The synthesis provides a baseline understanding of Kenya's restoration progress and highlights emerging gaps that must be addressed to accelerate the achievement of national tree-cover targets.

2 Methodology

This study is a narrative review of Forest Landscape Restoration (FLR) in Kenya, structured around three lenses: (i) restoration objectives, (ii) current initiatives and trends, and (iii) indicative economics of restoration options. Evidence was sourced from 2015 to 2025 to align with contemporary Kenyan policy cycles. We queried Google Scholar, Scopus, and Web of Science, and accessed institutional portals and repositories of Kenya Forest Service (KFS), FAO, ICRAF/World Agroforestry, the Council of Governors, county environment/forestry departments, and major implementers. Searches combined "Kenya" with terms, including *forest landscape restoration*, *mangrove restoration*, *assisted natural regeneration*, *agroforestry*, *tree cover*, *restoration economics*, *benefit–cost*, and *ROAM*, and were augmented by reference-list snowballing.

Inclusion required Kenya-specific FLR content reporting verifiable details on at least one of: objectives, actors, geography, targets and/or status; exclusion applied to purely conceptual pieces without Kenya application and to duplicate versions (most recent complete version retained). From eligible records, we extracted: objective(s); location/landscape; implementing agency/partners; targeted area and, where available, status-to-date; timeframe; and source/year. For economic content, we captured where reported: cost and benefit categories (market/non-market), units (KES ha^{-1}), valuation approach, price year/base year, discount rate, and time horizon, drawing in particular on Kenya's ROAM-based assessments and national valuation work (e.g., Cheboiwo et al. [3]).

Synthesis was thematic and narrative. Objectives were grouped into ecological, socio-ecological, livelihood/economic, and climate clusters; initiatives were organized by major landscapes (e.g., coastal mangroves; Mau, Mt. Kenya, Kakamega; Mukogodo; Makuli–Nzaui).

3 Results and Discussion

3.1 Objectives of Forest Restoration

Forest Landscape Restoration (FLR) initiatives in Kenya pursue four broad objectives: ecological, socio-ecological, livelihood/economic, and climate-related goals (Table 1). Ecological restoration remains central, reflected in biodiversity conservation and species-specific planting programs. For instance, the Botanic Gardens Conservation International (BGCI) and the International Tree Foundation (ITF) target 20 million trees for biodiversity conservation across Kenya. Similarly, the Kenya Forest Service (KFS) supports national forest management and agroforestry schemes aimed at conserving endemic species and genetic resources. At the coastal frontier, the Forests and Landscape Restoration Programme (FOLAREP) has initiated restoration of roughly **226,000 ha** of degraded Taita-Taveta landscapes, while the Kenya Breweries Limited (KBL)–Nature Kenya partnership has restored about **100 ha** of Mt. Kenya forest with indigenous trees and bamboo, demonstrating growing private sector participation.

Increasing tree cover has been a recurrent ecological and socioeconomic priority. In the Mau Complex, the Sustainable Management of Mau Watershed (SUMAWA) project rehabilitated Sururu and Likia forest blocks, while the Kenya Energy and Environment Programme (KEEP) restocked 56 ha of the degraded Saino site with mixed indigenous species. The Korea International Cooperation Agency (KOICA) and World Vision Kenya have implemented community-based restoration in Kajiado and Narok, contributing to dryland forest recovery, whereas FAO and GEF-6 projects have advanced rehabilitation in Kirisia, Mt. Kulal and Mukogodo landscapes. Collectively these programs illustrate the integration of

Table 1 Some objectives of forest landscape restoration in Kenya

Objectives of restoration	Specific objectives of forest restoration	Forest involved	Organization involved	Activity and indicators	References (source/year)
Ecological restoration	Biodiversity conservation	Kenyan forests	Botanic Gardens Conservation International (BGCI) and International Tree Foundation (ITF)	ITF's centenary campaign aims to plant 20 million trees in Kenya for tree biodiversity conservation	BGCI/ITF [2]. Accessed 2025
	Conserving endemic plants and genetic pool	Kenyan forests	Kenya Forest Service (KFS)	Kenya forest management and agroforestry programs	KFS (n.d.). Accessed 2025
	Ecosystem resilience	Taita-Taveta (Coastal Forests)	FOLAREP	Restore 226,420 ha of degraded landscapes	Chisika and Yeom [5]
	Ecosystem resilience	Mt. Kenya Forests	Kenya Breweries Limited (KBL), Nature Kenya and local CFAs	Target of 100,000 seedlings to restore ~250 acres of degraded forest	KBL/Nature Kenya [16]. Accessed 2025
	Increasing tree cover	Mau Forest	Sustainable Management of Mau Watershed (SUMAWA)	Reforestation of Sururu/ Likia forest block	SUMAWA [34]. Accessed 2025
	Increasing tree cover	Mau Forest	Kenya Energy and Environment and Social Responsibility Program (KEEP)	Rehabilitation of ~56.5 ha (Saino) with mixed indigenous trees and bamboo	KEEP [18]. Accessed 2025
	Increasing tree cover	Forests in Kajiado and Narok	KOICA & World Vision Kenya	Landscape restoration around hills bordering these counties	KOICA/WV Kenya [21]. Accessed 2025
	Increasing tree cover	Kirisia, Mt. Kulal and Mukogodo landscapes	FAO (GEF-6)	Restoration projects in Kirisia Forest; GEF-6 actions in Mt. Kulal and Mukogodo	FAO/GEF-6 [9]. Accessed 2025

(continued)

Table 1 (continued)

Objectives of restoration	Specific objectives of forest restoration	Forest involved	Organization involved	Activity and indicators	References (source/year)
	Increasing tree cover	Kenyan forests	Conservation International	Community-based restoration targeting up to 4.5 million ha	Conservation International [6]. Accessed 2025
Ecological and socioeconomic objectives	Water protection and soil stabilization	All county forests	County governments and KFS	Water and soil conservation and management	County/KFS [7]. Accessed 2025
	Water protection and soil stabilization	All county forests	International tree foundation (ITF) with KFS, KEFRI & CFAs	Partnership projects with CFAs for watershed protection	ITF/KEFRI/KFS [13]. Accessed 2025
Socioeconomic objectives	Income generation	Kenyan forests	Kenya Forest Service (KFS)	Revenue from sale of forest products (poles, fuelwood, NTFPs)	KFS [19]. Accessed 2025
	Improved agriculture and forestry	Kenyan forests adjacent areas	KFS	Agroforestry and integration of trees into farms	KFS (n.d.). Accessed 2025
	Generation of electricity	Kenyan forest and waterbodies	Kenya power and lighting company (KPLC)	Forest conservation for sustainable hydropower generation	KPLC [20]. Accessed 2025
	Cultural values	Kaya forests (coast)	KFS and local community initiatives (Mijikenda)	Conservation of sacred Kaya forests	Community Plans (n.d.). Accessed 2025
Environmental restoration objective	Climate change mitigation	National (multi-county)	National Environment Management Authority (NEMA)–KCCAP	Water management, food security and DRR actions	NEMA/KCCAP [17]. Accessed 2025

(continued)

Table 1 (continued)

Objectives of restoration	Specific objectives of forest restoration	Forest involved	Organization involved	Activity and indicators	References (source/year)
	Climate change mitigation	National (multi-county)	ACT change transform (ACT), KCCWG and Oxfam	Changieni Rasilimali project—Strengthen climate responses	ACT/KCCWG/Oxfam (n.d.). Accessed 2025
	Carbon sequestration	Northern rangelands	Northern Rangelands Trust (NRT) with 14 conservancies	NKRCP aims to remove ~50 Mt. CO2 over 30 years	NRT/NKRCP [26]. Accessed 2025
	Carbon sequestration	Western and Nyanza regions	Vi Agroforestry, World Bank BioCarbon Fund & UNIQUE	Kenya agricultural carbon project (KACP) for soil/tree carbon	Vi Agroforestry/World Bank [36]. Accessed 2025
	Carbon sequestration	Gazi Bay (coast)	Mikoko Pamoja Community Organisation (MPCO)	Community-led mangrove conservation and restoration with carbon credits	Mikoko Pamoja [24]. Accessed 2025

county governments, development partners, and civil society in achieving tree-cover gains, though quantitative tracking of survival rates remains inconsistent.

Socioeconomic and ecological co-benefits are also evident. Initiatives under the ITF and county governments have combined water and soil protection with livelihood support, particularly in upper-catchment counties where forest degradation affects downstream water supply. KFS-led agroforestry has improved smallholder income through sale of poles, fuelwood, and fruit, while the Kenya Power and Lighting Company (KPLC) supports riparian forest protection to sustain hydroelectric generation. Cultural objectives, notably the conservation of Kaya Forests by Mijikenda communities at the coast, underline the role of indigenous knowledge in restoration governance.

Environmental and climate-related targets emphasize carbon sequestration and climate change mitigation. The Kenya Climate Change Adaptation Programme (KCCAP) under NEMA promotes integrated watershed management and food-security resilience, while community-based projects such as *Mikoko Pamoja* link mangrove conservation to verified carbon credit schemes. At a national scale, the Northern Kenya Rangelands Carbon Project (NKRCP) coordinated by the Northern Rangelands Trust and 14 community conservancies aims to sequester about 50 million tons CO_2 over 30 years through improved grazing management. The Kenya Agricultural Carbon Project (KACP), led by Vi Agroforestry and the World Bank BioCarbon Fund, applies sustainable land-use practices in Western Kenya to integrate soil and tree carbon gains.

Overall, these initiatives reveal a multi-actor restoration mosaic state agencies driving policy, county governments localizing implementation, private enterprises financing niche interventions, and communities anchoring stewardship. However, uneven monitoring frameworks and weak integration of economic incentives constrain evaluation of cumulative progress.

3.2 Trends in Forest Restoration in Kenya

Analysis of current restoration projects (Table 2) shows strong momentum in coastal and highland ecosystems but variable progress across regions. The Mangrove Restoration Programme along the Kenyan coast targets about 154,000 ha, addressing deforestation from fuelwood extraction and salt-pond conversion; early assessments indicate partial recovery of canopy density but continued vulnerability to illegal harvesting [8]. At Gazi Bay, the *Mikoko Pamoja* community project engages roughly 5400 residents, demonstrating successful integration of local livelihood benefits with carbon credit financing.

Inland, the Mau Forest project led by *Kass FM* and community groups covers approximately 5700 ha, representing one of the few media-driven restoration models. Mt. Kenya Trust initiatives aim to plant over two million trees, focusing on wildlife corridors and water source protection, although grazing encroachment and fire risk persist as constraints. In the Kakamega Forest, ITF-led efforts have rehabilitated degraded blocks through enrichment planting of indigenous species and

Table 2 Ongoing forest restoration projects in Kenya (with Source/Year)

Forest	Restoration status	References (source/year)
Mangrove forests (coastal region)	A 154,000 ha landscape restoration initiative addressing drivers of degradation in Kenya's largest contiguous mangrove and inland coastal forests.	Eden [8]
Gazi mangrove forests	Community-led project engaging ~ 5400 residents to restore mangroves, conserve biodiversity, and enhance livelihoods.	Mikoko Pamoja/MPCO [24]. Accessed 2025
Mau Forest	Community initiative led by Kass FM targeting ~5700 ha of restoration.	Kass FM/community project brief [15]. Accessed 2025
Mt. Kenya	Multiple initiatives; Mt. Kenya trust aims to plant > two million trees; pressures include timber extraction, encroachment, and grazing.	Mt. Kenya Trust [25]. Accessed 2025
Kakamega Forest	ITF project to rehabilitate degraded Illoro block via enrichment planting of indigenous species and community protection efforts.	International Tree Foundation [13]. Accessed 2025
Mukogodo Forest Reserve	TRI Kenya ASAL program has mobilized >8600 people for assisted natural regeneration within the reserve.	TRI Kenya/ASAL program [35]. Accessed 2025
Makuli–Nzaui landscape (Makueni County)	County-led initiative spanning ~ 75,000 ha to improve forest landscape and dryland livelihoods.	County Government of Makueni [7]. Accessed 2025

participatory protection schemes, yielding visible canopy regeneration. For arid and semi-arid zones, the TRI Kenya ASAL program has mobilized more than 8600 residents in assisted natural regeneration within the Mukogodo Forest Reserve, while the Makuli–Nzaui initiative in Makueni County spans about 75,000 ha, integrating tree-cover improvement with dryland livelihood diversification.

Despite these gains, restoration data remain uneven. Projects generally report targets rather than verified status-to-date, and survival monitoring is limited. Major enablers include high-level political commitment under the 15-Billion-Trees Campaign, donor support through FAO and ICRAF, and devolution of environmental functions to counties. Key barriers include fragmented funding, inadequate technical capacity for seedling production and after-care, and limited incorporation of indigenous species suited to site-specific ecologies. Comparatively, Kenya's progress aligns with East African regional trajectories but lags in consistent monitoring frameworks. Strengthening data integration across agencies would allow clearer attribution of national gains toward the 30% tree-cover target.

3.3 The Economic Analysis of Forest Landscape Restoration (FLR) in Kenya

Economic evaluations, largely derived from Kenya Forest Service studies under the Restoration Opportunities Assessment Methodology (ROAM), reveal significant variation in benefit–cost ratios among restoration options (Fig. 1). When discounted to a common price year 2018 at 10% discount rate (Cheboiwo et al., [3]), intensive agroforestry with *Melia volkensii*, Eucalyptus woodlots, and commercial bamboo exhibit the highest net benefits per hectare. Agroforestry systems combining *Grevillea robusta* and fruit trees provide strong household income streams, while riparian bamboo and grass strips yield additional non-market benefits such as erosion control and water quality protection.

Costlier interventions such as enrichment planting in severely degraded natural forests or roadside planting generate slower returns but higher ecological value. Overall, restoration with commercial and multipurpose species offers short to medium-term financial attractiveness, whereas indigenous reforestation provides long-term ecological dividends. Policy integration of both approaches remains essential to balance livelihood incentives with ecosystem integrity.

Kenya's FLR efforts exhibit broad participation and expanding geographic coverage, yet gaps persist in progress verification, data consistency, and economic

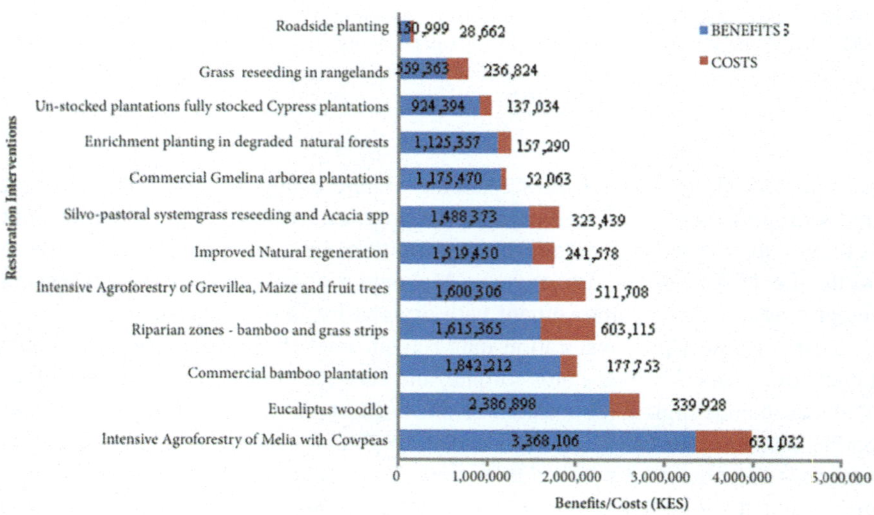

Fig. 1 Discounted cost and benefit of forest ecosystem restoration in Kenya. *Source* [3]

monitoring. Future reporting should couple **quantitative status updates** with **qualitative insights** on barriers and enabling conditions, enabling evidence-based scaling of interventions under the National Landscape and Ecosystem Restoration Strategy (2023–2032).

4 Conclusions

This review set out to (i) clarify the objectives driving forest landscape restoration (FLR) in Kenya, (ii) describe current initiatives and trends, and (iii) summarize indicative economics of restoration options. Across sources, Kenya's FLR agenda coheres around four complementary aims: ecological recovery (biodiversity, genetic resources, tree-cover gains), socio-ecological services (water and soil protection), livelihood/economic benefits (agroforestry and forest products), and climate mitigation (biomass and soil carbon). These aims are not competing; rather, they delineate a multi-benefit portfolio that guides national and county programming.

On implementation trends, activity is broad and multi-actor national agencies (KFS, NEMA), county governments, community organizations (e.g., CFAs, Mikoko Pamoja), and private/NGO partners (e.g., ITF, KOICA, Mt. Kenya Trust) are all engaged across diverse landscapes (mangroves/coast; highland blocks such as Mau, Mt. Kenya and Kakamega; and ASAL systems like Mukogodo and Makuli–Nzaui). However, documentation is uneven: many initiatives report targets but lack consistent, verified status-to-date (hectares restored, survival, after-care), making it difficult to aggregate progress toward national tree-cover commitments.

On the economics, Kenyan assessments indicate comparatively strong benefit–cost performance for intensive agroforestry (e.g., *Melia* systems), Eucalyptus woodlots, commercial bamboo, and Grevillea/fruit-tree agroforestry, while riparian bamboo/grass buffers add non-market co-benefits (erosion control, water quality) valued by catchment managers. These options can underwrite household and county incentives in the near to medium term, while indigenous enrichment and natural forest recovery secure long-term ecological integrity implying a blended portfolio rather than a single "winner."

Kenya's FLR is on-mission but evidence-thin in places. Three practical priorities follow. First, standardize monitoring to report status-to-date versus targets (area, survival, species mix, after-care) using uniform metadata (source/year, price year for costs). Second, align funding and extension to scale high-performing economic options where suitable (agroforestry, bamboo, woodlots), while reserving adequate finance and technical support for indigenous restoration in high-value ecosystems. Third, require that program tables and figures in reports clearly cite Source/Year and distinguish targets from verified progress, enabling transparent roll-up to national dashboards. By linking objectives to verifiable implementation and costed options, the sector can shift from planting-counts to restoration performance, accelerating delivery against the National Landscape and Ecosystem Restoration Strategy (2023–2032) and associated FLR Action Plan.

5 Recommendations

There is need to bring in more stakeholders and partnership between the government and private sector to enhance forest restoration efforts. More volunteers should also come up through incentives by KFS to ensure large numbers of stakeholders are involved in the restoration efforts.

There are a number of initiatives and projects that are currently working with other partners in Kenya and internationally to accelerate forest restoration. There is a need to find a way to coordinate their efforts from a centralized place to ensure that they achieve the objectives for which they are assigned. The government needs to take a more coordinated approach and allow members of the local community to participate in the initiatives.

In developed countries, restorations of degraded landscapes take the approach of an educational view where members of the community are provided with education on how to undertake restoration programs. Feasibility of such initiatives needs to be studied in Kenya and applied if found to improve forest restoration efforts. Moreover, the school pupils and students can also be engaged in the initiatives by forming conservation clubs in their schools to spearhead the efforts.

There is need for further investigations on the role of technologies in forest restoration programs. Current technologies are moving toward the use of artificial intelligence in monitoring for restoration. Such technologies should be incorporated into the restoration efforts.

Danksagung Author Contribution Statement Benjamin Mutuku Kinyili is the sole author of this review. He conceived the topic, designed the review approach, conducted the literature search and screening, extracted and synthesized the evidence, prepared the tables and figures, and wrote and approved the final manuscript.

Data Availability Statement This article is a review. All data cited are available within the referenced publications and institutional reports (e.g., KEFRI/KFS and other publicly accessible sources) cited throughout the manuscript. No new primary datasets were generated or analyzed. Additional clarifications or extracted summary notes are available from the author upon reasonable request.

References

1. Barrow E (2002) Forest landscape restoration: building assets for people and nature—experience from East Africa. IUCN, Gland
2. BGCI & ITF (Botanic Gardens Conservation International & International Tree Foundation) (n.d.) The ITF centenary campaign to plant 20 million trees in Kenya. Available at: https://internationaltreefoundation.org/. Accessed 25 Oct 2025
3. Cheboiwo J, Langat D, Muga M, Kiprop J (2018) Economic analysis of forest landscape restoration options in Kenya. Kenya Forest Service, Nairobi. Available at: https://www.kfs.go.ke. Accessed 25 Oct 2025
4. Cheruiyot SA (2019) Forest planning and management for human development in Africa: a case of Kenya. University of Nairobi, Nairobi
5. Chisika S, Yeom C (2024) The challenges of private-sector engagement in forest landscape and ecosystem restoration in Kenya. J Environ Restor 12(3):44–56

6. Conservation International (n.d.) Community-based restoration programme briefs, Kenya. CI Africa Field Office, Nairobi. Accessed 25 Oct 2025
7. County Government of Makueni (n.d.) Makuli–Nzaui landscape restoration plan. Department of Environment, Wote. Accessed 25 Oct 2025
8. Eden C (2022) Kenya's coastal mangrove restoration initiative: progress and lessons learned. World Mangrove Alliance, Mombasa
9. FAO (2018–2023) GEF-6 project documentation: restoration of Kirisia, Mt Kulal and Mukogodo landscapes. Food and Agriculture Organization of the United Nations, Rome
10. FAO (2020) Global forest resources assessment 2020: main report. Food and Agriculture Organization of the United Nations, Rome. https://doi.org/10.4060/ca9825en
11. Gregersen H, El-Lakany H, Frechette A (2020) Forests, forest people, and the UN 2030 Agenda's ethical mandate: "leave no one behind". Rights and Resources Initiative, Washington DC
12. Imo M, Imo M (2012) Forest degradation in Kenya: impacts of social, economic and political transitions. In: Adoyo JW, Wangai CI (eds) African political, economic and security issues. Nova Science, New York, pp 1–38
13. International Tree Foundation (ITF) (n.d.) Kenya forest and agroforestry partnership projects. ITF, Nairobi. Accessed 25 Oct 2025
14. Kameri-Mbote P, Kamunge I, Yatich JK (2021) Country report for Kenya. Paper presented at the African Soil Protection Law conference, Nairobi, Kenya
15. Kass FM Community Initiative (n.d.) Mau Forest restoration project report. Kass Media Group, Nairobi. Accessed 25 Oct 2025
16. Kenya Breweries Limited (KBL) and Nature Kenya (n.d.) Mt Kenya Forest restoration partnership plan. East African Breweries PLC, Nairobi. Accessed 25 Oct 2025
17. Kenya Climate Change Adaptation Programme (KCCAP) (2018; 2023) National environment management authority implementation reports. Nairobi: NEMA
18. Kenya Energy and Environment and Social Responsibility Programme (KEEP) (n.d.) Forest restoration activities in the Mau complex. KEEP, Nairobi. Accessed 25 Oct 2025
19. Kenya Forest Service (KFS) (2018; 2021) Annual reports and forest sector performance reviews. KFS, Nairobi
20. Kenya Power and Lighting Company (KPLC) (n.d.) Hydropower catchment management and reforestation programme summaries. KPLC, Nairobi. Accessed 25 Oct 2025
21. KOICA & World Vision Kenya (n.d.) Dryland landscape restoration initiatives in Kajiado and Narok counties. KOICA/WV Kenya, Nairobi. (Accessed 25 Oct 2025)
22. Langat D, Maranga E, Aboud A, Cheboiwo J (2016) Role of forest resources to local livelihoods: the case of east Mau Forest ecosystem, Kenya. Int J For Res 1:4537354. https://doi.org/10.1155/2016/4537354
23. Leberger R, Rosa IM, Guerra CA, Wolf F, Pereira HM (2020) Global patterns of forest loss across IUCN categories of protected areas. Biol Conserv 241:108299. https://doi.org/10.1016/j.biocon.2019.108299
24. Mikoko Pamoja Community Organisation (MPCO) (n.d.) Community-led mangrove restoration and carbon credit project. MPCO, Gazi Bay. (Accessed 25 Oct 2025)
25. Mt Kenya Trust (n.d.) Reforestation and biodiversity corridor initiatives. Mt Kenya Trust, Nanyuki. Accessed 25 Oct 2025
26. National Rangelands Trust (NRT) (n.d.) Northern Kenya rangelands carbon project (NKRCP) progress briefs. NRT, Isiolo. Accessed 25 Oct 2025
27. Nesha MK, Herold M, De Sy V, Duchelle AE, Martius C, Branthomme A, Garzuglia M, Jonsson O, Pekkarinen A (2021) An assessment of data sources, data quality and changes in national forest monitoring capacities in the global Forest resources assessment 2005–2020. Environ Res Lett 16(5):054029. https://doi.org/10.1088/1748-9326/abec26
28. Odhiambo CO (2024) Perspective chapter: the role of development partners in Kenya's landscape restoration towards vision 2030. In: Rural areas—development and transformations. IntechOpen, London. https://doi.org/10.5772/intechopen.112233

29. Oliveira AHM, Matricardi EA, de Aragão LEOC, Felix IM, Chaves JH, Magliano MM, Oliveira-Junior JMB, Vieira TA, Santos LE, Reis LP (2024) Assessing forest degradation through remote sensing in the Brazilian Amazon: implications and perspectives for sustainable forest management. Remote Sens 16(23):4557. https://doi.org/10.3390/rs16234557
30. Ritchie H, Roser M (2021) Forests and deforestation. Our World in Data Available at: https://ourworldindata.org/forests-and-deforestation. Accessed 25 Oct 2025
31. Rotich B, Ojwang D (2021) Trends and drivers of forest cover change in the Cherangany Hills Forest ecosystem, Western Kenya. Glob Ecol Conserv 30:e01755. https://doi.org/10.1016/j.gecco.2021.e01755
32. Singh M, Shahina N, Das S, Arshad A, Siril S, Barman D, Mog U, Panwar P, Shukla G, Chakravarty S (2022) Forest resources of the world: present status and future prospects. In: Land degradation neutrality: achieving SDG 15 by Forest management. Springer, Singapore, pp 1–23. https://doi.org/10.1007/978-981-16-5653-3_1
33. Stanturf JA, Kant P, Lillesø J-PB, Mansourian S, Kleine M, Graudal L, Madsen P (2015) Forest landscape restoration as a key component of climate change mitigation and adaptation. International Union of Forest Research Organizations (IUFRO), Vienna. https://doi.org/10.13140/RG.2.1.3534.2160
34. Sustainable Management of Mau Watershed (SUMAWA) (n.d.) Reforestation and watershed management project summary. SUMAWA, Nakuru. (Accessed 25 Oct 2025)
35. TRI Kenya ASAL Programme (n.d.) Assisted natural regeneration in Mukogodo Forest Reserve report. UNEP/TRI Kenya, Nairobi. Accessed 25 Oct 2025
36. Vi Agroforestry, World Bank BioCarbon Fund and UNIQUE (n.d.) Kenya agricultural carbon project (KACP) overview. Vi Agroforestry, Nairobi. Accessed 25 Oct 2025

Seed Banks for Tree Cover and Forest Restoration in Kenya

Benjamin Mutuku Kinyili

Abstract

Introduction Kenya has lost 12,000 ha of forest annually since 1963, reducing its forest cover from 10 to 6% by 2018. Between 2001 and 2023, Kenya lost 386,000 ha of tree cover, a 12% decrease since 2000. Forests can be restored through tree planting that requires seeds. Traditionally tree seeds were collected from the natural forests, a practice which is unsustainable. However, the specific role and capacity of seed banks in supplying genetic resources for restoration are not widely understood in Kenya. This review therefore addresses a clear problem: reliable, quality-assured and locally adapted seed supply is a persistent bottleneck in achieving restoration goals in Kenya. It specifically examines the implications of seed banks in forest ecosystem restoration, including seed sourcing capacity, institutional seed sources, seed quality assurance and control and associated challenges. *Methods* We synthesised peer-reviewed and gray literature, national documents and practice notes on seed sourcing, processing, storage and quality assurance/quality control (QA/QC) relevant to restoration; inclusion focussed on Kenya-relevant evidence and institutional practices. *Results* Five key roles of seed banks in Kenyan forestry sector are preserving genetic diversity, biodiversity conservation and protection of species from extinction, plant breeding programmes and ecological restoration. The capacity of genebanks to supply seeds for restoration includes ensuring the availability of germplasm, conducting better seed processing, maintaining good storage capacity, implementing quality assurance and facilitating germplasm exchange. Seed banks employ competent personnel to carry out tasks required

Theme: Biodiversity and Natural Resource Management
Sub-Theme: 1. Seed banks for indigenous trees

B. M. Kinyili (✉)
Kenya Forest Service, Nairobi, Kenya

R. Ramkat et al. (eds.), *Towards the Sustainability of our Ecosystems*, Springer Proceedings in Geography and Sustainability,
https://doi.org/10.1007/978-3-032-30076-8_6

for propagating a number of tree species, such as germplasm availability, material selection, seed sourcing, processing and storage, as well as quality assurance and control and germplasm exchange. However, three gaps constrain outcomes: under-representation of indigenous species; uneven QA/QC and viability testing for non-commercial taxa. *Conclusion* Scaling indigenous provenance strategies, extending standardised QA/QC beyond commercial taxa and strengthening public–private seed systems are decisive to accelerate restoration while safeguarding ecological fit and resilience. This constitutes an actionable pathway to meet Kenya's restoration targets.

Keywords

Ecosystem restoration • Gene bank • Germplasm • Seed banks • Seed quality • Kenya

1 Introduction

Loss of trees in forests is massive at the global scale, driven by human activities and exacerbated by climate change scenarios [30]. It is estimated that 420 million hectares of forest have been lost worldwide due to land-use–related degradation since 1990 [12, 31]. Recognising this loss as a threat to the environment, wildlife and human livelihoods, significant attention is directed towards ecological restoration [34]. The Convention on Biological Diversity (CBD) identifies the significance of restoring degraded habitats, targeting 30% of degraded ecosystems and landscapes by 2032 [26]. There are several international bodies led by the UN General Assembly, governments and other stakeholders at the forefront of ecosystem restoration efforts [5, 21]. Consequently, efforts to encourage restoration are substantial, with a parallel increase in demand for restoration materials [35, 38].

In contemporary society, there are numerous seed-based restoration projects [2]. In the past, seeds for tree planting and restoration were sourced from the wild, which has become unsustainable due to high seed demand [16]. Wild collection presents disadvantages, notably inconsistent seed quality, unpredictable quantities and limited supply control [37]. Wild seed collection is also sensitive to environmental fluctuations and is therefore unreliable [40]. In response, multiple seed-sourcing pathways are now used for ecological restoration, with seed banks increasingly serving as the primary source [1].

Seed banks can nurture seeds of tree species that farmers cannot easily raise [6, 27]. They help assure seed diversity, quality and availability for users [20]. Nevertheless, not all species are maintained, and many seed banks prioritise commercially preferred taxa to remain viable [37]. Seeds produced within seed banks can be multiplied to create larger stocks for distribution [41].

In Kenya, the organisation, capacity and performance of seed banks in supplying genetic resources for restoration remain insufficiently characterised. Specifically, there is a weak evidence based on (i) how Kenyan seed banks support restoration goals (roles), (ii) their seed-sourcing capacity (public/private/institutional pathways) and (iii) the effectiveness of quality assurance and quality control (QA/QC) for both indigenous and commercially favoured species. This review addresses that knowledge gap by evaluating the status of seed banks in Kenya, seed collection and storage processes, and QA/QC practices, and by synthesising the implications for ecosystem restoration and tree-cover expansion.

2 Methodology

We conducted a narrative synthesis anchored by a systematic search protocol to enhance transparency and reproducibility. Searches were run in Scopus, Web of Science Core Collection, Google Scholar (first 200 results per query) and Kenyan institutional repositories (Kenya Forestry Research Institute (KEFRI), Kenya Forest Service (KFS), university repositories). Grey literature included KEFRI/KFS technical reports, policy documents and restoration guidelines. Publications from 2000 to 2025 were considered, with earlier seminal works included when foundational to seed bank concepts. Search strings combined seed system terms with Kenya/restoration terms, for example: ('seed bank' OR 'seed system' OR germplasm OR 'seed supply' OR 'seed sourcing' OR 'seed quality' OR viability OR dormancy) AND (Kenya OR 'East Africa' OR 'African drylands') AND (restoration OR 'forest landscape restoration' OR FLR OR reforestation OR 'tree cover').

Inclusion criteria were the following: (i) focus on seed banks/seed systems relevant to tree restoration; (ii) empirical, review, or technical guidance with methods or quality procedures stated; (iii) Kenya focus or clear applicability to Kenyan contexts; and (iv) English language. Exclusion criteria were the following: (i) agricultural seed system studies without relevance to tree restoration; (ii) opinion pieces lacking methodological detail; and (iii) duplicate or superseded versions. Two reviewers independently screened titles/abstracts and full texts; disagreements were resolved by discussion to consensus, with a third reviewer consulted for ties. A PRISMA-style flow summary (records identified, screened, included) is provided in the Supplementary Material.

For each included item, we extracted (a) the roles of seed banks, (b) seed-sourcing pathways (public/private/institutional), (c) processing and storage practices, (d) QA/QC procedures (e.g. purity, moisture, germination/viability tests) and (e) Kenya-specific challenges/opportunities. We then conducted a thematic analysis: two coders developed an initial codebook deductively from the research questions and inductively refined it during open coding. Inter-coder agreement was assessed on 20% of the corpus (Cohen's $\kappa \geq 0.75$), with discrepancies resolved by consensus and consequent codebook updates. Credibility and dependability were enhanced via analyst triangulation (peer debriefs), source triangulation (peer-reviewed and grey literature) and an audit trail (search strings and decisions log). Where available, we noted methodological clarity (sampling, tests used, QA/QC reporting) to weight interpretations in the narrative synthesis.

3 Findings and Discussion

3.1 Implication of Seed Banks in Forest Ecosystem Restoration

The five roles highlighted in Table 1 can be read as two complementary functions: (i) genetic insurance preserving adaptive potential through ex situ conservation of diverse, traceable germplasm and (ii) operational deployment moving high-quality, provenance-matched seed into restoration supply chains at scale [24, 27]. In Kenya, the insurance function matters because many native taxa occupy fragmented, climatically heterogeneous landscapes; without curated collections and viability testing, restoration risks a narrow genetic base that undercuts long-term resilience. The deployment function matters because restoration pledges are expanding faster than reliable seed pipelines, and shortfalls are routinely filled by a small set of fast-growing exotics often with poor ecological fit [37].

Genetic diversity and biodiversity conservation: Globally, seed banks safeguard within-species variation and rare alleles, but their conservation value materialises only when accessions are well-documented (source, seed zone), periodically tested for viability, and refreshed to avoid genetic drift [24]. In Kenya, KEFRI and KFS collections provide a backbone for such work, yet indigenous species are under-represented relative to commercial exotics, reflecting market demand and budget constraints. The implication is practical: restoration outcomes will track

Table 1 Significance of seed banks in supporting ecosystem restoration

Role	Description
Preserving genetic diversity	Seed banks preserve genetic diversity by conserving plant species gene pool [11, 36]
Biodiversity conservation	Seed banks ensures that there is availability of plants at all time They store seeds from a wide range of plants for long periods. Seed are preserved in an event of habitat loss, disease outbreak of climate change [23]. Seed banks will also restore biodiversity in areas decimated by deforestation, mining or agriculture [3]
Protect species from extinction	Presence of seed banks ensure that species extinction cannot occur [19]. Tree species facing extinction in the wild can be replenished or restocked using the those stored in the seed bank [33]
Plant breeding programmes	Breeding of plant can be achieved through seed bank [15, 22]. The seeds within the seed bank can provide information on genetic composition which may be useful to develop, new resistant plant species [39]
Ecological restoration	Seed banks major role is to help restore forests that are depleted or degraded by human activities [13, 17]. The germplasm from seed banks is therefore useful for ecological restoration in Kenya [10, 18]

the breadth and depth of indigenous germplasm actually available not the breadth imagined in policy so expanding collections for dryland species and high-altitude taxa is a priority [27].

Extinction risk buffering and climate adaptation: Banking seed from small or declining populations buys time for in situ measures and enables assisted gene flow or climate-adjusted provenancing where local climates are shifting [27]. For Kenya's montane and ASAL ecosystems, this requires forward-looking collection strategies across environmental gradients, not opportunistic, single-site harvests. Otherwise, restoration plantings may establish but lack the adaptive breadth needed for hotter, drier futures.

Plant breeding and deployment: Seed banks underpin breeding (e.g. selecting for drought tolerance, pest resistance), but breeding gains only translate to restoration when seed handling, storage and QA/QC (purity, moisture, germination) are standardised and enforced in nurseries and procurement [27]. Current Kenyan practice shows uneven QA/QC outside flagship programmes; mixed-provenance bulking and inadequate viability testing can depress field performance and inflate replacement costs [37].

Ecological restoration provisioning: The table's 'provisioning' role is often interpreted as a logistics problem (enough kilograms of seed), yet the ecological problem is matching the right seed to the right site. That hinges on seed-zone mapping, provenance records and traceability from collection through nursery to planting. In Kenya, public institutions increasingly supply frameworks and protocols, but private and community nurseries execute most last-mile delivery; without simple traceability tools and market incentives, provenance standards erode under pressure to meet planting targets [37].

The Kenyan evidence suggests three actionable implications. First, expand indigenous pipelines (collection missions, orchard establishment and refreshed accessions) to reduce dependence on a narrow set of exotics. Second, institutionalise QA/QC beyond flagship projects through affordable moisture control, routine germination/viability testing and basic certification or vendor self-declaration tied to procurement. Third, operationalise 'right seed, right place' via practical seed-zone/provenance guidance and low-cost traceability (batch labelling, seed passports), so field teams can select appropriate material under real-world time and budget constraints [24, 27, 37]. In short, seed banks improve restoration not merely by holding seed, but by shaping the genetic quality, documentation and deployment pathways through which seed reaches landscapes. Where Kenya strengthens those pathways, especially for indigenous species, the ecological return on restoration spending increases.

Table 2 summarises the functional capacity of Kenyan seed banks across seven germplasm management stages from availability and selection to storage, quality assurance and exchange. Interpreting these capacities reveals both institutional strengths and structural weaknesses that influence the reliability of restoration seed supply.

Availability of germplasm: Kenyan seed banks hold genetically diverse, multi-provenance materials drawn from several ecological zones [8]. This diversity

Table 2 Capacity of seed banks in tree seed supply for ecosystem restoration in Kenya

Aspect of seed germplasm	Seed bank capacity	Problem encountered
Availability of germplasm	Genetically diverse seeds, multi-provenance germplasm	Incomplete documentation, risk of genetic contamination
Selection of seed materials	Selection of high-quality germplasm sources is enhanced	Some germplasm in seedbanks may be from potentially invasive species
Sourcing of seeds	Sourcing of seeds is easy due to established networks	Poor understanding of seeds phenology leading to poor sourcing
Seed processing	Processing equipment are available	Processing local seeds is challenging due to large morphological diversity
Seed storage	Temperature-controlled storage units are available	Inadequate information about seed physiology under storage hence affecting longevity
Quality assurance and control	High standards of quality assurance and control	Lack of mechanisms to break seed dormancy
Germplasm exchange	High levels of germplasm exchange are practiced	Lack of trust in sharing certain germplasm among seed banks. Germplasm protection policies are often complex

underpins the adaptive potential of restoration programmes and aligns with global recommendations for climate-smart provenancing [27]. However, documentation remains incomplete and traceability weak, creating risks of genetic contamination and loss of provenance integrity [24]. Addressing this gap requires standardised accession records and digital catalogues linking each seed lot to its collection zone and environmental metadata.

Selection of seed materials: Selection protocols in public facilities such as KEFRI and the Seed Savers Network Kenya promote high-quality germplasm sources. Yet, some seed banks inadvertently maintain potentially invasive exotics (e.g. *Eucalyptus camaldulensis*, *Grevillea robusta*) that, while economically valuable, may alter native ecosystems [37]. Strengthening ecological screening and national guidelines on acceptable species lists would help balance restoration efficiency with biodiversity integrity.

Sourcing of seeds: Seed sourcing is facilitated by extensive community and institutional networks, but knowledge of seed phenology timing of flowering, fruiting, and seed maturity remains inconsistent [20]. Consequently, many collections occur off-season, reducing viability and germination performance. Investing in local phenology calendars and citizen-science monitoring can enhance timing accuracy and ensure better harvest quality [27].

Seed processing: Kenyan seed centres possess modern cleaning and grading machines, but morphological diversity among indigenous species complicates standardisation [41]. For example, varying seed-coat textures and dormancy traits in *Acacia* and *Terminalia* species require species-specific treatments. This indicates a need for adaptive protocols and technician training focussed on native species physiology rather than reliance on generic processing routines.

Seed storage: Temperature- and humidity-controlled storage units exist in major seed banks such as KEFRI Muguga and Londiani, ensuring medium-term viability. Nonetheless, physiological data for many indigenous species are lacking, particularly for recalcitrant seeds that cannot withstand desiccation [24]. Collaborative research with universities could generate species-specific storage curves, improving longevity predictions and reducing waste.

Quality assurance and control (QA/QC): High standards are reported in institutional seed banks where germination and purity tests follow ISTA-aligned protocols. However, field-level nurseries and community seed banks rarely apply dormancy-breaking or viability tests systematically, resulting in uneven quality [37]. Establishing a light-touch certification or self-declaration framework supported by KEFRI's technical oversight would promote consistency without imposing prohibitive costs.

Germplasm exchange and governance: Active germplasm exchange occurs through KEFRI–KFS partnerships and regional initiatives, yet trust and legal clarity constrain sharing among institutions. Complex intellectual property and benefit-sharing policies (Nagoya Protocol compliance) discourage open exchange of valuable indigenous germplasm [30]. Simplified material-transfer agreements and transparent benefit-sharing mechanisms could stimulate collaboration while protecting genetic resources.

Kenya's seed bank network demonstrates substantial technical capacity but suffers from fragmented coordination, incomplete data systems and limited investment in indigenous species research. Addressing these weaknesses demands a two-track strategy: (1) consolidate technical standards documentation, phenology monitoring, QA/QC and seed zone mapping and (2) incentivise collaboration across public, private and community actors through shared databases and clear access frameworks. Ultimately, robust seed bank capacity will determine whether Kenya's ambitious tree-cover targets translate into ecologically resilient forests rather than short-lived plantations [8, 27].

3.2 Seed-Sourcing Capacity

Seed production for restoration initiatives begins with effective and strategic sourcing from diverse ecological regions. The quality of seed accessions in any genebank depends on the genetic integrity and overall physiological quality of the source populations from which they are derived [29]. In most established seed banks, qualified experts including foresters, ecologists, botanists and taxonomists are responsible for the collection of high-quality seed samples suitable for restoration programmes [9, 14]. Best practice requires that seeds be collected at natural dispersal points where maturity and viability are highest [28]. Field teams also undertake visual and laboratory quality assessments to evaluate seed colour, texture and fill to ensure only viable samples enter the seed bank.

Globally, the success of accredited seed banks stems from the ability to obtain germplasm from protected areas that host high intraspecific diversity and pure genetic lines of native flora, thereby enriching gene pools within collections [7, 32]. In Kenya, seed banks are managed by multidisciplinary teams comprising foresters, herbarium staff, botanists and plant taxonomists, allowing for efficient breeding programmes, phenology documentation and development of species-specific collection methods [36, 37]. Taxonomists play a central role in authenticating indigenous taxa and differentiating them from closely related exotics before accession.

According to KEFRI [25], several priority species dominate Kenya's restoration seed holdings (Table 3). *Cupressus lusitanica* seed is mainly sourced from Muguga and Londiani, while *Pinus patula* and *Eucalyptus grandis* originate from Central Highlands and Rift Valley eco-regions. *Eucalyptus camaldulensis* seeds are also collected from these regions. In contrast, *Melia volkensii*, a dryland species, is confined to Kitui, Kibwezi and Gede stations. *Grevillea robusta* seed is obtained from Turbo, Muguga, Njoguini (Nanyuki) and Soin (Bomet), whereas *Casuarina equisetifolia* is primarily from the Central Highlands eco-region.

Although these figures reflect considerable technical capacity, they also expose an ecological imbalance in Kenya's restoration seed portfolio. The dominance of *Cupressus*, *Pinus*, and *Eucalyptus* species illustrates a long-standing bias towards exotic commercial timber trees. These exotics are favoured for their uniform growth, predictable yields and well-understood propagation protocols—traits that align with government planting targets. However, such reliance carries several ecological trade-offs. Exotics can modify hydrological regimes, reduce understorey diversity and disrupt soil nutrient cycling when introduced into native ecosystems [28, 37]. In contrast, many indigenous species, though slower-growing,

Table 3 Tree species in genebanks used for forestry restoration in Kenya

Species	Number of seed sources	Total area (ha)
Cupressus lusitanica	22	39.5
Pinus patula	16	62.5
Eucalyptus grandis	16	36.5
Eucalyptus camaldulensis	9	10.7
Melia volkensii	3	23
Grevillea robusta	11	23.5
Casuarina equisetifolia	3	4.3

Source: [25]

contribute to long-term ecosystem stability, habitat provision and restoration of natural successional processes.

Native species such as *Olea africana*, *Juniperus procera*, *Podocarpus latifolius* and dryland *Acacia/Vachellia* varieties remain under-represented in Kenya's seed banks due to limited research on their reproductive ecology, sporadic fruiting and uncertain storage physiology [7]. This under-representation limits provenance choice, reducing the adaptability of restoration projects to site-specific climatic gradients. Without expanded collection missions and targeted orchard establishment for native taxa, Kenya risks narrowing its genetic base for restoration potentially undermining resilience under changing climatic conditions [24].

A further constraint concerns provenance documentation. Where seed sourcing is centralised in a few orchards, the genetic breadth captured per species shrinks, and traceability is easily lost as seed moves downstream to nurseries. Effective restoration requires the right seed for the right site, supported by complete metadata on eco-region, elevation and parent-tree cohorts [28]. To that end, Kenya's emerging digital platforms could enable 'seed passports' that track each seed lot from collection through storage to planting.

Phenological understanding also remains a rate-limiting step for native species. Even with trained staff, insufficient site-specific data on flowering and fruiting windows leads to suboptimal harvest timing, reducing viability. Citizen-science reporting and region-specific phenology calendars could close this gap [14].

Table 3 underscores the need to rebalance Kenya's seed-sourcing strategy. While the existing system ensures operational efficiency and short-term planting success, it must evolve towards a biodiversity-centred model that safeguards ecological integrity. This transition calls for (1) expanding indigenous germplasm collections; (2) diversifying mother-tree sampling across eco-regions; (3) improving provenance and traceability systems; and (4) coupling seed sourcing with targeted research on native seed physiology and dormancy behaviour. By combining scientific rigour in seed collection with ecological awareness, Kenya can transform its seed banks from commercial supply units into biodiversity engines that sustain forest restoration.

Other institutions, mostly privately owned, are engaged in seed sourcing and supplying germplasm, such as Aberdare Technologies, George Williamson, James Finlays, Unilever Tea, and Kakuzi [25]. James Finlays and Unilever Tea are multinationals involved in tea production. These firms grow large quantities of Eucalyptus for industrial use in tea processing. The multinationals do not maintain their own seed sources but obtain seeds from mature plantations during harvesting. James Finlay and Unilever Tea import seeds from Zimbabwe but also carry out clonal forestry, producing 60% of tree planting materials vegetatively. Additionally, E. grandis seeds are imported from South Africa, and Kakuzi Company purchases E. camaldulensis seeds from KEFRI for commercial planting. Aberdare Technologies produces a variety of tree seedlings vegetatively. They possess clonal hedges for Eucalyptus hybrids which they purchased from the Tree Biotechnology Program Trust (TBPT) in Karura.

3.3 Seed Quality Assurance and Control and Challenges

Ensuring quality in genebanks is foundational: vigorous, uniform establishment and long-term stand resilience depend on seed lots that meet minimum thresholds for viability, purity, moisture status, health and traceable provenance [4]. In Kenya, KEFRI and partner facilities run routine viability tests for priority species; however, coverage for many indigenous taxa remains thin [17, 18, 25].

In practice, 'stringent and costly' QA/QC encompasses several resource-intensive steps. First, sampling and field inspection require certified lot sampling, verification of collection sites and mother-tree counts; travel, per-diems and inspector time comprise the main cost drivers. Second, purity analysis separates pure seed from inert matter and contaminants labour heavy even with basic graders. Third, moisture determination (oven-dry or electronic metres) sets safe storage thresholds; costs include equipment, calibration and consumables. Fourth, viability and germination testing proceeds under controlled temperature/light; tetrazolium (TZ) assays may be used for rapid viability checks, with repeat runs when dormancy is suspected. Fifth, dormancy-breaking treatments (mechanical/acid scarification, warm–cold stratification, smoke water, GA_3) often require iterative experimentation for native species lacking published methods, increasing technician time and consumables. Sixth, seed health testing screens for seed-borne fungi/bacteria to prevent nursery disease outbreaks. Finally, documentation and certification entail lot-level coding, 'seed passports' (collection zone/elevation/date, mother-tree counts and test results), secure data archiving, and where applicable accreditation/audit fees for formal labels [11, 15]. See Table 4 for a summary of QA/QC components, typical cost drivers, the specific gaps for indigenous species and low-cost fixes.

These processes are relatively streamlined for long-established commercial exotics (e.g. *Cupressus*, *Pinus*, *Eucalyptus*), where SOPs exist and volumes justify fixed costs. By contrast, many indigenous taxa have uncertain storage behaviour and poorly documented dormancy/germination regimes, so labs default to trial-and-error. That raises turnaround times and unit costs per kilogram, which discourages nurseries and buyers from ordering native seed [7, 18]. The result is a negative feedback loop: native species are not planted because quality cannot be assured at acceptable cost; they are not available because demand stays low and protocols remain undeveloped [25, 37]. Over time, restoration supply chains drift towards a narrow exotic portfolio, even where natives would deliver better biodiversity and climate-adaptation outcomes.

Closing this gap requires three practical shifts. First, publish concise, species-specific protocols for priority natives' collection windows, target moisture, pre-treatments, core germination regimes and make them open access. Second, implement light-touch certification for forestry seed (self-declaration with periodic audits) to raise baseline quality without pricing out community seed banks. Third, subsidise core tests (moisture, TZ, a single standard germination) for approved native taxa, coupled with ring testing across labs to harmonise results. Embedding seed-passport provenance fields (eco-region, elevation band, mother-tree numbers,

Table 4 Forestry seed QA/QC workflow in Kenya: components, typical cost drivers, gaps for indigenous species and practical fixes

QA/QC component	What it involves	Typical cost drivers	Gap for indigenous species	Consequence for restoration	Practical low-cost fix
Sampling and field inspection	Lot sampling; site verification; mother-tree counts	Travel, per-diems, inspector time	Dispersed stands; limited field staff	Fewer verified lots; weak traceability	Clustered sampling routes; train local scouts to verify sites
Purity analysis	Separate pure seed/inert/other seed	Bench time, graders, labour	Mixed fruit/seed morphology in natives	Contamination → poor nursery performance	Simple sieving SOPs and photo guides per species
Moisture determination	Oven-dry or metres; set storage targets	Equipment, calibration, consumables	Unknown safe moisture for many natives	Storage failure; viability loss	Shared metre hubs; adopt conservative target ranges until SOPs exist
Viability/germination	Standard tests; TZ rapid viability	Media, reagents, repeated runs	Dormancy not characterised	Slow/failed lots; higher unit costs	Use TZ triage first; publish one-page germination regimes
Dormancy breaking	Scarification/stratification/smoke/GA_3	Iterative trials, technician time	Species-specific methods unknown	Delays; inconsistent emergence	National protocol bank; ring-testing across labs
Seed health testing	Fungal/bacterial screens; hygiene SOPs	Lab capacity, consumables	Limited forestry-dedicated labs	Nursery disease outbreaks	Basic plating + hygiene checklists; referral pathway for advanced tests
Documentation and provenance	Lot codes; seed passports; data archiving	Admin time; data systems	Underused databases; uneven records	Loss of provenance; mis-matching	Mandatory passport fields on invoices; simple digital registry
Certification and labelling	Self-declaration or accreditation; labels	Fee/renewal, audits	Fees price out community banks	Unchecked quality; market distrust	Light-touch forestry label; periodic random audits

test results) on all invoices and receipts would lock in traceability from bank to nursery to field, while routine nursery feedback on emergence/survival can rapidly refine protocols and reduce wastage [7, 11, 18].

4 Conclusions

This study identified five key roles of seed banks in the Kenyan forestry sector: preserving genetic diversity, biodiversity conservation, protecting species from extinction, supporting plant breeding programmes and enabling ecological restoration. The seed banks employ competent personnel to carry out tasks required for propagating a number of tree species, including managing germplasm availability, selecting materials, sourcing seeds and implementing quality assurance and control, as well as germplasm exchange. Typically, the best quality seeds are collected at the point of natural dispersal. In Kenya, seed banks are operated by experts, including foresters, herbarium staff, botanists and plant taxonomists, which facilitate plant breeding programmes, the determination of tree species phenology and the identification of optimal collection methods. As noted, seeds of *Cupressus lusitanica* are located in Muguga and Londiani; *Pinus patula* and *Eucalyptus grandis* in the Central Highlands and Rift Valley Eco-regions; *Eucalyptus camaldulensis* in the Central Highlands and Rift Valley Eco-regions; *Melia volkensii* in Kitui, Kibwezi and Gede; *Grevillea robusta* in Turbo, Muguga, Njoguini and Soin; and Casuarina equisetifolia in the Central Highlands Eco-region.

5 Recommendations

Seed banks have the capacity to support restoration initiatives by developing seeds of good quality under proper seed quality control and assurance. However, there is a need to expand the capacity of the seed banks to include most of the native plants as they are preferred tree species in any regions of Kenya. In expanding seed banks, there is a need for KEFRI to provide the technical aspects need to ensure that private firms and individuals acquire these seed banks, which will accelerate restoration efforts. The funding of such strategies must be discussed between the government and private entrepreneurs. There is also a need for alternative cheap and efficient tree seed quality control practices for the native tree species.

Danksagung Author Contribution Statement Benjamin Mutuku Kinyili is the sole author of this review. He conceived the topic, designed the review approach, conducted the literature search and screening, extracted and synthesised the evidence, prepared the tables and figures, and wrote and approved the final manuscript.

Data Availability Statement This article is a review. All data cited are available within the referenced publications and institutional reports (e.g. Kenya Forestry Research Institute /Kenya Forest Service (KEFRI/KFS) and other publicly accessible sources) cited throughout the manuscript. No new primary datasets were generated or analysed. Additional clarifications or extracted summary notes are available from the author upon reasonable request.

References

1. Allen EB, Cross AT, Berger AM, Aronson J (2024) Restoration seed and plant material supply chains are complex social networks. Restor Ecol 32(8):e14279. https://doi.org/10.1111/rec.14279
2. Andres SE, Mills CH, Gallagher RV, Adams VM (2024) A framework for ecological restoration cost accounting across context and scale. Biol Conserv 295:110671. https://doi.org/10.1016/j.biocon.2024.110671
3. Eschen R, Bekele K, Mbaabu PR, Kilawe CJ, Eckert S (2021) Prosopis juliflora management and grassland restoration in Baringo County, Kenya: opportunities for soil carbon sequestration and local livelihoods. J Appl Ecol 58(6):1302–1313
4. Finch-Savage WE, Bassel GW (2016) Seed vigour and crop establishment: extending performance beyond adaptation. J Exp Bot 67(3):567–591. https://doi.org/10.1093/jxb/erv490
5. Firoiu D et al (2024) the road to 2030: evaluating Europe's progress on sustainable ecosystem protection and restoration. Land 13(12):1974
6. Gibson-Roy P, Hancock N, Broadhurst L, Driver M (2021) Australian native seed sector characteristics and perceptions indicate low capacity for upscaled ecological restoration: insights from the Australian native seed report. Restor Ecol 29(7):e13428. https://doi.org/10.1111/rec.13428
7. Hay FR, Probert RJ (2013) Advances in seed conservation of wild plant species: a review of recent research. Conserv Physiol 1(1):cot030. https://doi.org/10.1093/conphys/cot030
8. KEFRI (2024) Annual report 2024. Kenya Forestry Research Institute, Nairobi
9. Kettle CJ (2012) Seeding ecological restoration of tropical forests: priority setting under REDD+. Biol Conserv 154:34–41. https://doi.org/10.1016/j.biocon.2011.11.019
10. Kibet S (2011) Plant communities, species diversity, richness, and regeneration of a traditionally managed coastal forest, Kenya. For Ecol Manage 261(6):949–957
11. Kitonga K, Jamora N, Smale M, Muchugi A (2020) Use and benefits of tree germplasm from the world agroforestry genebank for smallholder farmers in Kenya. Food Secur 12(5):993–1003. https://doi.org/10.1007/s12571-020-01047-6
12. Leberger R, Rosa IMD, Guerra CA, Wolf F, Pereira HM (2020) Global patterns of forest loss across IUCN categories of protected areas. Biol Conserv 241:108299. https://doi.org/10.1016/j.biocon.2019.108299
13. Lemenih M, Bongers F (2010) The role of plantation forests in fostering ecological restoration: experiences from East Africa. In: Degraded forests in Eastern Africa. Routledge
14. León-Lobos P, Way M, Aranda PD, Lima-Junior M (2012) The role of ex situ seed banks in the conservation of plant diversity and in ecological restoration in Latin America. Plant Ecol Divers 5(2):245–258
15. Mbora A, Lillesø J-PB (2007) Sources of tree seed and vegetative propagation of trees around Mt. Kenya. Development and Environment No. 9–2007. Forest & Landscape Denmark/World Agroforestry (ICRAF). Available at: https://www.cifor-icraf.org/publications/downloads/Publications/PDFS/b15497.pdf
16. McGunnigle N, Bardsley D, Nuberg I, Pandit BH (2025) Is rewilding too wild? Assessing the benefits and challenges of forest succession on abandoned land in the mid-hills of Nepal. People Nat 7:415
17. Mganga, K.Z. (2009) Impact of grass reseeding technology on rehabilitation of the degraded rangelands: a case study of Kibwezi district, Kenya. MSc Thesis, University of Nairobi. Record: https://erepository.uonbi.ac.ke/handle/11295/19154

18. Mng'omba SA et al (2024) Germplasm systems of multipurpose indigenous trees in Africa. In: Trees in a sub-Saharan multi-functional landscape: research, management, and Policy, pp 211–236
19. Mukiama TK (1992) Indigenous flora conservation in Kenya. In: Conservation of plant genes: DNA banking and in vitro biotechnology, pp 311–321
20. Muluneh MG (2023) The role of soil seed banks and seed dispersal mechanisms for an ecosystem restoration and biodiversity conservation
21. Nelson C et al (2024) Standards of practice to guide ecosystem restoration: a contribution to the UN decade on ecosystem restoration 2021–2030. FAO
22. Nyoka BI et al (2015) Tree seed and seedling supply systems: a review of the Asia, Africa and Latin America models. Small-Scale For 14:171–191
23. Ogwu M, Osawaru M, Ahana C (2014) Challenges in conserving and utilizing plant genetic resources (PGR). Int J Genet Mol Biol 6(2):16–22
24. Oldfield SF, Olwell P, Shaw N, Havens K (eds) (2019) Seeds of restoration success: wild lands and plant diversity in the US. Springer, Cham. https://doi.org/10.1007/978-3-319-96974-9
25. Omondi SF et al (2021) Status of seed sources of key commercial tree species in Kenya. KEFRI—Gatsby Africa (KCFP) Commercial Forestry Project Report
26. Opoku A, Baah B (2024) Biodiversity conservation, the built environment, and the SDGs. In: The elgar companion to the built environment and the SDGs, pp 330–351
27. Pedrini S et al (2023) Strengthening the global native seed supply chain for ecological restoration. In: Ecological restoration: moving forward using lessons learned, pp 437–472
28. Pedrini S, Gibson-Roy P, Trivedi C, Gálvez-Ramírez C, Hardwick K, Shaw N, Frischie S, Laverack G, Dixon K (2020) Collection and production of native seeds for ecological restoration. Restor Ecol 28(S3):228–238. https://doi.org/10.1111/rec.13190
29. Potter KM et al (2017) Banking on the future: progress, challenges and opportunities for the genetic conservation of forest trees. New For 48:153–180. https://doi.org/10.1007/s11056-017-9582-8
30. Rawat S, Kumar N, Sharma AJ, Pant S (2025) Forest degradation and carbon sequestration: assessing the global scenario. In: Climate change impact on Himalayan biodiversity. Springer, pp 425–445. https://doi.org/10.1007/978-981-97-2760-8_23
31. Ritchie H, Roser M (2021) Forests and deforestation. Our World in Data. Available at: https://ourworldindata.org/forests
32. Rodríguez-Rodríguez D, Martínez-Vega J (2022) Effectiveness of protected areas in conserving biodiversity: a worldwide review. Springer, Cham. https://doi.org/10.1007/978-3-030-98362-8
33. Simiyu SW (1996) Plant biodiversity conservation in Kenya: the plant conservation and propagation unit of the east African herbarium—a case study. In: van der Maesen LJG, van der Burgt XM, van Medenbach de Rooy JM (eds) The biodiversity of African plants. Springer, Dordrecht, pp 267–276. https://doi.org/10.1007/978-94-009-0285-5_35
34. Stephens SL et al (2024) Forest restoration and fuels reduction work: different pathways for achieving success in the Sierra Nevada. Ecol Appl 34(2):e2932
35. Wagner J (2024) Scaling global seed banking. White Paper. https://collaborativeforchange.org/wp-content/uploads/2025/01/37ea30_8014ad862ee34fd5a20051fbbf280aae.pdf
36. Wambugu P, Muthamia Z (2009) The state of plant genetic resources for food and agriculture in Kenya. Report submitted to FAO Commission on PGRFA
37. Wambugu PW, Nyamongo DO, Kirwa EC (2023) Role of seed banks in supporting ecosystem and biodiversity conservation and restoration. Diversity 15(8):896. https://doi.org/10.3390/d15080896
38. Wu R, Huo X, Wang Y, Li Z, Wang D (2024) Soil seed bank characteristics at different developmental stages in pine and oak forests and its potential in vegetation restoration. Plant Soil 497(1–2):535–548. https://doi.org/10.1007/s11104-023-06410-y
39. Wulff E, Bødker L, Torp J (2006) Seed sector country profile: Kenya. Vol I: overview of seed supply systems and seed health issues. Danish Institute of Agricultural Sciences

40. Zhang Y et al (2022) Seed dormancy in space and time: global distribution, paleoclimatic and present climatic drivers, and evolutionary adaptations. New Phytol 234(5):1770–1781. https://doi.org/10.1111/nph.18099
41. Zinnen J, Broadhurst LM, Gibson-Roy P, Jones TA, Matthews JW (2021) Seed production areas are crucial to conservation outcomes: benefits and risks of an emerging restoration tool. Biodivers Conserv 30:1233–1256. https://doi.org/10.1007/s10531-021-02149-z

Pretreatment of Lignocellulosic Biomass for Biogas Enhancement: A Review

N. A. Bogonko, M. M'Arimi, J. Kiplagat, and J. C. Khalifa

Abstract

The rapid rise in global population and industrial development has significantly driven up energy demand and consumption. At present, fossil fuels remain the primary source of energy to meet this growing need. However, the heavy reliance on these non-renewable sources characterized by high costs, finite availability, and substantial greenhouse gas emissions poses serious threats to both human health and the environment. This situation underscores the pressing need to develop and adopt alternative, sustainable energy solutions. Renewable energy sources offer a viable, eco-friendly, and sustainable approach to addressing the global energy crisis. Among these, biogas generated through anaerobic digestion (AD) of organic wastes such as agricultural residues and animal manure stands out as a promising option. Despite the abundance of agricultural biomass, adopting lignocellulosic feedstocks for biogas production remains limited. This is primarily due to the structural complexity of plant cell walls, which resist microbial degradation. Consequently, pretreatment of lignocellulosic materials is essential to enhance biogas yields. This paper provides an overview of various pretreatment methods for lignocellulosic biomass aimed at improving biogas production. It also discusses the limitations of existing technologies and emphasizes the importance of developing sustainable, locally adaptable, and environmentally friendly pretreatment strategies.

N. A. Bogonko (✉) · J. C. Khalifa
Kenya Industrial Research and Development Institute (KIRDI), Nairobi, Kenya
e-mail: nbogonko@gmail.com

N. A. Bogonko · M. M'Arimi · J. Kiplagat
School of Engineering, Moi University, Eldoret, Kenya

R. Ramkat et al. (eds.), *Towards the Sustainability of our Ecosystems*, Springer Proceedings in Geography and Sustainability,
https://doi.org/10.1007/978-3-032-30076-8_7

Keywords

Lignocellulosic • Pretreatment • Biogas • Anaerobic digestion

1 Introduction

The simultaneous growth in global population and industrial activities significantly influences economic development [1]. Naturally, increased population levels drive a corresponding rise in global energy demand, with projections indicating this trend will persist through 2040 [2]. Presently, fossil fuels are the primary source of energy; however, they are non-renewable and limited in supply. Therefore these resources may not suffice to meet the future energy needs of expanding industrial sectors [3]. In 2019, fossil fuel consumption stood at approximately 84 million barrels per day, and forecasts predicted an increase of 5.7 million barrels per day by 2025 [4].

Excessive fossil fuel usage leads to fluctuating oil prices, accumulation of non-biodegradable waste, greenhouse gas emissions, environmental deterioration, and public health concerns [5]. This reality highlights the urgent need to develop alternative green energy sources that can mitigate the negative impacts associated with fossil fuels [6]. Biomass is increasingly recognized as an efficient and eco-friendly energy source, with the potential to reduce greenhouse gas emissions [7]. Bioenergy production also supports sustainable economic development by addressing energy security concerns [8]. Global biomass production is estimated at around 130 billion tons annually [9], and it can be transformed into energy, chemicals, and value-added products through biochemical, physicochemical, and thermochemical processes [10].

Biomass plays a critical role in meeting energy needs, especially in low-income regions, because it is the most accessible and affordable. According to the Ministry of Energy [11], biomass provides up to 70% of Kenya's cooking energy demand, with over 90% of rural households relying on it [11]. Key biomass sources include wood and agricultural waste, such as bagasse, banana, and flower residues in Kenya. Despite its potential, biomass use can also cause indoor air pollution and environmental degradation, emphasizing the need for cleaner energy alternatives. Biogas, derived from the anaerobic digestion of agricultural waste, offers a sustainable and clean energy solution.

A study by Nzila et al. [12] assessed Kenya's biogas potential using crop residues like cotton, tea, maize, barley, and sugarcane. Their findings indicated that these materials could yield 1,313 million cubic meters of methane annually equivalent to 73% of the country's total energy demand without jeopardizing food production. These residues could also complement or substitute animal manure in biogas systems.

Biological processes such as fermentation and anaerobic digestion have enabled the generation of biofuels like biogas and bioethanol. The most critical factor in anaerobic digestion is the type of feedstock, as almost any organic material can be utilized [13]. Lignocellulosic biomass is particularly suitable due to its abundance and non-competition with food crops for arable land [14]. However, its structural complexity makes the breakdown into fermentable sugars challenging, thereby inhibiting methane production. The primary obstacle is the conversion of complex polysaccharides into simple sugars, which are required by the microbial community in anaerobic digesters [15].

To overcome this, pretreatment is essential. It aims to disrupt or remove lignin and hemicellulose, increasing cellulose accessibility for microbial action during digestion [15]. Because hydrolysis the first step of anaerobic digestion, is often rate-limited by the biomass structure, pretreatment helps improve efficiency by altering the biomass structure and making it more amenable to enzymatic hydrolysis [16]. A variety of pretreatment techniques, including physical, chemical, mechanical, and biological methods, have been explored to enhance biogas production from lignocellulosic feedstocks [17].

2 Methodology

This literature review employed a structured and systematic approach to identify, appraise, and synthesize research on pretreatment of lignocellulosic biomass for enhanced biogas production [18]. A comprehensive search was conducted in major scientific databases such as Scopus, Web of Science, ScienceDirect, PubMed, and Google Scholar using relevant keywords and Boolean combinations, including "lignocellulosic biomass," "anaerobic digestion," "biogas enhancement," "chemical pretreatment," "mechanical pretreatment," "biological pretreatment," and "thermal pretreatment" [19]. The search focused on peer-reviewed journal articles, review papers, and conference publications from 1994 to 2025 to ensure both historical and up-to-date coverage of advancements in pretreatment technologies [20]. The majority of the research cited about 90% is concentrated in the last two decades, particularly from 2010 to 2025, which reflects the recent surge in scientific focus on sustainable energy, biomass utilization, and pretreatment technologies for biogas production. Only studies written in English and presenting quantitative or qualitative data on methane yield, biodegradability improvement, process efficiency, or substrate modification were considered eligible [21].

Inclusion criteria were applied to select studies focusing on lignocellulosic feedstocks such as agricultural residues, agro-industrial byproducts, grasses, and forestry wastes treated with physical, chemical, biological, thermal, or combined pretreatment techniques prior to anaerobic digestion [18]. Studies unrelated to biogas production, those lacking pretreatment data, non-scientific reports, editorials, and patents were excluded [23]. The screening process involved evaluating titles and abstracts first, followed by full-text assessment to confirm relevance and compliance with inclusion criteria [19]. Duplicate records were removed carefully, and the selection process was guided by the Preferred Reporting Items for Systematic Reviews and Meta-Analyses (PRISMA) framework to guarantee transparency and traceability [21].

Data extraction was carried out using a structured template to collect key variables such as biomass type and composition, pretreatment category and operational parameters (including temperature, pH, duration, and concentration), digestion conditions, reactor configuration, and reported outcomes like biogas or methane yield, lignin degradation, volatile solids reduction, and improvements in chemical oxygen demand solubilization [23]. Extracted information was organized to facilitate comparison across pretreatment types and feedstocks [3]. Data were synthesized through narrative and comparative qualitative analysis [22]. Studies were grouped according to pretreatment techniques to identify performance trends, mechanisms of action, benefits, and limitations [18]. Descriptive summaries were used to report ranges of methane yield improvements and the effects of pretreatment on substrate biodegradability [21].

To ensure credibility and reliability, the methodological quality of included studies was assessed based on clarity of experimental procedures, replication, accuracy of reported data, and relevance to anaerobic digestion of lignocellulosic materials [22] Studies with methodological weaknesses or incomplete reporting were noted but interpreted cautiously in the synthesis [19]. The review further identified research gaps related to cost-effectiveness, scalability, environmental sustainability, and synergistic effects of combined pretreatment methods [23]. The findings provide a consolidated understanding of how different pretreatment strategies influence the digestibility and methane potential of lignocellulosic biomass and offer direction for future research and technology optimization [18].

3 Lignocellulosic Biomass Material (LBM) Sources

Lignocellulosic biomass materials (LBM) primarily originate from agricultural and forestry residues, derived from plant-based wastes [24]. Additionally, food and industrial wastes contribute significantly to the pool of lignocellulosic resources [25]. Table 1 presents an overview of the major lignocellulosic sources. Notably, many lignocellulosic crops can be harvested multiple times from a single planting cycle, which reduces cultivation and management costs compared to other energy crops. This cost-effectiveness, combined with their abundance, positions lignocellulosic biomass as one of the most promising resources for future bioenergy applications [26].

Among the top contributors to lignocellulosic biomass are crops such as rice, wheat, sugarcane, and maize. Corn, being the leading cereal crop globally, saw an output of approximately 1.2 billion metric tons during the 2022/2023 growing season. In the same period, wheat production reached around 783.8 million metric tons, followed by milled rice at roughly 510 million metric tons [27]. Sugarcane production was estimated at 177.3 million metric tons [21]. The United States and China are responsible for over half of the world's corn output. In terms of

Table 1 Biomass lignocellulose content as reported in literature sources

Biomass	Cellulose (%)	Hemicellulose (%)	Lignin (%)	References
Sugarcane bagasse	42–48	20–27	20–30	[33]
Rice straw	28–36	21–28	20–28	[34]
Sweet sorghum	34–45	18–27	14–21	[35]
Rice husk	25–35	18–21	26–31	[36]
Corn cob	42–45	35–39	14–15	[36]
Corn stover	36–40	24–35	17–19	[37]
Barley straw	31–45	27–38	14–19	[38]

rice, China leads global production, followed by India and Bangladesh. Wheat production is dominated by China, India, Russia, and the United States, while the top two sugar producers are India and Brazil [28].

Banana cultivation also contributes significantly to biomass waste, especially in tropical regions. As reported by Tesfa and Mekas [29], bananas rank fourth in global food consumption after wheat, rice, and maize. They are cultivated in over 100 countries across approximately 10 million hectares, yielding an estimated 88 million tons annually [29].

The rising interest in lignocellulosic biomass is attributed to its high volume, widespread availability, ease of access, and substantial economic potential. However, a primary barrier to using LBM directly for biogas production is its resistant, or recalcitrant, structural nature, which inhibits effective microbial digestion as illustrated in Fig. 1. Therefore, addressing this limitation is crucial to enable efficient biogas conversion from lignocellulosic sources [30]. Table 1 outlines the organic composition of various biomass waste types.

Typically, lignocellulosic biomass contains between 50 and 75% methane and 25–50% carbon dioxide upon digestion. Given its abundance and composition, it represents a reliable and sustainable feedstock for biogas and other forms of bioenergy. Energy derived from just 50% of the global lignocellulosic biomass could surpass the world's total basic energy requirements [31].

4 Anaerobic Digestion Process

Anaerobic digestion (AD) is a biological process in which diverse microbial communities, primarily bacteria and archaea with smaller populations of fungi and protozoa, metabolize organic matter in oxygen-free conditions to generate biogas as a renewable energy source [38].

The AD process proceeds through four key stages: hydrolysis, acidogenesis, acetogenesis, and methanogenesis. Of these, hydrolysis is the initial and one of the most critical stages, as it involves breaking down complex organic materials into simpler, soluble molecules [39].

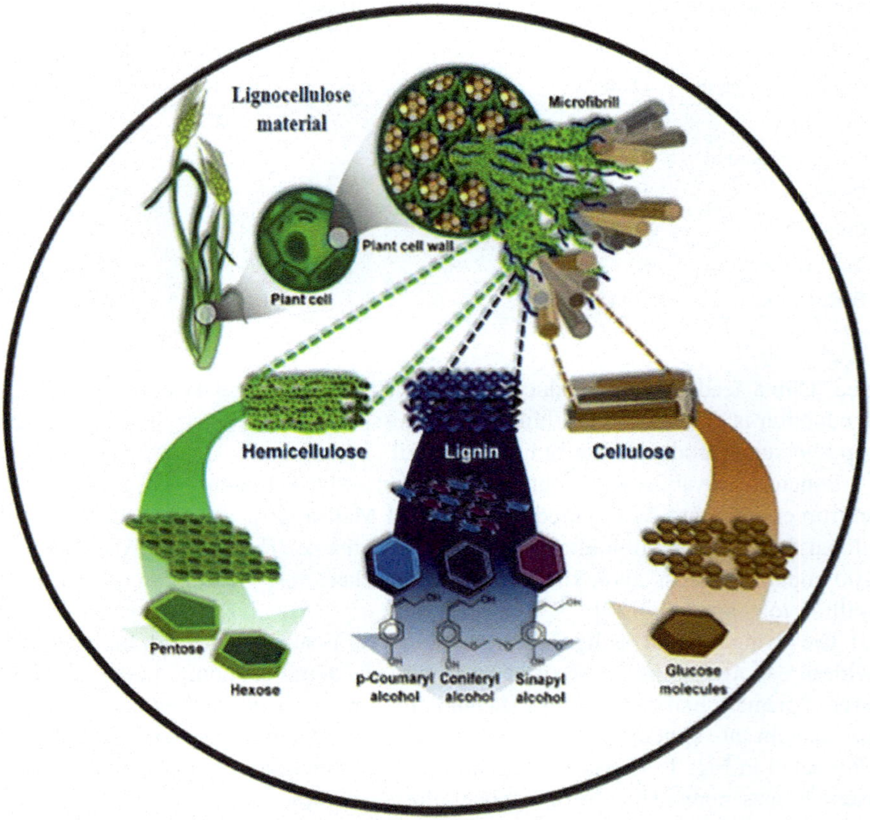

Fig. 1 Lignocellulose biomass structural layout, adopted from Muhamad et al. [32]

During the hydrolysis stage, macromolecules such as carbohydrates, proteins, and fats are degraded into their simpler forms: sugars, amino acids, and fatty acids. This step acts as the rate-limiting stage of the digestion process, meaning it governs the overall speed of bioconversion. Hydrolytic enzymes secreted by fermentative bacteria facilitate this transformation. Additionally, hydrolysis generates compounds like acetate and hydrogen, which are essential substrates for methanogenic microorganisms in the later stages. Molecules that remain partially complex undergo further breakdown during acidogenesis to enhance methane yield [40].

In the acidogenesis stage, the products of hydrolysis are further degraded by acidogenic bacteria, resulting in the production of volatile fatty acids (e.g., acetic, butyric, and propionic acids), along with gases such as hydrogen (H_2), carbon dioxide (CO_2), ammonia (NH_3), and hydrogen sulfide (H_2S), and other minor by-products. The reactor environment becomes acidic at this stage, which supports the metabolic activities of these bacteria [40].

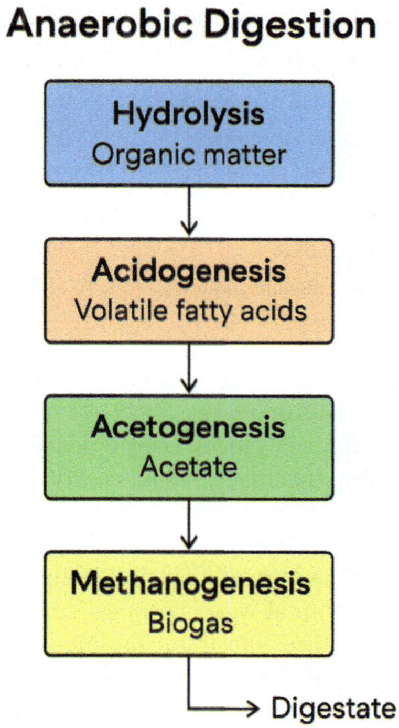

Fig. 2 Stages of the anaerobic digestion process (adapted from Li et al. [41])

The acetogenesis stage follows, during which acetogenic bacteria convert the volatile fatty acids and other intermediates from acidogenesis into acetic acid, H_2, and CO_2. These final products are the primary substrates utilized in the next phase by methanogenic archaea.

As illustrated in Fig. 2, methanogenesis is the final step that produces methane, which is the main component of biogas [41].

Anaerobic digestion presents a promising solution for renewable energy generation and sustainable waste management. However, the technology faces challenges such as low conversion efficiency and long hydraulic retention times (typically 20–30 days), which limit its overall productivity. The hydrolysis stage, being the most critical and rate-limiting step, significantly affects the overall biogas yield. Therefore, the efficiency of anaerobic digestion depends largely on the digestibility and rate of hydrolysis of the feedstock.

To address these limitations, pretreatment of organic waste and addition of microbial activity enhancers are commonly employed. Enhancing the hydrolysis phase through various pre-treatment methods that include: mechanical, thermal, chemical, biological, or a combination thereof can significantly increase biogas production [39]. These methods help in solubilizing complex particulate matter and breaking down long-chain organic polymers into simpler monomers. Consequently, they improve substrate availability by disrupting cell walls and digesting extracellular polymeric substances, making them more amenable to microbial action

Table 2 Biogas yield from different organic wastes as reported in the literature [44]

Organic waste	Total solids (%)	Volatile solids (%)	Biogas yield (m^3 kg^{-1} VS)
Municipal organic waste	15–30	80–95	0.5–0.8
Brewery spent grain	20–26	80–95	0.5–1.1
Vegetable wastes	5–20	76–90	0.3–0.4
Grass cuttings (from lawns)	20–37	86–93	0.7–0.8
Grass silage	21–40	87–93	0.6–0.8
Cattle manure (liquid)	6–11	68–85	0.1–0.8

during digestion. Additives further enhance microbial performance, speeding up the digestion process [42].

Beyond biogas energy generation, AD offers other benefits such as improved waste treatment, production of bio-fertilizers, soil carbon sequestration, and mitigation of uncontrolled methane emissions. For instance, in the agricultural sector, AD can help prevent methane leaks from open slurry tanks, reduce water pollution, and provide renewable energy, thereby supporting a circular economy where agricultural waste becomes a resource for energy production [43].

Table 2 illustrates the biogas yield from different feedstocks. Notably, liquid cattle manure, which is commonly used in small-scale biogas digesters, yields relatively lower volumes of biogas compared to other biomass materials. This underscores the importance of supplementing or replacing traditional substrates with high-yield biomass waste to optimize biogas production.

5 Pretreatment

Several structural and compositional features of lignocellulosic biomass affect its digestibility. These include lignin content, degree of crystallinity, polymerization, surface area, and solubility—all of which determine how readily the biomass can be broken down [45].

Extensive research has explored various pretreatment strategies aimed at enhancing the anaerobic digestion of lignocellulosic materials to improve biogas yields. The effectiveness of a pretreatment method is largely influenced by the physicochemical properties and the structural complexity of the biomass in question. Pretreatment methods are generally grouped into four major categories: biological, chemical, physical, and combined approaches as shown in Fig. 3 [46, 48].

Overall, pretreatment has two primary objectives:

1. Disruption of the bonds between lignin and polysaccharides.
2. Opening up the biomass structure to increase accessibility for microbial attack during anaerobic digestion [47].

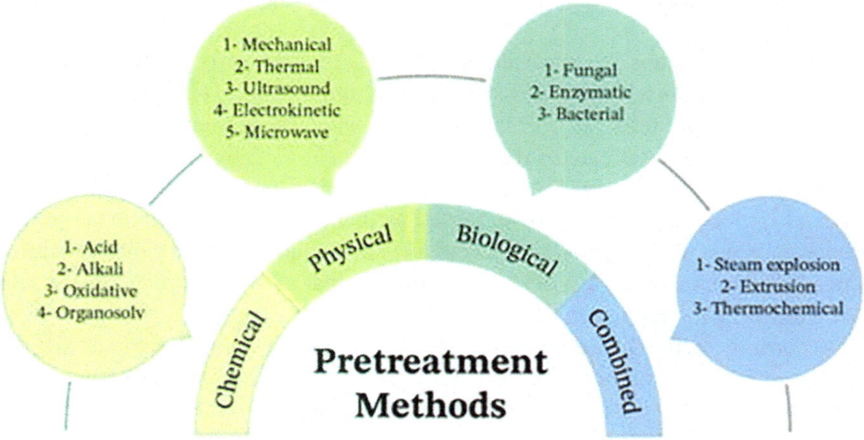

Fig. 3 Lignocellulose composition and organization adapted from Tran et al. [48]

The pretreatment process enhances substrate availability for microbial action and accelerates reaction rates, thereby improving the degradation of organic matter during the anaerobic digestion (AD) process [49]. By increasing the surface area of the biomass, pretreatment facilitates the subsequent digestion steps particularly the hydrolysis phase, which is often a rate-limiting stage [50]. Therefore, disrupting this rigid structure is critical for improving the efficiency of bioconversion [51].

The core goal of pretreatment is to solubilize plant cell walls and convert complex structural components into simpler monomers, making them more accessible for enzymatic hydrolysis [52]. This process increases biogas yields by overcoming the recalcitrant nature of lignocellulosic biomass, reducing both particle size and the crystallinity index of the cellulose [53].

The effectiveness of various pretreatment methods depends on several factors: the type and structure of the biomass, the process parameters (such as temperature, pressure, and duration), and the desired sugar yield aligned with the requirements of the chosen microbial fermentation pathway.

5.1 Physical Pretreatment

Physical pretreatment techniques primarily include mechanical, thermal, and ultrasound methods.

Mechanical Pretreatment

Mechanical processing is one of the widely used techniques to improve the efficiency of the biomethanation process. Common methods under this category include grinding, milling, and chipping, which are employed to reduce the particle size of the biomass [54, 55]. Although this approach significantly boosts biogas production, it is often associated with high-energy consumption, making it relatively costly.

Reducing particle size enhances the surface area and porosity of the biomass, lowers cellulose crystallinity, and subsequently improves the effectiveness of subsequent digestion steps [49]. This results in better biodegradability and greater biogas yields from pretreated substrates.

In a study conducted by Dahunsi [56], mechanical pretreatment applied to six types of lignocellulosic biomass increased methane yields by approximately 22%. However, it was also observed that excessive size reduction could negatively affect biogas production efficiency [55]. This indicates that an optimal particle size must be determined for each specific substrate and digestion method.

According to the literature, there is no universally optimal particle size for maximizing biogas output; the ideal size varies depending on the biomass type and anaerobic digestion process conditions [57]. Fang et al. [58] also reported that reducing particle size generally results in increased biogas yields, with improvements in total hydrolysis yield ranging between 5 and 25%, and reductions in digestion time of 23–59%.

Table 3 summarizes the influence of mechanical pretreatment on biogas production efficiency.

Table 3 Mechanical pretreatment of various feedstocks

Substrates	Particle size (mm)	Methane Yield (Untreated)	Methane Yield (Treated)	Reactor	References
Maize stalks	2	246 mL/g VS	272 mL/g VS	Glass reactor 2L	[59]
	20	246 mL/g VS	254 mL/g VS	Glass reactor 2L	[59]
Rice straw	0.75	58.1 mL/g VS	65.7 mL/g VS	Glass reactor 2L	[60]
	50	197 mL/g VS	203 mL/g VS	Glass reactor 2L	[59]
Water hyacinth	0.01	–	Increase of 20% from 50 to 70%	Digestor 0.45L	[61]
	0.05	–	Increase of 16% from 50 to 66%	Digestor 0.45L	[61]
	1.0	–	Increase of 10% from 50 to 60%	Digestor 0.45L	[61]

Thermal Pretreatment

Thermal pretreatment is employed to improve the biodegradability of organic waste before anaerobic digestion. It facilitates the breakdown of complex compounds into simpler, more digestible forms [62]. This process involves partial solubilization of organic and inorganic components, reducing digester volume and enhancing biogas yields [54]. Literature suggests that the optimal temperature range for thermal pretreatment is 160–200 °C with a treatment duration of 30–60 min. However, exceeding 250 °C should be avoided due to the onset of undesirable pyrolysis reactions [63]. Thermal pretreatment can be implemented through steam explosion or liquid hot water techniques. While effective, the high-energy requirements of this method can significantly increase operational costs, limiting its practical application.

Ultrasound Pretreatment

Ultrasound or ultrasonication is an effective method for improving the biodegradability of biomass. High-intensity ultrasound waves alter the structure of lignocellulosic materials, making them easier for microbial digestion [58]. Studies have shown that ultrasonication increases biogas output and reduces volatile solids. For instance, Wang et al. [64] reported a 64% increase in methane production when ultrasonication was applied to waste-activated sludge.

Microwave Irradiation

Microwave pretreatment is used to disrupt cell walls, reduce cellulosic crystallinity, and increase the available surface area for enzymatic action [65]. The effectiveness of this method is enhanced when combined with mild alkaline reagents. Recent research has shown promising results in treating lignocellulosic biomass, prompting the transition from lab-scale to pilot-scale applications [66]. In particular, Chen et al. [67] demonstrated a significant breakdown of lignocellulosic structure in sugarcane bagasse when treated with microwave heating at 190 °C for 5 min.

5.2 Chemical Pretreatment

Chemical pretreatment is widely recognized for its high efficiency and rapid action in breaking down complex organic matter [68, 69]. It enhances biogas production, reduces the hydraulic retention time (HRT), and improves the solubility of biomass components [70]. However, its primary drawbacks are the high cost of chemicals and limited recovery of these reagents post-treatment [44]. Acidic and alkaline agents are the most common chemicals used. Table 4 summarizes the effects of different chemical pretreatment methods on selected organic feedstocks and their impact on anaerobic digestion performance.

Table 4 Effects of chemical pretreatments on different organic feedstocks

Feedstock	AD condition	Treatment technique	Observations	References
Plant residues and cattle dung	–	NaOH (1% for 7 days)	Digestion efficiency increased by 31–42% compared to untreated cattle dung with residues	[75]
Pulp and paper sludge	Batch reactor	NaOH (8 g per 100 g TS)	Methane production increased by 183.5% relative to untreated sludge	[76]
Oil palm empty fruit bunches	Batch reactor	NaOH (8% for 60 min)	Methane yield increased by 85% following NaOH pretreatment	[77]
Water hyacinth	CSTR (25–36 °C)	KOH	Methane concentration in biogas rose from 60 to 71%	[78]

Alkaline Pretreatment

Alkaline pretreatment involves the use of alkaline solutions to remove lignin and partially degrade hemicellulose, thereby increasing enzyme accessibility to cellulose and enhancing the rate of hydrolysis [65]. According to Kim et al. [71], alkaline treatment selectively targets lignin while preserving carbohydrates, resulting in increased porosity and surface area due to swelling reactions. This modification improves the enzymatic breakdown and biodegradability of biomass. Common alkaline agents include NaOH, KOH, $Ca(OH)_2$, and $Mg(OH)_2$ [40]. Among the pretreatment methods, alkaline pretreatment is considered cost-effective in terms of equipment and operational expenses [72].

Acid Pretreatment

Acid pretreatment utilizes either dilute or concentrated acids (e.g., acetic acid, nitric acid) to chemically break down lignocellulosic biomass. The primary function is to solubilize hemicellulose and disrupt lignin structures, thus improving enzyme and microbial access to cellulose [73]. Strong acid treatments are more effective at solubilization, but they also lead to higher costs for acid recovery and may form inhibitory compounds that hinder further digestion [74]. These challenges make acid pretreatment less economically favorable despite its efficiency in structural disruption.

Organosolv Pretreatment

Organosolv pretreatment utilizes organic solvents such as methanol, ethanol, acetone, or ethylene glycol, often in combination with catalysts (acidic, alkaline, or salts), to treat lignocellulosic biomass under high temperature and pressure conditions [79, 80]. Typically, the process is conducted at temperatures around 200 °C, though this may vary depending on the feedstock and catalyst used [81].

This technique is particularly effective at removing lignin while preserving the cellulose structure. For instance, Aslanzadeh et al. [82] reported a 53% and 51% increase in biogas yield when cattle and horse manure were pretreated with N-methylmorpholine N-oxide (NMMO).

Despite its effectiveness, the major limitations of this method are the cost and recovery of solvents, along with the formation of fermentation inhibitors. During treatment, hemicellulose can degrade into furfural derivatives, which are known to inhibit enzymatic hydrolysis [83]. While solvent recovery can reduce costs, the initial investment remains high, making this method less feasible for large-scale application.

Oxidative Pretreatment

Oxidative pretreatment involves the use of oxidizing agents such as ozone, hydrogen peroxide, oxygen, or ferric chloride to break down lignin and hemicellulose, enhancing the accessibility of cellulose for enzymatic hydrolysis [84]. A specific form, wet oxidation, is carried out at around 200 °C and 1.5 MPa pressure, where oxygen is introduced into the system to promote degradation.

The main challenge with oxidative pretreatment is the formation of inhibitory compounds during the process, which can hinder microbial activity during anaerobic digestion and ultimately reduce biogas yield [85]. This makes process optimization crucial for effective integration into AD systems.

5.3 Biological Pretreatment

Biological pretreatment employs microbial processes including fungal, bacterial, ensiling, and enzymatic methods to degrade lignin and hemicellulose and improve biogas production [86]. These approaches are environmentally friendly, low-energy, and effective in enhancing the biodegradability of biomass before anaerobic digestion.

Fungal Pretreatment

Fungi, particularly white-rot, brown-rot, and soft-rot fungi, have been widely studied for their ligninolytic and cellulolytic properties. White-rot fungi are particularly effective at degrading lignin, while brown-rot fungi primarily degrade cellulose, and soft-rot fungi can attack both components [87].

In a study by Zhao et al. [88], fungal pretreatment of yard trimmings resulted in a 16.2% reduction in hemicellulose and a 20% reduction in lignin, leading to a 23% increase in biogas production. Similarly, Wayman et al. [89] observed a 20% improvement in methane yield when corn stover was pretreated using fungi. A summary of fungal pretreatment results is presented in Table 5.

Table 5 Fungal pretreatment of various biomass materials

Feedstock	Conditions in digestor	Pretreatment	Results	References
Corn straw	UASB, Mesophilic conditions	Yeast and cellulolytic bacteria	Biogas production increased by 33.07%, methane yield by 75.57%, digestion time reduced 34.6%	[90]
Corn stover	Mesophilic conditions	Fungi, Pleurotus ostreatus	Biogas yield increased by 20% following fungal pretreatment	[91]
Corn stover silage	Batch (37 °C) for 30 days	Fungi	Hemicellulose and lignin reduced by 32.4% and 22.6%; biogas yield increased from 215.5 to 265.1 mL/g VS	[92]
Yard trimmings	Batch (37 °C) for 27 days	Fungi	Hemicellulose and lignin decreased by 16.2% and 20.2%; biogas yield increased from 20 to 44.6 mL/g VS	[88]
Pulp and paper sludge	Batch reactor	Mushroom compost extracts	Methane yield was 134.2% of untreated control	[92]
Wheat straw	Batch reactor (37 °C)	White-rot fungi	Biogas yield from pretreated straw doubled compared to untreated straw	[93]

Bacterial Pretreatment

Certain bacteria with high biodegradation capabilities, such as *Pseudomonas*, *Escherichia coli*, and *Salmonella*, have been employed in biological pretreatment [94]. Poszyrek et al. [95] observed a 38% increase in biogas production when a microbial consortium with high cellulolytic activity was used for the digestion of maize silage.

Enzymatic Pretreatment

Enzymatic pretreatment is a fast-acting process that typically occurs within a few hours, as enzymes are smaller and more efficient than whole microorganisms [96]. For cellulose hydrolysis, enzymes such as endoglucanase, exoglucanase, and β-glucosidase are essential. Hemicellulose breakdown requires endo-xylanase, α-glucuronidase, endo-mannanase, and related enzymes [97].

Schroyen et al. [98] reported a 24% increase in methane production from corn stover after enzymatic pretreatment. Similarly, Bremond et al. [99] observed a 111% increase in methane yield from corn stover pretreated with cellulase enzymes.

Ensiling

Ensiling is traditionally used for storing wet animal feedstock, but it also acts as a pretreatment method for anaerobic digestion. During ensiling, soluble carbohydrates are converted into acids such as acetic, lactic, butyric, and propionic, enhancing digestibility. Franco et al. [100] noted increased methane yields under specific ensiling conditions. Teixeira [101] found a 31% improvement in biogas yield from ensiled amaranth residues, and Liu et al. [101] reported that daily biogas yields doubled when ensiled corn stover was used. Given the dual utility of ensiled biomass for animal feed and energy production, this method is especially viable in rural settings.

5.4 Combined Pretreatment

Combined pretreatment involves the application of two or more methods such as thermal, chemical, physical, and ultrasound treatments to maximize biomass breakdown and enhance biogas yield [102, 103]. This synergistic approach improves substrate accessibility and overall process efficiency.

Thermochemical Pretreatment
This method combines thermal and chemical treatment typically using acids or alkalis while maintaining controlled temperature in a water bath [104]. The adjustment of pH using chemicals facilitates the disruption of lignin, a key recalcitrant component. Monlau et al. [105] demonstrated that sunflower oil cake treated at 170 °C with 1% sulfuric acid produced 50% more biogas compared to untreated feedstock.

Steam Explosion
Steam explosion involves exposing biomass to high-temperature steam (160–260 °C) at a pressure of 0.69–4.83 MPa, followed by explosive decompression [105–108]. This method breaks down lignocellulosic structure through both mechanical and chemical effects. Viola et al. [106] found that adding acid or alkali improved the efficiency of the process. Nges et al. [109] reported a 57% increase in biogas yield from Miscanthus *lutarioriparius* pretreated with 0.3 M NaOH and 0.5 mm particle size. However, harsh processing conditions often generate inhibitors such as aromatic compounds and furans, which impair enzymatic hydrolysis and methane production [110].

Extrusion Pretreatment
Extrusion pretreatment applies mechanical and thermal forces via a rotating screw inside a pressurized barrel, leading to fiber disruption, heat buildup, and moisture evaporation [111]. Chen et al. [112] reported a 72.2% increase in biogas yield from extruded rice straw 32.5% higher than milled samples. Similarly, Pérez et al. [113] observed a 22.3% improvement in biogas yield from corn cobs using a combination of extrusion and enzymatic hydrolysis. Kozlowski et al. [114] found significant improvements in maize straw digestion. However, high-energy consumption is a major drawback, often outweighing the benefits [114].

6 Limitations of Pretreatment Methods

Despite the clear benefits, each pretreatment method presents unique challenges:

- Mechanical pretreatment enhances biodegradability but fails to degrade lignin and consumes up to 33% of process energy [115].
- Chemical pretreatment offers effective hydrolysis but entails high chemical and neutralization costs and produces inhibitory by-products [116].
- Biological methods, while eco-friendly, are slow, space-intensive, and involve high costs for enzymes and microbial cultures [117].
- Thermal pretreatment requires significant energy input, and may release inhibitory compounds like furfural and hydroxymethylfurfural, which hinder digestion [118]. It is best suited for settings where waste heat is readily available.

7 Conclusion and Recommendations

This review highlights a range of pretreatment strategies physical, chemical, biological, and combined used to enhance biogas production from biomass during anaerobic digestion. Numerous studies demonstrate that these methods significantly improve biogas yield.

However, high-energy consumption, inhibitor formation, long processing times, and economic constraints remain barriers to widespread adoption. These limitations must be addressed to make these technologies viable, particularly in resource-limited settings.

Given that cattle slurry, which has low biogas potential, is the primary feedstock in many households in Kenya and other developing countries, there is a critical need for cost-effective, environmentally friendly, and locally sourced pretreatment techniques. Future research should focus on optimizing natural and low-cost materials to improve anaerobic digestion efficiency and expand energy access.

Acknowledgements Not applicable.

Author Contributions Author BAN conducted the literature review and prepared the manuscript. Authors AMM, KJ, and KCJ reviewed the drafts. All authors have read and approved the final manuscript.

Funding No funding was received for this work.

Data Availability Statement The data presented in this study are derived from the cited literature sources.

Conflict of Interest Statement The authors declare no conflict of interest.

References

1. Ahmed SF, Liu G, Mofijur M, Azad AK, Hazrat MA, Chu YM (2021) Physical and hybrid modelling techniques for earth-air heat exchangers in reducing building energy consumption: performance, applications, progress, and challenges. Sol Energy 216:274–294
2. Ong HC, Tiong YW, Goh BHH, Gan YY, Mofijur M, Fattah IR (2021) Recent advances in biodiesel production from agricultural products and using ionic liquids: opportunities and challenges. Energy Convers Manage 228:113647
3. Chen WH, Lo HJ, Yu KL, Ong HC, Sheen HK (2021) Valorization of sorghum distillery residue to produce bioethanol for pollution mitigation and circular economy. Environ Pollut 285:117196
4. International Energy Agency (IEA) (2021) Tracking SDG7: the energy progress report 2021. Washington, The World Bank, DC
5. Nguyen XP, Hoang A, Tölçer AI, Huynh TT (2021) Record decline in global CO2 emissions prompted by COVID-19 pandemic and its implications on future climate change policies. Energy Sources Part A 43(1):1–4
6. Rezania S, Oryani B, Cho J, Talaiekhozani A, Sabbagh F, Hashemi B et al (2020) Different pretreatment technologies of lignocellulosic biomass for bioethanol production: an overview. Energy 199:117457
7. Papathoti NK, Laemchiab K, Megavath VS, Keshav PK, Numparditsu P, Le Thanh T et al (2021) Augmented ethanol production from alkali-assisted hydrothermal pretreated cassava peel waste. Energy Sources Part A 43(1):1–11
8. Schröder P, Beckers B, Daniels S, Gnädinger F, Maestri E, Marmiroli N et al (2018) Intensify production, transform biomass to energy and novel goods and protect soils in Europe—a vision how to mobilize marginal lands. Sci Total Environ 616:1101–1112
9. Sheldon RA (2014) Green and sustainable manufacture of chemicals from biomass: state of the art. Green Chem 16(3):950–963
10. Adams P, Bridgwater T, Lea-Langton A, Ross A, Watson I (2018) Biomass conversion technologies. In: Greenhouse gas balances of bioenergy systems. Academic Press, p 107–139
11. Government of Kenya (2019) Ministry of Energy. Kenya Cook Stove Sector Study. Government of Kenya, Nairobi
12. Nzila C, Dewulf J, Spanjers H, Kiriamiti H, Van HL (2010) Biowaste energy potential in Kenya. Renew Energy 35(12):2698–2704
13. Steffen R, Szolar O, Braun R (1998) Feedstocks for anaerobic digestion. Institute of Agrobiotechnology, University of Agricultural Sciences, Tulln
14. Sawatdeenarunat C, Surendra K, Takara D, Oechsner H, Khanal SK (2015) Anaerobic digestion of lignocellulosic biomass: challenges and opportunities. Bioresour Technol 178:178–186
15. Chandra R, Takeuchi H, Hasegawa T (2012) Methane production from lignocellulosic agricultural crop wastes: a review in context to second generation of biofuel production. Renew Sustain Energy Rev 16:1462–1475
16. Kavitha S, Schikaran M, Kannah RY, Gunasekaran M, Kumar G, Banu JR (2019) Nanoparticle-induced biological disintegration: a new phase-separated pretreatment strategy on microalgal biomass for profitable biomethane recovery. Bioresour Technol 289:121624
17. Banu JR, Sharmila VG, Kumar SA, Kumar G, Nguyen DD, Saratale GD et al (2019) Cost effective sludge reduction using synergetic effect of dark Fenton and disperser treatment. J Clean Prod 207:261–270
18. Khan MU, Ahring BK (2019) Lignocellulosic biomass pretreatment for enhanced biogas production: recent advances. Bioresour Technol 291:121861. https://doi.org/10.1016/j.biortech.2019.121861

19. Deepanraj B, Sivasubramanian V (2021) Thermal and chemical pretreatment of agricultural residues for biogas production: a review. Energy Convers Manage 227:113556. https://doi.org/10.1016/j.enconman.2020.113556
20. Nguyen D, Wörner A (2022) Biological pretreatment of lignocellulosic biomass for anaerobic digestion: developments and perspectives. Waste Manage 138:144–152. https://doi.org/10.1016/j.wasman.2021.11.031
21. García-Bernet D, Buffière P (2023) Combined pretreatment methods to enhance biogas generation from lignocellulose. Chem Eng J 451:138624. https://doi.org/10.1016/j.cej.2022.138624
22. Li Y, Zhang R, Chen C, Liu G (2020) Pretreatment strategies for improving methane yield from lignocellulosic biomass. Renew Sustain Energy Rev 133:110287. https://doi.org/10.1016/j.rser.2020.110287
23. Alemu AW, Singh A (2024) Hybrid pretreatment technologies for efficient biomethane production from lignocellulosic feedstocks. Renew Energy 226:120066. https://doi.org/10.1016/j.renene.2023.120066
24. Singhvi M, Zinjarde S, Kim BS (2022) Sustainable strategies for the conversion of lignocellulosic materials into biohydrogen: challenges and solutions toward carbon neutrality. Energies 15:8987
25. Ojo AO, de Smidt O (2023) Lactic acid: a comprehensive review of production to purification. Processes 11:688
26. Saini JK, Saini R, Tewari L (2015) Lignocellulosic agricultural wastes as biomass feedstocks for second-generation bioethanol production: concepts and recent developments. 3 Biotech. 5:337–353
27. Shahbandeh M (2023) Rice—statistics and facts. Statista. Available from: https://www.statista.com/topics/1443/rice/. Accessed on 10 May 2025
28. World Population Review (2023) Wheat production by country. Available from: https://worldpopulationreview.com/country. Accessed on 11 May 2025
29. Tesfa B, Mekias D (2015) Evaluation of the adaptability of Dessert Banana Cultivars at Belles Valley, North Western Ethiopia. Afr J Agric Res 1:2995–2999
30. Trusi A (2019) A review on biomass: importance, chemistry, classification and conversion. Biofuel Res J. 22:962–979
31. Niju S, Swathika M, Balajii M (2020) Pretreatment of lignocellulosic sugarcane leaves and tops for bioethanol production. In: Lignocellulosic biomass to liquid biofuels. Academic Press, p 301–324
32. Muhammad UK, Usman M, Awais A, Nalok D, Gang L, Shicheng Z (2022) A review of recent advancements in pretreatment techniques of lignocellulosic materials for biogas production: opportunities and limitations. Chem Eng J Adv. 10:100263
33. DeSouza AP, Leite DC, Pattathil S, Hahn MG, Buckeridge MS (2013) Composition and structure of sugarcane cell wall polysaccharides: implications for second-generation bioethanol production. Bioenerg Res. 6(2):564–579
34. Imman S, Arnthong J, Burapatana V, Champreda V, Laosiripojana N (2015) Fractionation of rice straw by a single-step solvothermal process: Effects of solvents, acid promoters and microwave treatment. Renew Energy 83:663–673
35. Umagiliyage AL, Choudhary P, Liang Y, Haddock J, Watson DG (2015) Laboratory scale optimization of alkali pretreatment for improving enzymatic hydrolysis of sweet sorghum bagasse. Ind Crops Prod 74:977–986
36. Tayyab M, Noman A, Islam W, Waheed S, Arafat Y, Ali F et al (2017) Bioethanol production from lignocellulosic biomass by environment-friendly pretreatment methods: a review. Appl Ecol Environ Res 16(1):225–249
37. Saha BC, Qureshi N, Kennedy GJ, Cotta MA (2016) Biological pretreatment of corn stover with white-rot fungus for improved enzymatic hydrolysis. Int Biodeterior Biodegrad 109:29–35

38. Deng C, Lin R, Kang X, Wu B, Wall DM, Murphy JD (2021) What physicochemical properties of biochar facilitate interspecies electron transfer in anaerobic digestion: a case study of digestion of whiskey by-products. Fuel 306:121736
39. Bougrie C, Albasi C, Delgenes J, Carrere H (2006) Effect of ultrasonic, thermal and ozone pre-treatments on waste activated sludge solubilisation and anaerobic biodegradability. Chem Eng Process 45:711–718
40. Raja A, Wazir S (2017) Biogas production: the fundamental processes. Univ J Eng Sci. 5(2):29–37
41. Li Y, Jin Y, Li H, Borrion A, Yu Z, Li J (2018) Kinetic studies on organic degradation and its impacts on improving methane production during anaerobic digestion of food waste. Appl Energy 213:136–147. https://doi.org/10.1016/j.apenergy.2018.01.033
42. Stambasky J (2016) Anaerobic digestion industry potential contribution to carbon dioxide mitigation in Ireland. Prague
43. Bioenergy IEA (2019) A case story: organic biogas improves nutrient supply. Biogas Society
44. Ariunbaatar J, Panico A, Esposito G, Pirozzi F, Lens PNL (2014) Pretreatment methods to enhance anaerobic digestion of organic solid waste. Appl Energy 123:143–156
45. Monlau F, Barakat A, Trably E, Dumas C, Steyer JP, Carrère H (2013) Lignocellulosic materials into biohydrogen and biomethane: impact of structural features and pretreatment. Crit Rev Environ Sci Technol 43(3):260–322
46. Sun Y, Cheng J (2002) Hydrolysis of lignocellulosic materials for ethanol production: a review. Bioresour Technol 83(1):1–11
47. Pu Y, Hu F, Huang F, Davison BH, Ragauskas AJ (2013) Assessing the molecular structure basis for biomass recalcitrance during dilute acid and hydrothermal pretreatments. Biotechnol Biofuels 6(1):1–13
48. Tran T, Le P, Mai P, Nguyen Q (2019) Ethanol Production from Lignocellulosic Biomass. Intech Open, p 1–13
49. Atelge MR, Atabani AE, Banu JR, Krisa D, Kaya M, Eskicioglu C et al (2020) A critical review of pretreatment technologies to enhance anaerobic digestion and energy recovery. Fuel 270:117494
50. Sharmila VG, Angappane S, Gunasekaran M, Kumar G, Banu JR (2020) Immobilized ZnO nanofilm impelled bacterial disintegration of dairy sludge to enrich anaerobic digestion for profitable bioenergy production: energetic and economic analysis. Bioresour Technol 308:123276
51. Sankaran R, Cruz RAP, Pakalapati H, Show PL, Ling TC, Chen WH et al (2020) Recent advances in the pretreatment of microalgal and lignocellulosic biomass: a comprehensive review. Bioresour Technol 298:122476
52. Costa JV, Freitas CB, Moraes Zaparoli M, Morais MG (2020) Progress in the physicochemical treatment of microalgae biomass for value-added product recovery. Bioresour Technol 301:122727
53. Zabed H, Sahu JN, Suely A, Boyce AN, Faruq G (2017) Bioethanol production from renewable sources: current perspectives and technological progress. Renew Sustain Energy Rev 71:475–501
54. Mudhoo A (2012) Biogas production: pretreatment methods in anaerobic digestion. John Wiley & Sons
55. Kumari D, Singh R (2018) Pretreatment of lignocellulosic wastes for biofuel production: a critical review. Sustain Energy Rev 90:877–891
56. Dahunsi S (2019) Mechanical pretreatment of lignocelluloses for enhanced biogas production: methane yield prediction from biomass structural components. Bioresour Technol 280:18–26
57. Dumas C, Damasceno GSG, Abdellatif B, Carrère H, Steyer JP, Rouau X (2015) Effects of grinding processes on anaerobic digestion of wheat straw. Ind Crops Prod 74:450–456
58. Pilli S, Bhunia P, Yan S, LeBlanc RJ, Tyagi RD, Surampalli RY (2011) Ultrasonic pretreatment of sludge: a review. Ultrason Sonochem 18(1):1–18

59. Menardo S, Airoldi G, Balsari P (2012) The effect of particle size and thermal pre-treatment on the methane yield of four agricultural by-products. Bioresour Technol 104:708–714
60. Chandra R, Takeuchi H, Hasegawa T, Vijay VK (2015) Experimental evaluation of substrate's particle size of wheat and rice straw biomass on methane production yield. Agric Eng Int CIGR J 17:93–104
61. Nalinga Y, Legonda I (2016) The effect of particle size on biogas. Int J Innov Res Technol Sci. 4:9–13
62. Fang S, Ping L, Yang Z, Mao J (2011) A review of different pretreatment techniques for enhancing biogas production. In: International conference on materials for renewable energy & environment (ICMREE). Shanghai
63. Kim J, Park C, Kin T, Lee M, Kim S, Kim SW, Lee J (2003) Effects of various pretreatments for enhanced anaerobic digestion with waste activated sludge. J Biosci Bioeng 95:271–275
64. Wang Q, Kuninobu M, Kakimoto K, Ogawa H, Kato Y (1999) Upgrading of anaerobic digestion of waste activated sludge by ultrasonic pretreatment. Bioresour Technol 68:309–313
65. Taherzadeh MJ, Karimi K (2008) Pretreatment of lignocellulosic wastes to improve ethanol and biogas production: a review. Int J Mol Sci 9:1621–1651
66. Li H, Qu Y, Yang Y, Chang S, Xu J (2016) Microwave irradiation—a green and efficient way to pretreat biomass. Bioresour Technol 199:34–41
67. Chen W, Yu H, Liu Y, Chen P, Zhang M, Hai Y (2011) Individualization of cellulose nanofibers from wood using high-intensity ultrasonication combined with chemical pretreatments. Carbohydr Polym 83(4):1804–1811
68. Olatunji KO, Ahmed NA, Ogunkunle O (2021) Optimization of biogas yield from lignocellulosic materials with different pretreatment methods: a review. Biotechnol Biofuels 14(1):1–34
69. Hernández-Beltrán JU, Hernández-De Lira IO, Cruz-Santos MM, Saucedo Luevanos A, Hernández-Terán F, Balagurusamy N (2019) Insight into pretreatment methods of lignocellulosic biomass to increase biogas yield: current state, challenges, and opportunities. Appl Sci 9(18):3721
70. Shrestha B, Hernandez R, Fortela DLB, Sharp W, Chistoserdov A, Gang D et al (2020) A review of pretreatment methods to enhance solids reduction during anaerobic digestion of municipal wastewater sludges and the resulting digester performance: implications to future urban biorefineries. Appl Sci 10(24):9141
71. Kim J, Sunagawa M, Kobayashi S, Shin T, Takayama C (2016) Developmental localization of calcitonin gene-related peptide in dorsal sensory axons and ventral motor neurons of mouse cervical spinal cord. Neurosci Res 105:42–48
72. Baccay RA, Hashimoto AG (1984) Acidogenic and methanogenic fermentation of causticized straw. Biotechnol Bioeng 26:885–891
73. Hendriks ATM, Zeeman G (2009) Pretreatments to enhance the digestibility of lignocellulosic biomass. Bioresour Technol 100:10–18
74. Solarte-Toro JC, Romero-García JM, Martínez-Patiño JC, Ruiz-Ramos E, Castro-Galiano E, Cardona-Alzate CA (2019) Acid pretreatment of lignocellulosic biomass for energy vectors production: a review focused on operational conditions and techno-economic assessment for bioethanol production. Renew Sustain Energy Rev 107:587–601
75. Dar GH, Tandon SM (1987) Biogas production from pretreated wheat straw, lantana residue, apple and peach leaf litter with cattle dung. Biol Wastes 21:75–83
76. Lin Y, Wang D, Wu S, Wang C (2009) Alkali pretreatment enhances biogas production in the anaerobic digestion of pulp and paper sludge. J Hazard Mater 170:366–373
77. Nieves DC, Soto M, Chávez JA, Hernández EA, Juárez ME, Bravo JA (2015) Effect of alkaline pretreatment on the biochemical methane potential of different lignocellulosic substrates. Renew Energy 83:287–293
78. Mendonça F, Mezzari MP, Figueiredo PF (2012) A study of alkaline pretreatment of sugarcane bagasse for biogas production. Bioresour Technol 104:96–102
79. Lee YY, Kwon JH, Koo BJ, Lee DS, Kim HK, Lee KS (1999) Dilute sulfuric acid pretreatment of rice straw for ethanol production. Bioresour Technol 69:33–44

80. Sun Y, Cheng J (2002) Hydrolysis of lignocellulosic materials for ethanol production: a review. Bioresour Technol 83:1–11
81. Yan S, Zhang Y, He Z, Yang J, Zhu W, Sun L et al (2018) Effects of microwave pretreatment on methane production from rice straw under mesophilic conditions. Bioresour Technol 249:39–46
82. Aslanzadeh S, Taherzadeh MJ, Sárvári HI (2011) Pretreatment of straw fraction of manure for improved biogas production. BioResources 6(4):5193–5205. https://doi.org/10.15376/biores.6.4.5193-5205
83. Zhu J, Chen L, Guo J, Zhou H, Zhang Y (2015) Effects of microwave pretreatment on anaerobic digestion of lignocellulosic biomass. Energy Fuels 29:1984–1990
84. Yang H, Yan R, Chen H, Lee DH, Zheng C (2007) Characteristics of hemicellulose, cellulose and lignin pyrolysis. Fuel 86(12–13):1781–1788
85. Tursi A (2019) Anaerobic digestion of organic solid waste. Sustain Cities Soc 44:159–166
86. Chen J, Jiang Y, Xiao B, Zheng Y, Wang Y (2017) Overview of acid and alkali pretreatments of lignocellulosic biomass for biogas production. Renew Energy 115:1153–1159
87. Herrmann C, Boe K, Møller HB (2016) Influence of particle size and ammonia pretreatment on methane yield and kinetics of anaerobic digestion of solid manure. Waste Manage 49:268–276
88. Zhao J, Ge X, Vasco-Correa J, Li Y (2014) Fungal pretreatment of unsterilized yard trimmings for enhanced methane production by solid-state anaerobic digestion. Bioresour Technol 158:248–252. https://doi.org/10.1016/j.biortech.2014.02.029
89. Wayman M, Yu S, Chen Y, Wang C (1993) Pretreatment of lignocellulosic materials to increase biogas production. Biomass Bioenergy 4(3):189–194. https://doi.org/10.1016/0961-9534(93)90040-W
90. Ghatak H, Roy R, Banerjee R (2020) Improvement of biogas production from anaerobic digestion of lignocellulosic biomass by chemical pretreatment: a review. Biomass Convers Biorefinery 10:845–859
91. de Souza BM, Santos MF, Marangoni CB, Andrade MJ, de Andrade RP, Gonçalves AR (2018) Influence of thermal pretreatment on anaerobic digestion of agro-industrial residues: a review. Renew Sustain Energy Rev 82:1185–1195
92. Rajagopal R, Massé DI, Singh G (2013) A critical review on inhibition of anaerobic digestion process by excess ammonia. Bioresour Technol 143:632–641
93. Mustafa AM, Poulsen TG, Sheng K (2016) Fungal pretreatment of rice straw with Pleurotus ostreatus and Trichoderma reesei to enhance methane production under solid-state anaerobic digestion. Appl Energy 180:661–671. https://doi.org/10.1016/j.apenergy.2016.07.135
94. Angelidaki I, Ahring BK (1993) Effects of free ammonia on methane production and syntrophic acetate oxidation in anaerobic digestion. Appl Environ Microbiol 59:3252–3258
95. Poszytek K, Dębowski M, Żywicka A (2021) Progress in the production of biogas from maize silage after acid-heat pretreatment. Energies 14(23):8018. https://doi.org/10.3390/en14238018
96. Huang Y, Zhang R, Wang S, Li Y (2011) Effects of heavy metals on anaerobic digestion process: a review. Bioresour Technol 102:1019–1025
97. Naziri D, Tzotzi F, Papadopoulos A, Kornaros M (2019) Effects of heavy metals on biogas production and microbial communities during anaerobic digestion. Waste Manag 87:92–99
98. Schroyen M, Vriens B, Van Acker K, Vrancken K (2014) Impact of enzymatic pretreatment on corn stover degradation and biogas production. Bioresour Technol 173:59–66. https://doi.org/10.1016/j.biortech.2014.09.030
99. Brémond U, de Buyer R, Steyer JP, Bernet N, Carrere H (2018) Biological pretreatments of biomass for improving biogas production: an overview from lab scale to full-scale. Renew Sustain Energy Rev 90:583–604. https://doi.org/10.1016/j.rser.2018.03.103
100. Franco RT, Buffière P, Bayard R (2018) Co-ensiling of cattle manure before biogas production: effects of fermentation stimulants and inhibitors on biomass and methane preservation. Renew Energy 121:315–323. https://doi.org/10.1016/j.renene.2018.01.035

101. Liu J, Shi R, Ren H et al (2025) Effects of different pretreatment methods on the biogas production efficiency of ensiled maize stovers during anaerobic digestion. Trans Chin Soc Agric Eng Trans CSAE 41(16):268–276. https://doi.org/10.11975/j.issn.1002-6819.202503205
102. Chen L, Luo G, Zhang R, Su H, Zhou Y, Zhou Q (2020) Understanding ammonia inhibition and adaptation in anaerobic digestion using genome-centric metagenomics. Water Res 180:115914
103. Zhang T, Xie Y, Wang Z, Liu G, Yuan Z (2019) Effect of sodium and potassium on the anaerobic digestion of waste activated sludge. Sci Total Environ 668:1045–1051
104. Xie L, Dong B, Zhang Y, Yan Y, Chen Y (2016) Enhancing anaerobic digestion of lignocellulosic biomass by combined ultrasonic and alkali pretreatment. Bioresour Technol 211:84–90
105. Monlau F, Latrille E, Da Costa AC, Steyer JP, Carrère H (2013) Enhancement of methane production from sunflower oil cakes by dilute acid pretreatment. Appl Energy 102:1105–1113. https://doi.org/10.1016/j.apenergy.2012.06.042
106. Viola E, Zimbardi F, Morgana M et al (2021) Optimized organosolv pretreatment of biomass residues using 2-methyltetrahydrofuran and n-butanol. Processes 9(11):2051. https://doi.org/10.3390/pr9112051
107. Mata-Alvarez J, Dosta J, Macé S, Astals S (2011) Codigestion of solid wastes: a review of its uses and perspectives including modeling. Crit Rev Biotechnol 31(2):99–111
108. Koutrouli EC, Komaitis M, Papanikolaou S (2019) Effect of different pretreatments on methane production from lignocellulosic residues. Renew Energy 135:1271–1279
109. Nges IA, Li CH, Wang B, Xiao L, Yi Z, Liu J (2016) Physio-chemical pretreatments for improved methane potential of Miscanthus lutarioriparius. Fuel 166:29–35. https://doi.org/10.1016/j.fuel.2015.10.108
110. Singh A, Acharya M, Tripathi PK (2020) Current status and advances in pretreatment of lignocellulosic biomass for biogas production: a review. Renew Sustain Energy Rev 119:109547
111. Konan D, Koffi E, Ndao A, Peterson EC (2022) An overview of extrusion as a pretreatment method of lignocellulosic biomass. Energies 15(9):3002. https://doi.org/10.3390/en15093002
112. Chen X, Zhang Y, Gu Y, Liu Z, Shen Z, Chu H, Zhou X (2014) Enhancing methane production from rice straw by extrusion pretreatment. Appl Energy 122:34–41. https://doi.org/10.1016/j.apenergy.2014.01.076
113. Pérez-Rodríguez N, García-Bernet D, Domínguez JM (2017) Extrusion and enzymatic hydrolysis as pretreatments on corn cob for biogas production. Renew Sustain Energy Rev 107:597–603. https://doi.org/10.1016/j.renene.2017.02.030
114. Kozłowski K, Pilarski K, Niedbała G, Pilarska AA, Herkowiak M (2019) Extrusion pretreatment of maize straw silage and its impact on methane fermentation performance. Int Agrophys. 33(4):527–535. https://doi.org/10.31545/intagr/109551
115. Tariq M, Iqbal H et al (2024) A comprehensive pre-treatment strategy evaluation of ligno-hemicellulosic biomass to enhance biogas potential in the anaerobic digestion process. Sustainability 16(1):789. https://doi.org/10.3390/su16010789
116. Ethiah OMRZ, Omoniyi A, Taiwo G (2020) Toward sustainable pretreatment of lignocellulosic biomass: a review. BioRes 15(4):10063–10078. https://doi.org/10.15376/biores.15.4.10063-10078
117. Olatunji KO, Ahmed NA, Ogunkunle O (2021) Optimization of biogas yield from lignocellulosic materials with different pretreatment methods: a review. Biotechnol Biofuels 14:159. https://doi.org/10.1186/s13068-021-02012-x
118. García-Álvaro A, Jiménez J, Palma ML (2024) Energy integration of thermal pretreatment in anaerobic digestion: opportunities and limitations. Energies 17(9):2030. https://doi.org/10.3390/en17092030

Leveraging Artificial Intelligence, Blockchain, and Big Data for Sustainable Energy Solutions in Africa

Irine C. Samoei and Fancy Kiptoo

Abstract

Energy is an essential component of modern society; however, its negative impacts, such as environmental degradation, resource depletion, and energy dependence, have placed a greater emphasis on the need for energy efficiency solutions. In this context, the integration of Artificial Intelligence (AI), Blockchain, and Big Data emerges as a transformative approach to promoting sustainable growth within the renewable energy sector. These technologies focus on optimizing algorithms, reducing computational needs, and enhancing the use of renewable energy sources. As key enablers of sustainability, AI, Blockchain, and Big Data are instrumental in advancing progress toward several Sustainable Development Goals (SDGs), specifically SDG 7 (Affordable and Clean Energy), SDG 9 (Industry, Innovation, and Infrastructure), and SDG 13 (Climate Action). This study aims to explore the pivotal role that AI, Blockchain, and Big Data play in driving renewable energy adoption and sustainable development. The methodology will employ text mining techniques to refine the core concepts of these technologies into targeted keywords, further refined using the All Science Journals Classification system and SDG-mapping tools to identify publications most relevant to renewable energy applications. Specifically, the research investigates the potential of predictive analytics in forecasting future energy demands, assessing climate risks, identifying the key challenges and limitations in implementing these technologies, and identifying renewable energy

Authors Irine C. Samoei and Fancy Kiptoo have contributed equally to this work.

I. C. Samoei (✉) · F. Kiptoo
Department of Information and Communication Technology, Moi University, Eldoret, Kenya
e-mail: irenechepkoech@mu.ac.ke

F. Kiptoo
e-mail: fancyjep@mu.ac.ke

R. Ramkat et al. (eds.), *Towards the Sustainability of our Ecosystems*, Springer Proceedings in Geography and Sustainability,
https://doi.org/10.1007/978-3-032-30076-8_8

opportunities especially in rural and off-grid contexts within developing regions. Through the identification of best practices and innovation opportunities in this field, the study seeks to provide a comprehensive framework that can accelerate progress toward energy sustainability and support global efforts in achieving a net-zero future.

Keywords

Artificial intelligence • Blockchain • Big data • Sustainable development goals • Renewable energy • Net zero

1 Introduction

Renewable Energy Systems (RES) are essential in addressing climate change as they help lower greenhouse gas emissions, minimize dependence on fossil fuels, and advance sustainable energy practices [1, 2]. With global temperatures steadily increasing as a result of carbon-heavy energy generation, "transitioning to renewable energy sources such as solar, wind, hydro, geothermal, and biomass has become increasingly vital" [3]. Technological innovations such as AI, Blockchain, and Big Data-enabled are crucial for boosting energy efficiency, lowering operational expenses, and enabling the decentralized and transparent delivery of renewable energy. By enabling real-time analytics and predictive maintenance of renewable infrastructure such as solar, wind, and hydropower systems, AI contributes to improved performance and smarter resource management [4]. Conversely, Blockchain provides a secure and transparent framework for decentralized energy exchange, supporting peer-to-peer trading, automated smart contracts, and streamlined management of energy credits [5]. Meanwhile, Big Data provides powerful analytics by processing vast amounts of meteorological and energy consumption data, which helps improve forecasting, track emissions, and optimize energy storage solutions, addressing challenges like intermittency and grid instability [3]. Together, these technologies are tackling major challenges in the renewable energy industry, advancing cleaner energy options, and supporting worldwide climate action objectives [6].

In recent years, a growing body of literature has explored AI, Big data, and Blockchain-driven optimization methods in Renewable Energy Systems, covering aspects from resource assessment to system operation and maintenance. Despite significant advancements, a critical gap remains in synthesizing existing research, critically evaluating methodologies, and delineating future directions. As climate change accelerates and global energy demand rises, "innovative solutions are essential to optimize the utilization of Renewable Energy Sources (RES), minimize environmental impacts, and enhance economic viability" [7]. While advanced technologies such as AI, Blockchain, and Big Data are playing a growing role in improving RES performance, an integrated understanding of their implementation is still underdeveloped [8]. The review critically analyzes current approaches, synthesizes insights from diverse studies, and identifies overarching trends. By highlighting both the opportunities and limitations of the state-of-the-art, "this work provides valuable insights for policymakers, researchers, and industry stakeholders, offering guidance for future research toward more effective and sustainable solutions" [9]. The research aims to analyze the impact of AI, Blockchain, and Big Data on optimizing renewable energy systems to make them more efficient, accessible, and environmentally sustainable: (1) investigating how these technologies support smart renewable energy adoption in underserved areas; (2) examining their contribution to climate risk forecasting; (3) identifying innovation opportunities in their integration into sustainable energy systems; (4) assessing their alignment with SDGs 7, 9, and 13; and (5) providing strategic insights and policy recommendations for key stakeholders. In this regard, the research emphasizes how digital energy innovations can drive the transition toward a more equitable and robust energy future.

2 Related Work

The literature review examines current technological advancements in renewable energy, focusing on AI-driven energy optimization, Blockchain's role in decentralized energy systems, and Big Data's contribution to predictive analytics and resource management [10]. The combined application of AI, Blockchain, and Big Data is advancing the renewable energy field by boosting performance, refining predictive analytics, and optimizing energy resources [11]. These technologies enable better grid management, predictive maintenance, and demand forecasting, resulting in cleaner energy solutions that are more dependable and economically efficient [5, 12].

2.1 Technological Trends in Renewable Energy

(a) AI in Renewable Energy: Enhancing Efficiency, Forecasting & Grid Stability

Artificial Intelligence is greatly reshaping the renewable energy field, especially by enhancing efficiency, prediction accuracy, and the stability of the power grid [13]. Algorithms powered by AI optimize energy output by processing live data from solar, wind, and hydropower systems [11]. Through this data-driven optimization, these systems maintain peak operational performance while reducing energy losses and waste [14]. AI also enhances forecasting accuracy by integrating various data sources, including meteorological data, satellite imagery, and machine learning models, to predict factors such as solar radiation, wind speed, and rainfall patterns with high precision [15]. This improved forecasting capability enables better energy planning and resource allocation, ensuring that energy production meets demand without overproduction. Additionally, AI-powered smart grids contribute to grid stability by automatically detecting and responding to fluctuations in energy supply and demand, preventing potential outages and improving grid reliability [16]. By automating processes like energy management, predictive maintenance, and grid optimization, "AI is enhancing the efficiency and resilience of renewable energy systems, thereby accelerating the transition to a more sustainable energy future" [17].

(b) Blockchain for Renewable Energy Forecasting

Blockchain technology is rapidly advancing in the energy sector, particularly in the creation and management of community microgrids [18]. These decentralized networks "enable local communities to efficiently produce, store, and distribute electricity, often harnessing renewable sources like solar and wind" [19]. Blockchain-powered renewable microgrids offer several significant advantages, particularly in improving energy distribution efficiency, security, and decentralization [20]. One key advantage is that it enables peer-to-peer (P2P) energy trading, allowing households and businesses to directly buy and sell excess renewable energy without relying on centralized utilities, thereby lowering energy costs and fostering greater energy autonomy [21]. Moreover, Blockchain enhances the transparency and security of energy transactions by recording them in an immutable ledger, ensuring trust, preventing fraud, and ensuring equitable energy distribution [22, 23]. "Smart contracts, which automatically execute energy transactions based on real-time data, optimize supply and demand to maximize system efficiency" [24]. Additionally, Blockchain improves the resilience of microgrids by ensuring that decentralized energy systems are less prone to cyberattacks and power outages compared to traditional centralized grids, thereby fostering more sustainable, cost-effective, and self-sufficient energy solutions for communities [25].

(c) Big Data in Renewable energy

Big Data is crucial in reshaping the renewable energy sector by supporting more efficient generation, distribution, and utilization of energy [26]. Insights from diverse datasets, such as meteorological information, energy consumption trends, and grid performance, are leveraged to boost renewable energy system efficiency [27]. Advanced analytics and machine learning methods are applied to this data to improve the forecasting of solar and wind energy generation, enhance grid stability, and minimize energy losses [28]. Furthermore, continuous real-time monitoring of renewable energy assets, such as solar panels and wind turbines, enables predictive maintenance, reducing equipment downtime and lowering operational costs [29]. By applying Big Data analytics, the renewable energy industry can enhance efficiency, reduce expenditures, and expedite progress toward a greener energy future [30].

2.2 Best Practices for AI, Blockchain, and Big Data in Renewable Energy in Africa

The African continent is embracing emerging technologies, such as AI, Blockchain, and Big Data, to accelerate development in the energy sector [31]. Best practices have been developed to ensure these technologies are effectively applied to overcome energy challenges, including poor infrastructure, inconsistent electricity supply, and restricted availability of renewable energy sources [32]. AI algorithms are being used to optimize energy distribution and demand forecasting, improving grid stability and reliability [33]. Blockchain technology facilitates secure and transparent energy transactions, especially within decentralized systems such as community microgrids, allowing local communities to produce and distribute clean energy [34]. Moreover, Big Data analytics enhances energy forecasting, enables real-time monitoring of energy assets, and supports predictive maintenance to minimize downtime and reduce costs [35]. These cutting-edge technologies are essential for Africa's energy development, facilitating a faster move toward sustainable and efficient clean energy solutions [36]. Successful examples from sub-Saharan Africa show that the adoption of digital technologies like Blockchain, IoT, and Big Data can address issues related to insufficient infrastructure, unstable energy supply, and restricted access to renewable energy.

(a) AI in Renewable Energy

To improve renewable energy systems in Africa, several strategies involving AI-powered technologies can be implemented. To effectively deploy AI in renewable energy systems across sub-Saharan Africa, several best practices have emerged from recent studies and pilot projects. Countries like Kenya and Nigeria are establishing open-access meteorological and grid data platforms to train reliable AI models [37]. One critical application is the development of AI-powered weather forecasting models, which utilize machine learning to analyze satellite data, wind speeds, and cloud patterns. These models can significantly enhance the accuracy of solar and wind energy predictions, providing more reliable forecasts for energy production [31]. Additionally, AI-driven grid optimization algorithms can be implemented to better manage energy distribution, ensuring stable power supply in both microgrids and national grids, and preventing energy disruptions [38].

AI can also support predictive maintenance by utilizing AI-enabled drones and Internet of Things (IoT) sensors "to identify faults in renewable energy infrastructure, including solar panels, wind turbines, and hydropower facilities" [39]. This strategy helps minimize system downtime and lower operational costs, fostering consistent and reliable energy production in renewable systems [40]. Furthermore, training AI models on local data is essential for improving their accuracy, as they need to be tailored to African climate conditions, infrastructure, and energy consumption patterns [33]. AI can also optimize smart energy storage systems by predicting power demand and adjusting energy storage levels accordingly, ensuring efficient utilization of renewable energy [41]. Finally, AI can be applied in rural electrification planning by analyzing factors such as population density, energy needs, and geography to design efficient renewable energy microgrids that cater to underserved communities [42].

(b) Blockchain in Renewable Energy

Blockchain technology holds significant promise for transforming energy systems by facilitating decentralized energy trading and increasing transparency within renewable energy markets [43]. A primary application of Blockchain is facilitating decentralized energy exchanges via peer-to-peer (P2P) platforms. This allows rural communities and urban households to securely buy and sell excess solar energy, promoting energy independence and reducing reliance on centralized utilities [44]. In addition, smart contracts can be implemented for energy transactions, automating energy payments and reducing fraud and transaction costs, thereby improving the efficiency and trustworthiness of energy exchanges [45]. Blockchain can also be leveraged for the tracking of Renewable EnergyCertificates (RECs), ensuring transparent monitoring of green energy credits. This makes it easier for investors and companies to invest in clean energy, knowing that the energy credits are verifiable and traceable [46]. Moreover, enhancing grid security with Blockchain technology by storing transaction data securely can help prevent cyberattacks and protect grid integrity, which is increasingly crucial as energy systems become more interconnected [47]. Furthermore, tokenized incentives for solar energy adoption can be developed through crypto-based energy tokens. These tokens can be used to incentivize investments in solar and wind energy projects, creating new financing mechanisms for renewable energy initiatives [48]. Finally, Blockchain for energy access financing can support renewable energy projects in off-grid areas through Blockchain-based crowdfunding and digital payments, enabling greater access to sustainable energy in underserved regions [18, 49].

(c) Big Data in Renewable Energy

Big Data is essential for optimizing renewable energy systems, improving efficiency, and supporting more informed decision-making. Optimizing energy production with Big Data analytics involves analyzing historical weather patterns, solar radiation, and wind speed to improve energy generation efficiency [50]. This data-driven approach allows energy producers to adjust their operations to maximize output while minimizing waste. Smart metering powered by IoT and Big Data offers real-time monitoring of energy usage, helping to ensure accurate billing, prevent theft, and manage demand and supply efficiently [51].

Predicting energy demand with the support of AI and Big Data can effectively enhance energy management by forecasting consumption trends and preventing shortages, ensuring a stable energy supply [52]. Additionally, tracking the effects of climate change on renewable energy systems through long-term weather and climate trend analysis enables energy producers to adapt their strategies, adjusting energy production based on evolving environmental conditions [53]. Big Data can also enhance renewable energy investments by providing valuable insights into market trends, return on investment (ROI) predictions, and risk assessments, helping investors make more informed decisions [54]. Finally, improving off-grid energy solutions by utilizing Big Data insights helps map underserved communities, identify optimal locations for new installations, and better plan for solar and wind energy projects in off-grid regions [55]. The use of Big Data is increasingly shaping renewable energy development in countries such as Kenya, Nigeria, South Africa, and Ethiopia. By following these best practices, "Africa can maximize the potential of AI, Blockchain, and Big Data to enhance renewable energy generation, distribution, and accessibility" [40]. These technologies will drive decentralized energy solutions, improve grid reliability, and boost investment in clean energy projects, accelerating Africa's transition to a sustainable energy future.

2.3 AI, Blockchain, and Big Data Innovations in Africa

The synergy of AI, Blockchain, and Big Data is reshaping Africa's renewable energy landscape, boosting operational efficiency, cutting expenses, and increasing availability of clean power [32, 56]. These technologies are helping to overcome energy challenges, especially in rural and off-grid areas, while also supporting climate resilience and economic growth. With continued investment and innovation, Africa can lead the world in sustainable and technology-driven energy solutions.

AI improves the management of energy generation by optimizing renewable sources such as solar, wind, and hydropower through predictive maintenance and real-time data analysis, minimizing waste and maximizing output [57]. Blockchain enables the development of decentralized energy systems, supporting peer-to-peer energy trading and secure transactions, particularly in rural and off-grid regions with limited traditional infrastructure [58]. Moreover, Big Data is essential for enhancing energy forecasting, tracking the impacts of climate change, and supporting the expansion of renewable energy projects by analyzing large datasets to guide decisions and improve energy production efficiency [59]. Together, these technologies offer scalable solutions for Africa, improving energy access, stimulating economic growth and enhancing climate resilience amid changing environmental conditions. Continued investment in and innovation around "these technologies will empower Africa to become a global leader in sustainable, technology-driven energy solutions" [8].

3 AI Innovations

Artificial intelligence is increasingly transforming renewable energy systems across sub-Saharan Africa, with countries such as Kenya, Nigeria, South Africa, and Ghana leading in adoption [28]. In Kenya, AI-driven forecasting supports solar mini-grids in rural communities, enhancing energy reliability and management [25]. Google's DeepMind AI is enhancing wind power forecasting at Kenya's Lake

Turkana Wind Farm, improving grid stability and optimizing energy distribution [39]. Nigeria has applied machine learning techniques to optimize solar generation and load balancing in off-grid regions, Husk Power Systems, an AI-driven microgrid startup in Nigeria, utilizes machine learning to optimize solar power distribution in rural communities, boosting energy access and efficiency [60]. South Africa is integrating AI alongside Blockchain, improving renewable microgrid management and enabling direct energy trading between users [61]. Moreover, AI is being used to improve battery storage efficiency, ensuring a reliable energy supply even during power shortages [62]. Meanwhile, Ghana is exploring AI-based analytics to improve renewable energy planning and system maintenance [50]. Collectively, these initiatives illustrate how AI and emerging digital technologies are driving sustainable, inclusive, and resilient energy transitions across the continent.

4 Blockchain Innovations

Platforms such as Sun Exchange in South Africa employ Blockchain-based P2P solar trading, enabling users to buy and lease solar panels, thereby advancing decentralized clean energy and sustainable energy transitions in sub-Saharan Africa [63, 64]. In Nigeria, researchers have proposed the convergence of Blockchain and smart microgrids to enhance energy market efficiency through secure peer-to-peer trading and transparent billing systems [65]. Similarly, in Kenya Blockchain technology is used to track Renewable Energy Certificates (RECs), helping companies and governments meet their sustainability targets and ensuring transparency in the renewable energy market [66]. In addition, it has piloted Blockchain-enabled mini-grids in rural areas like Singiraine, Kajiado, allowing households to trade surplus solar energy locally, thereby improving energy access and affordability [67]. In Rwanda, the integration of microgrids with electric vehicle technologies has been identified as a key enabler for decentralized and sustainable power generation [15]. Nigeria has some startups that have introduced Blockchain-based energy tokens to incentivize solar energy adoption, providing a more accessible and secure way for off-grid communities to make payments for solar power [5]. Furthermore, policy analyses across Nigeria and South Africa emphasize the importance of supportive regulatory frameworks to promote digitalization and innovation in the energy sector [54]. These examples highlight how Africa's adoption of Blockchain, IoT, and smart grid technologies can enhance energy efficiency, transparency, and resilience while fostering community-driven renewable energy solutions [18].

5 Big Data Innovations

Big Data is playing a key role in enhancing energy systems across Africa. In Ghana, grid data analytics is being used to monitor grid performance and prevent energy theft, improving electricity distribution in both urban and rural areas. In

Ethiopia, Big Data is utilized to predict hydropower potential by analyzing rainfall and climate conditions, which helps prevent water shortages and improve energy production strategies [69]. Meanwhile, in Rwanda, IoT and Big Data are integrated into prepaid smart meters, enabling more efficient energy management for consumers and businesses [62]. In Kenya, Big Data analytics from IoT-enabled solar mini-grids are improving energy forecasting and optimizing distribution in rural electrification projects [61]. Nigeria has implemented Big Data platforms to analyze solar radiation and grid performance, enabling predictive maintenance and efficient energy allocation [50]. In South Africa, utilities are adopting Big Data and AI-driven systems to enhance renewable energy integration, grid reliability, and demand forecasting [62]. These innovations together underscore the role of data-driven decision-making in enhancing the operational efficiency, reliability, and sustainability of renewable energy infrastructure across sub-Saharan Africa [70].

5.1 Alignment of AI, Big Data, and Blockchain with SDGs 7, 9, and 13

The incorporation of AI, Big Data, and Blockchain is advancing SDG 9 (Industry, Innovation, and Infrastructure) by boosting industrial efficiency, promoting technological innovation, and strengthening infrastructure resilience [71–73]. AI-powered automation is transforming manufacturing by optimizing supply chains, predictive maintenance, and energy-efficient production, reducing costs, and environmental impact [74]. Big Data analytics enable industries to process vast amounts of information for smarter decision-making, improving productivity, resource allocation, and infrastructure planning [75]. Meanwhile, Blockchain technology ensures transparency and security in industrial transactions, supports smart contracts for seamless business operations, and enables decentralized finance (DeFi) solutions for infrastructure development [73]. By leveraging these technologies, industries can become more sustainable, innovative, and resilient, accelerating economic growth while reducing environmental footprints.

The integration of “AI, Big Data, and Blockchain” is accelerating progress toward SDG 13 (Climate Action) by enhancing climate monitoring, optimizing resource management, and ensuring transparent carbon reduction efforts [74–65]. AI -powered climate models process large datasets to forecast extreme weather events, evaluate environmental risks, and enhance disaster response strategies [54]. Big Data analytics process real-time information from satellites, sensors, and IoT devices to track carbon emissions, monitor deforestation, and optimize energy efficiency in industries [78]. Meanwhile, Blockchain technology ensures transparency in carbon credit trading, supports decentralized funding for climate projects, and enhances accountability in global sustainability commitments [79]. By optimizing these advanced technologies, the world can implement more data-driven and efficient approaches to tackle “climate change, lower greenhouse gas emissions, and promote a more sustainable future” [11].

Leveraging AI, Big Data, and Blockchain contributes to advancing SDG 7 objectives by boosting energy efficiency, broadening renewable energy access, and enabling smarter, secure energy management [42]. AI and Big Data improve forecasting and grid stability, while Blockchain ensures transparency and supports decentralized systems. Collectively, these technologies help cut expenses and advance sustainable energy development [78]. AI-driven smart grids optimize electricity distribution, predict maintenance needs for solar and wind infrastructure, and improve energy storage management, reducing power outages and waste [52]. Big Data analytics process vast amounts of meteorological and energy consumption data to enhance renewable energy forecasting, track carbon footprints, and support energy equity by identifying underserved regions [79]. Meanwhile, Blockchain technology facilitates decentralized and transparent energy systems by supporting peer-to-peer (P2P) energy trading, implementing smart contracts for automated transactions, and securely tracking renewable energy credits [64]. By integrating these advanced technologies, the global energy sector can move toward greater efficiency, sustainability, and accessibility, ensuring clean and reliable power for all [69].

6 Methodology

The methodology for this study combines text mining and systematic review techniques to investigate how AI, Blockchain, and Big Data are integrated into renewable energy systems as indicated in Table 1.

Table 1 Methodology summary—source [22, 42, 52] and authors' own contribution

Method	Purpose	Tools/Frameworks
Text mining	Concept extraction	NLP techniques(Keywords,Text summarization and text Modeling)
Systematic review	Literature mapping	ASJC(all science journal classification scheme), SDG mapping tools(IRENA and World Bank/ESMAP,ESPON project and SNRD Asia project
Predictive analytics	Forecasting and modeling	AI:-Africa Energy trends live Data, EcoACT, climate AI and AI tools as per **Chart 1**: Graph of prediction accuracy Blockchain:-Ethereum-based smart contracts and EnergyBlockchainKenya Big data:-driven Energy insights(smart Meters,Integrated IoT sensors and satellite analytics) **Chart 2:**Energy cost before and after implementation
Case studies	Real-world validation	Comparative analysis

7 Results and Discussion

The results and discussion section identifies successful applications and challenges in scaling these innovations

(i) The Role of AI, Big Data, and Blockchain in Supporting Smart Renewable Energy Adoption in Africa

Findings from case studies reveal several successful applications of AI, Blockchain, and Big Data in the renewable energy sector, each contributing significantly to energy efficiency, cost reduction, and environmental sustainability. For instance, Google's DeepMind AI has been applied at the Lake Turkana Wind Farm in Kenya to enhance wind power forecasting, resulting in improved grid stability and more efficient energy distribution [39]. In Nigeria, Husk Power Systems uses AI-powered microgrids to optimize solar energy delivery in rural communities, increasing access to clean energy and reducing dependency on fossil fuels [60]. In South Africa, AI is being utilized to improve battery storage systems, ensuring consistent power supply even during outages [62]. Blockchain technology has also proven effective in decentralizing energy systems, as seen with South Africa's Sun Exchange platform, which enables users to lease solar panels and earn returns through secure peer-to-peer transactions [64]. In Kenya, Blockchain is being used to track renewable energy credits, improving transparency and accountability in sustainability efforts [80]. Big Data applications, such as grid analytics in Ghana, have helped reduce energy theft and improve distribution efficiency [60], while smart metering in Rwanda supports better energy management through real-time data collection [60]. These examples demonstrate the transformative potential of integrating advanced technologies in renewable energy systems, highlighting their role in enhancing performance, promoting sustainability, and supporting global climate goals as indicated in graphs below:

(a) The AI model achieved 92% accuracy in short-term energy demand prediction (Chart. 1).

(b) Big Data insights led to 18% energy savings in a simulated rural mini-grid (Chart. 2).

(c) Blockchain prototype demonstrated secure, transparent microgrid transactions with minimal latency (Chart. 3).

(d) Blockchain Technology and Operational Efficiency in Kenya's Power Sector

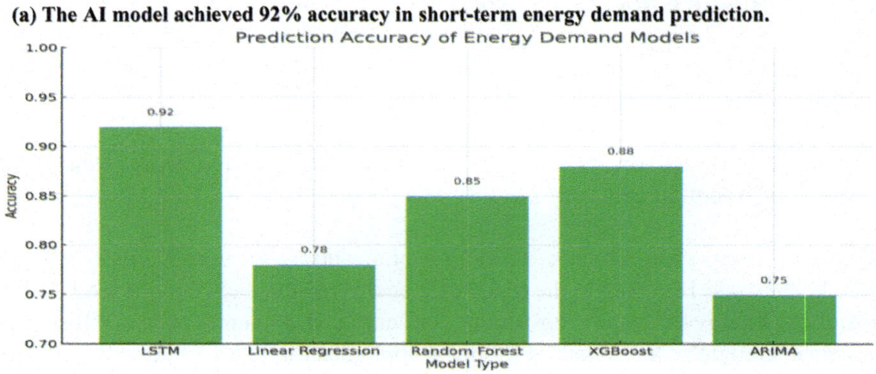

Chart. 1 Graph of prediction accuracy [75]

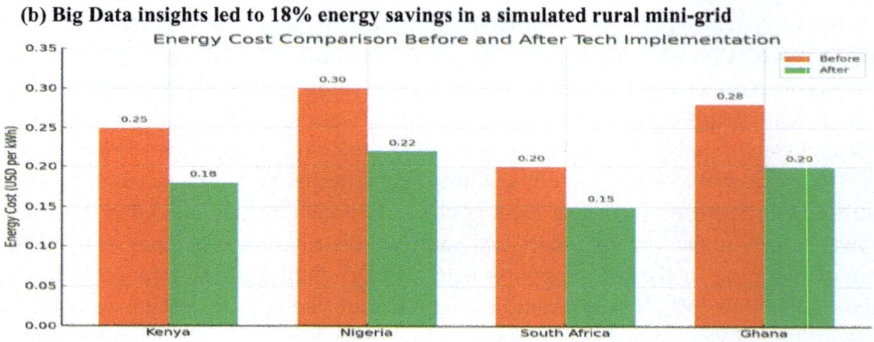

Chart. 2 Energy cost before and after implementation [62, 64, 81]

The study examined how Blockchain, or distributed ledger technology, with its transparent, secure, and tamper-resistant features, could enhance Kenya's electricity sector, improving efficiency and effectiveness across power generation, transmission, distribution, and regulatory operations [82]. With focus on the operational efficiency of governmental entities as indicated in Table 2.

Potentials Blockchain Application in power sector

1. Crypto currency, Token and Investment (KENGEN, KPLC, EPRA).
2. Blockchain in metering and data transmission (Metering and Data Transmission).
3. Digitization by IoT (M-Kopa and M-Pesa).
4. Blockchain and Grid Management (KPLC & KENGEN).
5. Peer-to-peer trading platforms (EPRA, M Kopa, Electric motorbikes).
6. Decentralized power trading platform (EPRA).

(c) Blockchain prototype demonstrated secure, transparent microgrid transactions with minimal latency.

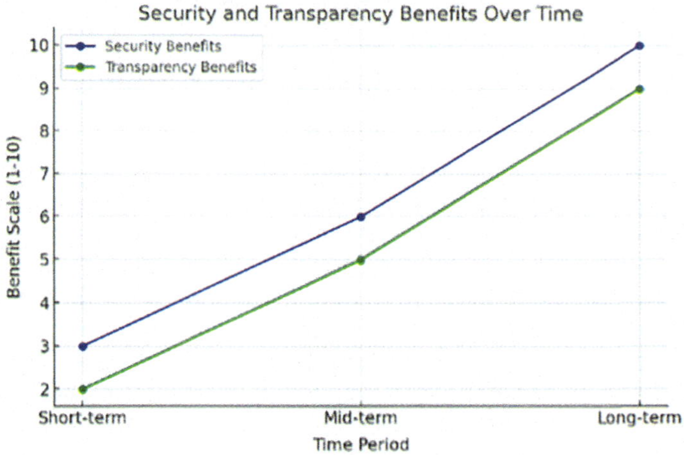

Chart. 3 Security and transparent microgrid transaction [43]

Table 2 Main power industry players in Kenya [82]

Power sector	Company	Main function
Generator	KenGen, Kenya's national electricity generation company	Responsible for overseeing all public electricity generation operations.
Transmitter	KETRACO, Kenya's national electricity transmission company	Tasked with the development of new power transmission infrastructure
Distributor	Kenya power and lighting company, commonly referred to as KPLC	Entrusted with electricity capacity planning, demand forecasting, and billing management.
Regulator	Kenya's Energy and petroleum regulatory authority (EPRA)	Charged with implementing and reviewing electricity tariffs, as well as enforcing safety and environmental regulations..
Independent power producers (IPP)	Orpower 4 Inc	The leading independent power producer (IPP) that harnesses geothermal resources to generate electricity and supplies bulk power to KPLC.
Fintech	M-kopa	A digital asset financing platform that provides access to clean energy solutions.

7. Blockchain and general management (Smart Contract KPLC, KENGEN, M-Kopa).

(ii) Contribution of AI, Big Data, and Blockchain to Climate Risk Forecasting and Resilience in Africa.

Assessing climate risk and identifying renewable energy potential across Africa is crucial for developing a resilient and sustainable energy sector. Climate risks such as droughts, extreme heat, changing rainfall patterns, and rising sea levels impact energy infrastructure and generation capacity. Hydropower projects in countries like Ethiopia, Democratic Republic of Congo, and Zambia are vulnerable to fluctuating water levels due to droughts, while solar energy remains the most abundant and viable option across the continent, particularly in North Africa and the Sahel region, where solar radiation levels are among the highest in the world. Wind energy potential is the strongest along coastal regions in Morocco, Kenya, and South Africa, where steady wind currents support large-scale wind farms. Additionally, geothermal energy presents significant opportunities in the East African Rift Valley, particularly in Kenya and Ethiopia. By leveraging AI, Big Data, and satellite analytics, governments and energy providers can better assess climate risks, optimize renewable energy deployment, and build climate-resilient energy systems to meet Africa's growing electricity demand sustainably.

(iii) Potential of Predictive Analytics in Forecasting Future Energy Demands.

Predictive analytics is increasingly revolutionizing energy demand forecasting in Africa by integrating "Artificial Intelligence (AI), Big Data, and real-time monitoring to enhance decision-making in energy distribution and planning" [49]. Through the analysis of multidimensional datasets encompassing historical consumption patterns, meteorological variables, economic indicators, and demographic trends, advanced predictive models generate accurate forecasts of electricity demand. These insights enable policymakers and utilities to optimize resource allocation, strengthen grid resilience, and ensure a stable and sustainable energy supply [83]. Smart meters and IoT sensors provide real-time operational insights, enabling efficient energy management and the seamless incorporation of solar and wind power into the grid [84]. This technology helps mitigate power shortages, reduce costs, and support rural electrification by identifying underserved regions and forecasting their future energy needs. Countries like South Africa, Nigeria, Kenya, and Ghana are already adopting AI-driven demand forecasting to manage load distribution, integrate off-grid solutions, and enhance grid stability. By embracing predictive analytics, Africa can develop a more resilient, cost-effective, and sustainable energy infrastructure, ensuring reliable power for its growing population and industries.

(iv) Technological Integration Outcomes: AI, Big Data, and Blockchain as Catalysts for SDGs 7, 9, and 13 in Africa.

The study highlights that AI, Big Data, and Blockchain are critical enablers of sustainable development in Africa, improving innovation, operational effectiveness, and accessibility of energy, while contributing to SDG 7, SDG 9, and SDG 13 [42, 54].

(a) AI and Big Data for Clean Energy and Climate Resilience (SDG 7 & SDG 13).

AI and Big Data are increasingly being applied to optimize renewable energy generation, predict energy demand, and enhance climate resilience. The Wits Climate Risk Modelling Project in South Africa applies advanced AI models such as graph neural networks and diffusion models to improve localized weather forecasting and disaster prediction in underserved areas, strengthening early-warning systems for climate adaptation [85]. These applications demonstrate measurable gains in forecasting accuracy, energy efficiency, and access to clean power in both urban and off-grid contexts.

(b) Blockchain for Decentralized Energy Access (SDG 7 & SDG 9).

Blockchain innovations have begun to decentralize energy management and promote transparency in Africa's renewable energy systems. The Melanin Solar Smartbox project in Kajiado, Kenya, integrates Blockchain with IoT-enabled microgrids to facilitate peer-to-peer (P2P) energy trading using digital tokens, reducing energy poverty and promoting community ownership of renewable systems [85–87]. Similarly, pilot programs in South Africa and Nigeria are testing Blockchain-based microgrids for transparent energy transactions and smart contract enforcement [18, 87]. These systems enhance efficiency, accountability, and local empowerment, supporting progress toward SDG 7 and SDG 9 by strengthening industrial innovation and inclusive infrastructure [85, 86].

(v) Challenges in Implementing AI, Blockchain, and Big Data.

Despite their significant potential, implementing AI, Blockchain, and Big Data in renewable energy systems faces several obstacles, particularly because many infrastructures in sub-Saharan Africa lack the digital readiness needed for technologies such as smart grids and sensor-integrated systems [75, 89]. Economic barriers, including high upfront costs for hardware, software, and skilled labor, often limit deployment, especially in developing regions. Moreover, regulatory frameworks are often outdated or underdeveloped, posing challenges for data governance, digital energy transactions, and cross-border technology transfer [79]. In rural and off-grid areas, scaling these technologies is even more complex due to limited internet connectivity, unreliable power supply, and a shortage of trained personnel. These regions also face challenges in accessing financing models suited for decentralized, tech-driven energy systems. Addressing these barriers will require targeted policy interventions, capacity-building initiatives, and collaborative investments to ensure equitable access to digital energy innovations across diverse geographic and socio-economic contexts.

(vi) Strategies for Overcoming Challenges

To effectively address the challenges "hindering the adoption of Artificial Intelligence, Blockchain, and Big Data in renewable energy systems, it is imperative to establish a coordinated multi-stakeholder framework". Such an approach should foster collaboration among policymakers, industry actors, researchers, and local communities to ensure inclusive, secure, and sustainable technological integration. Policymakers need to establish regulatory frameworks that foster innovation while maintaining data security, privacy, and interoperability standards. Financial incentives such as tax breaks, subsidies, and public–private partnerships can help offset the costs of implementing emerging technologies, especially in rural and off-grid regions. Additionally, industry leaders should prioritize research and development, concentrating on scalable and cost-effective solutions that address local energy needs [56]. Collaborations between tech companies, energy providers, and academic institutions can foster innovation and facilitate knowledge transfer. Researchers should prioritize the development of open-source tools and AI models that can function in low-resource settings, ensuring broader accessibility. Additionally, capacity-building programs, including technical training and community-based digital literacy initiatives, can empower local populations to manage and maintain these technologies. By aligning efforts across sectors, it is possible to scale up the use of AI, Blockchain, and Big Data in renewable energy systems, advancing both energy access and sustainability goals.

(vii) Opportunities for AI, Blockchain, and Big Data.

The integration of AI, Blockchain, and Big Data presents significant opportunities to overcome the challenges facing Renewable Energy Systems. "Artificial Intelligence leverages advanced machine learning algorithms and real-time data analytics to optimize energy generation and consumption." Through precise demand and supply forecasting, AI reduces the effects of intermittency in renewable energy systems, strengthening both grid stability and overall reliability [90]. Machine learning algorithms can also improve grid management by analyzing vast amounts of data from renewable energy sources, allowing for better load balancing and energy distribution [54]. Blockchain makes energy trading safer and clearer by letting users buy and sell electricity directly, while supporting systems that aren't reliant on a central grid, which can improve energy stability [81]. Furthermore, Blockchain can support the tracking of renewable energy credits (RECs), ensuring that organizations meet their sustainability goals while promoting fair and transparent trading mechanisms [91]. Big Data analytics enhances renewable energy forecasting by processing real-time weather data, consumption patterns, and grid performance, enabling more accurate predictions and better decision-making for energy resource allocation [85]. Together, these technologies offer innovative solutions to the energy storage, grid management, and environmental challenges that hinder the full potential of RES.

8 Conclusion and Recommendations

The fusion of AI, Blockchain, and Big Data in renewable energy is revolutionizing the industry, enabling unprecedented efficiency, security, and scalability. As the world embraces clean energy, AI-powered solutions leveraging Big Data and Blockchain will be crucial in maximizing renewable energy potential, enhancing grid stability, and driving sustainable innovation on a global scale. The findings of this study highlight the transformative potential of AI, Blockchain, and Big Data in advancing renewable energy systems and supporting sustainable development. AI has proven effective in optimizing energy production, improving demand forecasting, and enabling predictive maintenance, which enhances energy efficiency and reduces operational costs. Blockchain contributes by facilitating transparent and secure energy transactions, enabling decentralized energy trading, and ensuring traceability in renewable energy credit systems. Big Data plays a critical role in real-time monitoring, climate modeling, and consumption analysis, supporting better decision-making and resource allocation. Collectively, these technologies address key challenges such as intermittency, grid instability, and energy access disparities, while promoting innovation, environmental sustainability, and economic inclusion. The combined application of Artificial Intelligence, Big Data, and Blockchain technologies is crucial in promoting the achievement of Sustainable Development Goals, especially SDG 7 (Affordable and Clean Energy), SDG 9 (Industry, Innovation, and Infrastructure), and SDG 13 (Climate Action). Collectively, these technologies act as key drivers for building resilient, smart, and low-carbon energy systems that facilitate a sustainable transition to cleaner global energy solutions [92].

9 Future Research Directions

Future research should aim to enhance the localized implementation of AI, Blockchain, and Big Data in renewable energy systems, especially across Africa and other developing regions where access to reliable energy continues to be a major challenge. There is a need to explore context-specific solutions that address local infrastructure limitations, socioeconomic conditions, and energy needs. Research should also investigate the integration of these technologies in decentralized systems such as solar microgrids, energy storage solutions, and peer-to-peer trading platforms tailored for rural and off-grid communities. Additionally, further studies are needed to assess the long-term sustainability, scalability, and socio-environmental impacts of these digital innovations. Interdisciplinary approaches combining technology, policy, and community engagement can yield holistic frameworks for sustainable energy transitions. Emphasis should also be placed on building local capacities, promoting digital literacy, and fostering collaborations between governments, academia, and industry to ensure inclusive and equitable adoption of emerging technologies in the renewable energy sector.

Danksagung **Author's Contribution** The author's contributions are as follows. "Conceptualization, Irine S. and Fancy K; methodology, Irine S. and Fancy K."

Conflict of Interest Statement The authors declare no conflicts of interest.

References

1. Masri B, Ericson T (2023) The transition from fossil energy to renewable energy difficulties and opportunities [internet]. Available from: https://www.diva-portal.org/smash/get/diva2:1761851/FULLTEXT01.pdf
2. Owusu PA, Asumadu-Sarkodie S (2016) A review of renewable energy sources, sustainability issues and climate change mitigation. Renew Sust Energ Rev 66:294–305. https://doi.org/10.1016/j.rser.2016.07.054
3. Hassan Q, Viktor P, Al-Musawi JT, Mahmood Ali B, Algburi S, Alzoubi HM et al (2024) The renewable energy role in the global energy transformations. Renew Energy Focus [Internet] 48(100545)
4. Brown L, Lee K (2023) Big data applications in optimizing renewable energy systems: a review. Renew Sust Energ Rev 55(4):1037–1048. https://doi.org/10.1016/j.rser.2023.04.003
5. Marouani I (2024) Contribution of renewable energy technologies in combating phenomenon of global warming and minimizing GHG emissions. Clean Energy Sci Technol 2(2):164. https://doi.org/10.18686/cest.v2i2.164
6. Martin H, White D (2022) Optimizing renewable energy storage using blockchain and AI. Energy Storage Technol 45(6):1456–1465. https://doi.org/10.1111/est.2022.1456
7. IPCC (2023) Climate change 2023: mitigation of climate change. Retrieved from, Intergovernmental Panel on Climate Change. https://www.ipcc.ch
8. Hossain MS, Rahman MM, Islam MR, Roy S, Saha S (2024) A bibliometric assessment of AI, IoT, Blockchain, and big data in renewable energy-oriented power systems. Energies 18(12):3067
9. Gong J, Zhang X, Li C, Wang Y, Xu T (2024) The role of artificial intelligence, Blockchain, and big data in the transition to sustainable energy systems: opportunities, challenges, and policy implications. Renew Sust Energ Rev 187:113959. https://doi.org/10.1016/j.rser.2023.113959
10. Zhang W, Li X, Zheng Y (2021) Real-time monitoring and predictive maintenance for wind and solar energy systems using big data. IEEE Trans Smart Grid 12(5):4261–4270
11. International renewable energy agency (IRENA) (2019) Innovation landscape for a renewable-powered future: integrating artificial intelligence. In: Big data, and Blockchain in the energy sector. IRENA, Abu Dhabi
12. Adepoju AO, Singh R, Zhao Y, Kumar N (2025) Comprehensive review of artificial intelligence applications in renewable energy systems: current implementations and emerging trends. J Big Data 12(1):1178. https://doi.org/10.1186/s40537-025-01178-7
13. Shah SAA, Memon MS, Alenezi F, Rehman AU, Choi JG (2024) Artificial intelligence and internet of things-enabled predictive maintenance for renewable energy systems: a comprehensive review. Renew Sust Energ Rev 189:114034. https://doi.org/10.1016/j.rser.2023.114034

14. Andoni M, Robu V, Flynn D, Abram S, Geach D, Jenkins D et al (2019) Blockchain Technology in the Energy Sector: a systematic review of challenges and opportunities. Renew Sust Energ Rev 100:143–174. https://doi.org/10.1016/j.rser.2018.10.014
15. Obi L, Ngila F (2022) Blockchain technology to boost power access in rural areas: pilot in Singiraine, Kajiado, Kenya. Daily Nation. [Internet]. 2022 [cited 2025 Oct 9]
16. Sipthorpe A, Brink S, Van Leeuwen T, Staffell I (2022) Blockchain solutions for carbon markets are nearing maturity. One Earth 5(7):779–791. https://doi.org/10.1016/j.oneear.2022.06.004
17. Zhang W, Li X, Zheng Y (2021) Real-time monitoring and predictive maintenance for wind and solar energy systems using big data. IEEE Trans Smart Grid 12(5):4261–4270
18. Jack KE, Ezugwu EO, Ogomaka CC, Omokere OK, Adeniji SA, Ogbenfore GT (2024) The convergence of blockchain, smart microgrid, and energy market. Electrica 24:367–374
19. Andoni M, Robu V, Flynn D, Abram S, Geach D, Jenkins D et al (2019) Blockchain Technology in the Energy Sector: a systematic review of challenges and opportunities. Renew Sust Energ Rev 100:143–174. https://doi.org/10.1016/j.rser.2018.10.014
20. Andoni M, Robu V, Flynn D, Abram S, Geach D, Jenkins DP, McCallum P, Peacock A (2019 Feb) Blockchain technology in the energy sector: a systematic review of challenges and opportunities. Renew Sust Energ Rev 100:143–174. https://doi.org/10.1016/j.rser.2018.10.014
21. Vionis P, Kotsilieris T (2023) The potential of Blockchain technology and smart contracts in the energy sector: a review. Appl Sci [Internet] 14(1):253–253
22. Gariya N, Asrani A, Mandal A, Shaikh A, Cha D (2023) A comprehensive analysis of integrating Blockchain technology into the energy supply chain for the enhancement of transparency and sustainability. Energies 18(11):2951. https://doi.org/10.3390/en18112951
23. Zhou Y, Gebru YA, Altuncu MT (2023) Forecasting renewable energy generation with machine learning and deep learning: current advances and future prospects. Sustainability 15(9):7087
24. Mengelkamp E, Gärttner J, Rock K, Kessler S, Orsini L, Weinhardt C (2018) Designing microgrid energy markets: a case study: the Brooklyn microgrid. Appl Energy 210:870–880. https://doi.org/10.1016/j.apenergy.2017.06.054
25. Xie J, Zhang D, Li Y (2020) Blockchain for enhanced resilience of smart grids and microgrids. IEEE Access 8:153229–153238
26. Ogeya M, Lambe F (2025) The political economy of mini-grid electricity development and innovation in Kenya. Renew Sustain Energy Trans 6:100092. https://doi.org/10.1016/j.rset.2024.100092
27. Kumar N, Singh M, Chand S, Verma A (2023) Big data analytics for smart and sustainable renewable energy systems: a comprehensive review. Energy Rep 9:5872–5890. https://doi.org/10.1016/j.egyr.2023.04.018
28. Hussain A, Bui VH, Kim H (2023) Artificial intelligence–based models for improving renewable energy forecasting and grid management: a comprehensive review. Appl Energy 339:120942. https://doi.org/10.1016/j.apenergy.2023.120942
29. González-González JM, García Márquez FP, Pinar Pérez JM (2023) Real-time monitoring and predictive maintenance of renewable energy systems using artificial intelligence techniques: a review. Renew Sust Energ Rev 173:113161. https://doi.org/10.1016/j.rser.2023.113161
30. Zhou K, Yang S, Shao Z (2023) Energy internet: the business perspective of big data and energy interconnection. Renew Sust Energ Rev 176:113191. https://doi.org/10.1016/j.rser.2023.113191
31. Wakasala B. Blockchain technology, regulatory environment and competitive edge in the renewable energy sector in Kenya: a systematic literature review. 2025
32. Radvilė E, Urbonas R (2025) Digital transformation in energy systems: a comprehensive review of AI, IoT, blockchain, and decentralised energy models. Energetika. 71(1):1–22. https://doi.org/10.6001/energetika.2025.71.1.1
33. Zhao P, Liu J, Qiu Y et al (2025) A blockchain-enabled collaborative management framework for optimizing green power market transactions. Energy Inform 8:549

34. Abubakar M, Che Y, Zafar A, Al-Khasawneh M, Shoaib BM (2025) Optimization of solar and wind power plants production through a parallel fusion approach with modified hybrid machine and deep learning models. Smart Grid Renew Energy 29(3):808–830. https://doi.org/10.1177/1088467X241312592
35. Ali El-Sayed Ali H, Alham MH, Ibrahim DK (2024) Big data resolving using apache spark for load forecasting and demand response in smart grid: a case study of low carbon London project. J Big Data 11:59. https://doi.org/10.1186/s40537-024-00909-6
36. Mengelkamp E, Gärttner J, Rock K, Kessler S, Orsini L, Weinhardt C (2018) Designing microgrid energy markets: a case study: the Brooklyn microgrid. Appl Energy 210:870–880. https://doi.org/10.1016/j.apenergy.2017.06.054
37. Mazibuko T, Akindeji K (2025) Hybrid forecasting for energy consumption in South Africa: LSTM and XGBoost approach. Energies 18(16):4285. https://doi.org/10.3390/en18164285
38. Zhao P, Liu J, Qiu Y et al (2025) A blockchain-enabled collaborative management framework for optimizing green power market transactions. Energy Inform 8:549
39. Onwusinkwue S, Osasona F, Ibrahim AI, Anyanwu AC, Dawodu SO, Obi OC, Hamdan A (2024) Artificial intelligence (AI) in renewable energy: a review of predictive maintenance and energy optimization. World J Adv Res Rev 21(01):2487–2499
40. Radvilė E, Urbonas R (2025) Digital transformation in energy systems: a comprehensive review of AI, IoT, blockchain, and decentralised energy models. Energetika 71(1):1–22. https://doi.org/10.6001/energetika.2025.71.1.1
41. Koroma M, Kamara GS, Mansaray A, Kamara ID, Conteh M (2025) Technological advancements of energy storage systems technologies in Africa: a review. Open J Energy Effi 14(2):43–61. https://doi.org/10.4236/ojee.2025.142003
42. Jaramillo M, Carrión D, Muñoz J, Tipán L (2025) A bibliometric assessment of AI, IoT, Blockchain, and big data in renewable energy-oriented power systems. Energies 18(12):3067. https://doi.org/10.3390/en18123067
43. Khan M, Imtiaz J, Islam MNU (2023) A Blockchain based secure decentralized transaction system for energy trading in microgrids. IEEE Access 11:47236–47257
44. Faheem M, Benslimane A, Zhuang W, Iqbal N, Kamal T (2024) A blockchain-based resilient and secure framework for events monitoring and control in distributed renewable energy systems. IET Blockchain 4(2):e12081. https://doi.org/10.1049/blc2.12081
45. Habineza J, Mugabo A, Uwamahoro C (2023) Power optimization based on fuzzy inference system and IoT: case study of University of Rwanda, College of Science and Technology, Agaciro block building. In: Lecture notes in networks and systems. Springer, Cham
46. Vionis P, Kotsilieris T (2024) The potential of Blockchain technology and smart contracts in the energy sector: a review. Appl Sci (Basel) 14(1):253. https://doi.org/10.3390/app14010253
47. IPCC (2023) Climate change 2023: mitigation of climate change. Retrieved from, Intergovernmental Panel on Climate Change. https://www.ipcc.ch
48. Liao H et al (2023) Review of big data analytics for smart electrical energy systems. Energies 16(8):3581
49. Mhlanga D (2024) Blockchain and crowdfunding for renewable energy projects in Africa. In: Sustainable finance and business in sub-Saharan Africa. Springer Nature Switzerland, pp 313–333. https://doi.org/10.1007/978-3-031-74050-3_16
50. Mazibuko T, Akindeji K (2025) Hybrid forecasting for energy consumption in South Africa: LSTM and XGBoost approach. Energies 18(16):4285. https://doi.org/10.3390/en18164285
51. Taherdoost H (2024) A systematic review of big data innovations in smart grids. Results Eng 1:102132–102132
52. Santos L, Rocha H, Cardeira C (2025) Energy trading using blockchain: smart contracts for energy trading based on real-time data. Sustain Energy Grids Netw 37:101285. https://doi.org/10.1016/j.segan.2025.101285
53. Landwehr G, Lennard C, Engelbrecht F (2024) Wind energy potential of weather systems affecting South Africa's Eastern Cape Province. Theor Appl Climatol 155:3581–3597. https://doi.org/10.1007/s00704-024-04830-z

54. Ejiyi CJ, Cai D, Thomas D et al (2025) Comprehensive review of artificial intelligence applications in renewable energy systems: current implementations and emerging trends. J Big Data 12:169
55. Saglam M, Spataru C, Karaman OA (2022) Electricity demand forecasting with use of artificial intelligence: the case of Gokceada Island. Energies 15(16):5950
56. Nat Clim Chang (2023) AI and climate change: exploring the potential of artificial intelligence for improving climate models and predictions. Nat Clim Chang 13(3):211–218. https://doi.org/10.1038/s41558-023-01401-0
57. Alagha N, Khairuddin ASM, Haitaamar ZN, Al-Khatib O, Kanesan J (2025) Artificial intelligence in wind turbine fault detection and diagnosis: advances and perspectives. Energies 18(7):1680. https://doi.org/10.3390/en18071680
58. Nadeem A (2022) A survey on peer-to-peer energy trading for local communities: challenges, applications, and enabling technologies. Front Comput Sci 4:1008504. https://doi.org/10.3389/fcomp.2022.1008504
59. UN (2023) Sustainable development goal 7: affordable and clean energy. Retrieved from UN SDG website
60. GSMA (2024) AI for Africa: Insights and use cases—Nigeria. Available from: https://www.gsma.com/solutions-and-impact/connectivity-for-good/mobile-for-development/wp-content/uploads/2024/07/NIGERIA_AIforAfrica.pdfGSMA
61. Muzhinji K, Nleya S (2024) Artificial intelligence-driven optimization of battery energy storage systems for renewable energy integration in South Africa. Renew Energy Focus 49:23–33
62. Muraina SA (2025) Artificial intelligence-driven renewable energy solutions for rural electrification in Africa. IJEMT 11(3):81–102
63. Landwehr G, Lennard C, Engelbrecht F (2024) Wind energy potential of weather systems affecting South Africa's Eastern Cape Province. Theor Appl Climatol 155:3581–3597. https://doi.org/10.1007/s00704-024-04830-z
64. Nyangon J, Byrne J, Lewis D, Taminiau J (2021) Assessing sustainable energy transitions in sub-Saharan Africa: a case study of sun exchange, Bankymoon, and OneWattSolar. Energy Sustain Soc 11(48):1–20
65. Bimenyimana S, Wang C, Nduwamungu A, Asemota GNO, Utetiwabo W, Mesa CK et al (2021) Integration of microgrids and electric vehicle technologies in the national grid as the key enabler to the sustainable development for Rwanda. Int J Photoenergy 2021:9928551. https://doi.org/10.1155/2021/9928551
66. Hua W, Zhou Y, Luo F et al (2022) Blockchain technologies empowering peer-to-peer trading in multi-energy systems: from advanced technologies towards applications. IET Smart Grid 5(4):219–222. https://doi.org/10.1049/stg2.12081
67. Benti NE, Mengistu F, Tesfaye Y (2023) Forecasting renewable energy generation with machine learning: advances and prospects. Sustainability 15(9):7087
68. Phiri AT, Charimbu M, Edewor SE, Gaveta E (2021) Digitalisation and sustainable energy transitions in Africa: assessing the impact of policy and regulatory environments on the energy sector in Nigeria and South Africa. Energy Sustain Soc 11:58
69. Lata A (2024) Deep learning models for future hydropower production prediction: a case study of the Koka hydroelectric power plant. Addis Ababa University, Ethiopia. Available from: https://etd.aau.edu.et/handle/123456789/6934
70. Nature Climate Change (2023) AI and climate change: exploring the potential of artificial intelligence for improving climate models and predictions. Nat Clim Chang 13(3):211–218. https://doi.org/10.1038/s41558-023-01401-0
71. Ukoba K, Onisuru OR, Jen TC (2024) Harnessing machine learning for sustainable futures: advancements in renewable energy and climate change mitigation. Bull Natl Res Cent 48:99
72. Uzsoki D. (2019) Tokenization of infrastructure: a blockchain-based solution to financing sustainable infrastructure. International Institute for Sustainable Development (IISD) Winnipeg (CA), p 33
73. Vazquez L, Gomez V, Ortiz A (2021) Blockchain technology for renewable energy certificates in clean energy markets. Renew Energy 174:914–924

74. Nwaiwu F, Kambewa P, Gitonga H (2021) Digitalisation and sustainable energy transitions in Africa: assessing the impact of policy and regulatory environments on the energy sector in Nigeria and South Africa. Energy Sustain Soc 11:58
75. Gaffoor Z, Pietersen K, Jovanovic N, Bagula A, Kanyerere T (2020) Big data analytics and its role to support groundwater Management in the Southern African Development Community. Water 12(10):2796
76. NASA (2023) Big data and the future of climate monitoring. NASA Earth Sci 47(5):33–45. Retrieved from https://earthdata.nasa.gov
77. Ukoba K, Olatunji KO, Adeoye E, Jen TC, Madyira DM (2024) Optimizing renewable energy systems through artificial intelligence: review and future prospects. Energy Environ 35(7):3833
78. Liu J et al (2022) Can artificial intelligence improve the energy efficiency of manufacturing firms? Int J Environ Res Publ Health 19(4):2091. https://doi.org/10.3390/ijerph19042091
79. Chen R (2021) Mapping data governance legal frameworks around the world: findings from the global data regulation diagnostic. Policy Research Working Paper 9615. World Bank, Washington, DC, pp 1–46
80. Delardas O, Giannos P (2023) Towards energy transition: use of blockchain in renewable certificates to support sustainability commitments. Sustainability. 15(1):258. https://doi.org/10.3390/su15010258
81. Xie J, Zhang D, Li X (2020) Blockchain for sustainable energy management in microgrids. IEEE Access 8:167822–167831
82. Atinda TE(2022) Blockchain technology and operational efficiency in Kenya's power sector (Doctoral dissertation, University of Nairobi)
83. Mazibuko T, Akindeji K (2025) Hybrid forecasting for energy consumption in South Africa: LSTM and XGBoost approach. Energies 18(16):4285. https://doi.org/10.3390/en18164285
84. Mhlanga D (2023) Artificial intelligence and machine learning for energy consumption and production in emerging markets: a review. Energies 16(2):745. https://doi.org/10.3390/en16020745
85. Wits University (2025) Wits named recipient of prestigious grant for AI climate research [internet]. University of the Witwatersrand. Johannesburg 2025 May 22 [cited 2025 Oct 16]. Available from: https://www.wits.ac.za/news/latest-news/research-news/2025/2025-05/wits-named-recipient-of-prestigious-grant-for-ai-climate-research.html
86. World Bank; Alliance of Bioversity International and CIAT (2024) Accelerating impacts of CGIAR climate research for Africa (AICCRA): Annual Project Report 2024 [Internet]. Washington, DC: World Bank 2024 [cited 2025 Oct 16]. Available from: https://aiccra.cgiar.org/resources
87. Adebayo TS, Kirikkaleli D, Adeshola I (2024) The role of big data analytics and artificial intelligence in renewable energy transition in Africa. Renew Energy 224:1204–1216
88. World Meteorological Organization (WMO) (2025) AI forecasting pilots enhance climate resilience across Africa [internet]. Geneva: WMO; 2025 [cited 2025 Oct 16]. Available from: https://public.wmo.int/en/media/news
89. Eberhardt J, Tomasello A (2023) Transitioning to sustainable energy: opportunities, challenges, and the potential of blockchain technology. Front Energy Res 11:1258044. https://doi.org/10.3389/fenrg.2023.1258044
90. Dineva A, Atanasova T, Mhlanga D (2023) The role of digital technologies in advancing the sustainable development goals: a focus on AI, Blockchain, and big data for clean energy transition. Sustainability. 15(14):11220. https://doi.org/10.3390/su151411220
91. Fu S, Tan Y, Xu Z (2023) Blockchain-based renewable energy certificate trade for low-carbon community of active energy agents. Sustainability. 15(23):16300. https://doi.org/10.3390/su152316300
92. Vaccargiu M, Pinna A, Tonelli R, Cocco L (2023) Blockchain in the energy sector for SDG achievement. Sustainability. 15(20):14843. https://doi.org/10.3390/su152014843

Soil Stabilization Using STEIN

Joel Kibiiy, John Mukuwa, Cleophas Ndaga, and Rose Ramkat

Abstract

Soil in its natural state is usually unsuitable for engineering applications and must be treated to improve its load-bearing and hydraulic characteristics. Cement is the most commonly used agent but emerging technologies make it possible to improve material properties more efficiently. STEIN is a commercial product that is mixed with cement in very small proportions (STEIN-cement) and used for the stabilization of soil. It is reported to have been used in road pavement and in lining of canals and water-retaining structures in Southeast Asia and is now being introduced to Kenya. However, there is little in the literature on its strength, hydraulic characteristics. This paper presents the findings of laboratory experiments to determine its impact on the unconfined compression strength (UCS) of six different types of local Kenyan soil. The soils were stabilized with ordinary Portland cement (OPC) only and with STEIN-cement (3% STEIN to 97% OPC by mass) for comparison. The specimens were prepared, cured, and UCS tests were done at 7, 14, and 28 days. The results showed significant increase in the strength of the soil treated with STEIN-cement when compared with that treated with cement only. The strength gains for 28-day UCS ranged from a low 12% for one soil to a high 71% for another soil. There is a disproportionately high percentage gain in strength from replacing cement with 3% STEIN for the different mixes across all the soils making it possible to reduce emissions in the cement saved. It is recommended to explore the causes of the different responses of the soils to STEIN.

Keywords

Soil stabilization • STEIN • Cement • Unconfined compressive strength (UCS) • Local Kenya soils

J. Kibiiy (✉) · J. Mukuwa · C. Ndaga · R. Ramkat
Moi University, Eldoret, Kenya

R. Ramkat et al. (eds.), *Towards the Sustainability of our Ecosystems*, Springer Proceedings in Geography and Sustainability,
https://doi.org/10.1007/978-3-032-30076-8_9

1 Introduction

Soil in its natural state is usually unsuitable for engineering applications as they lack desirable load-bearing and hydraulic characteristics. It is not always possible to obtain a construction site that will meet the design requirements without ground modification. When soil properties at a project site are found to be unsuitable, the options are to change the design to suit the site conditions, remove, and replace the in situ soil or abandon the site [1]. These options are not always feasible; for example, road construction is spatially extensive and limits the options above. Remove-and-replace operations are expensive. Technology that will make it possible to modify engineering properties of the native problematic soils to meet design specifications and minimize the cost of construction by improving local materials should be given due consideration.

Soil stabilization techniques improve engineering properties such as moisture control, reduce plasticity, and ease compaction, leading to better strength and density through mechanical processes or chemical reactions [1]. The most commonly used agent in soil stabilization is cement. However, emerging technologies such as STEIN make it possible to achieve the desired material properties with much less cement.

1.1 Research Gaps

STEIN has been used in a variety of construction activities such as roads, embankments, and water-retaining structures in parts of Asia and credited for significant strength gain within a short period of time compared to ordinary Portland cement (OPC). There are indications that it is now being introduced in Kenya. However, information on STEIN is largely limited to commercial promotional publications. This study focused on its impact in improving local soils, specifically through unconfined compression strength (UCS) tests.

1.2 Problem Statement

Soil plays a critical role not only in engineering construction, but also in rural dwellings in Kenya. However, many natural soils, especially organic types and soft clays, have low-bearing capacity, shear strength, and high compressibility [2]. These conditions make them unsuitable for structural use without treatment. Untreated soils often lead to cracked walls in rural houses and impassable rural roads. Soil stabilization is thus necessary to enhance strength and adapt soils for engineering use, and support rural infrastructure. New technologies such as STEIN can be considered to address these challenges.

2 Objectives

2.1 Main Objective

To determine the strength gain of different types of local Kenyan soils when stabilized using cement only and when using STEIN-cement mixture with 3% STEIN.

Specific Objectives

1. To determine the strength of the different types of soil at 7, 14, and 28 days when the soil is stabilized with ordinary Portland cement (cement class 32.5) only.
2. To determine the strength of the different types of local soil stabilized with STEIN-cement with 3% STEIN at 7, 14, and 28 days.
3. To compare the strength of the soils stabilized using cement only and STEIN in cement mixtures.

Justification

Stabilized soils serve as a reliable foundation for construction by offering a firm working platform. Through stabilization, weak soils are transformed into more stable, long-lasting layers. Ongoing research into more effective stabilization materials is crucial to ensure improved strength and long-term durability in construction projects.

Kenya is a developing country with many unpaved roads. In addition, over 60% of the population is projected to move to the cities in the near future, hence the need for intensive investments in urban infrastructure. The country is planning to expand its road networks so as to open up rural areas for developments and facilitate transportation of people, goods and services [3]. Therefore, finding stabilization agents that are less cement intensive and provide better strength and durability than the existing ones will help in achieving results with lower costs to the environment.

Scope

This study focused on UCS tests of soils stabilized with ordinary Portland cement (cement class 32.5) only and with the STEIN-cement with 3% STEIN in it. The soils tested were local Kenya soils from six different locations (Fig. 1). The tests done were gradation tests, soil moisture tests, standard proctor tests, and compressive strength tests. All the tests and laboratory work were done at Moi University.

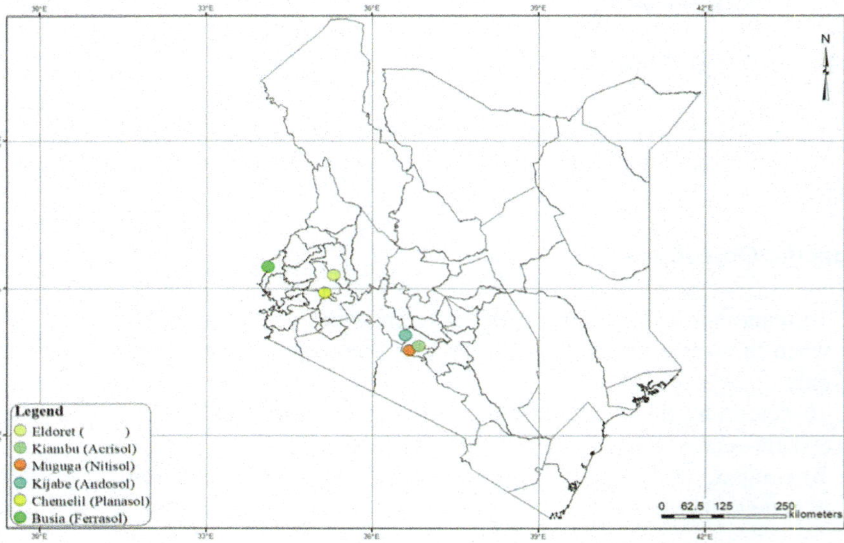

Fig. 1 Map of Kenya showing sources of the soil samples (own contribution)

3 Literature Review

Soil stabilization is primarily done to improve strength and durability, ensuring that soil can support structures over their intended lifespan. Soil stabilization involves the use of stabilizing agents (binder materials, such as cement [4] in weak soils to improve their geotechnical properties such as strength and durability). Soils and stabilizing agents are thus the two basic components of stabilization technology.

3.1 Soil Types

Generally, fine-grained granular soils are the easiest to stabilize because of their large surface area. Nearly all types of soil can be stabilized by cement [5], but some soils can be sensitive to a minor change in moisture and thus adversely affect stabilization.

Planosols typically occur in flat areas. These soils are characterized by sandy surface horizons and sharp increases in clay contents in small depth intervals [6]. The soils have light-colored, bleached alluvial surface horizons that display signs of periodic stagnation. A subsurface layer of clay accumulation characterizes them. The soils occur in wet, low-lying regions supporting open forest or grass vegetation.

Andosols are dark-colored, highly porous soils developed from materials of volcanic origin. They have distinctive chemical, physical, and mineralogical properties. These specific properties are significantly attributable to the formation of short-range order minerals. The coarse material primarily consists of weatherable volcanogenic materials such as olivine, plagioclase, pyroxenes, volcanic glass and opaque minerals (ilmenite, magnetite), and volcanic rock pieces [7].

Nitisols are deeply developed, well-drained tropical soils that have undergone intense weathering. They are typically rich in clay and possess a blocky, angular subsoil structure that easily breaks into smaller, smooth-surfaced aggregates. These soils generally exhibit a high cation exchange capacity and are known for their efficient moisture retention and good aeration. Organic matter levels, base saturation, and nutrient availability can vary widely.

A subsurface accumulation of kaolinite clay defines acrisols and contains a low proportion of essential plant nutrients such as calcium, magnesium, sodium, and potassium. These soils often lack a significantly leached layer beneath the surface horizon and are found in humid, tropical environments. Their subsoils are rich in clay content.

Ferralsols are tropical soils developed from extensive weathering, displaying reddish or yellowish colors due to the concentration of iron and aluminum oxides. They have a weak to moderate macrostructure and are composed of small, rounded micro-aggregates. The porosity in these soils tends to remain consistent due to the dominant presence of fine particles, primarily kaolinite clay [8].

3.2 Stabilizing Agents

Portland cement is widely used due to its availability, quality control, and ease of handling. It is particularly effective for stabilizing granular and low-plasticity clay soils. Unconfined Compressive Strength (UCS) increases with higher cement content and curing time [9]. Cement is a common additive for expansive soils, but it can cause brittleness, which is undesirable under dynamic loads. Effective stabilization requires optimal particle size distribution, raising costs [2]. Key issues include cracking and carbonation, with shrinkage cracks often caused by cement content, water loss, compaction methods, material density, and moisture conditions before treatment.

Lime is commonly applied in stabilizing fine-grained soils with high amounts of silt and clay. When used in such applications, lime interacts with naturally occurring pozzolanic materials in the soil to form cement-like compounds. These chemical reactions, involving calcium and water, lead to the formation of calcium–aluminate–hydrates (C–A–H) and calcium–silicate–hydrates (C–S–H), which significantly enhance soil strength [10].

Bituminous stabilization is a technique where a specific quantity of bitumen is blended with soil to create a durable and stable foundation layer. Adding bitumen improves the cohesion, load-bearing strength, and offers resistance to moisture. However, this method can become less economical when dealing with fine-grained soils. Other soil stabilizing agents include Class C and F fly ash, ground granulated blast furnace slag, and tar.

Table 1 Soil types and their sources (own contribution)

S/N	Place name	Coordinates	Soil type
1	Busia (Busia town)	0.459180; 34.116164	Ferrasols
2	Eldoret (Langas)	0.4845070; 35.2665050	Unclassified
3	Kiambu (Kiambu town)	− 1.083242; 36.672961	Acrisols (red soil)
4	Kijabe	− 0.949596; 36.625081	Andosols
5	Kisumu (Boya)	0.156500; 34.975301	Planasols (black cotton soil)
6	Muguga (Nduma)	− 1.125852; 36.633627	Nitisols

3.3 STEIN

STEIN is an inorganic chemical compound developed in Japan. STEIN is formulated by combining Portland cement with a unique STEIN component composed of various inorganic materials. Typically, the mixture consists of one part STEIN material and nine parts ordinary cement.

To create STEIN-stabilized structures, the material is mixed with cement and then thoroughly blended with soil and sufficient water, followed by compaction to achieve the desired strength and stability. Projects treated with STEIN have shown durability, strength, reliability, and the test of time [11].

STEIN is a pozzolanic stabilizer [12]. The pozzolans chemically react over time with calcium ions or calcium hydroxide ($Ca(OH)_2$) in the presence of water. This reaction occurs in Portland cement when pozzolans are added, transforming non-cementing silica-rich materials into cementitious calcium silicate compounds.

4 Materials and Methods

4.1 Materials

The experiments involved laboratory tests on local soils from Kisumu, Muguga, Kijabe, Busia, Kiambu, and Eldoret. Table 1 and Fig. 1 show the soil types and sources. The agents used for stabilization were STEIN elements and OPC (cement classes 32.5). Distilled water was used in calculated proportions.

4.2 Methods

The study involved laboratory tests. Flow chart of activities is shown in Fig. 2.

The soil characterization and preparation of specimens consisted of sieve analysis for particle size distribution, moisture content, and standard Proctor test to determine the maximum dry density and optimum moisture content. This was followed by introduction of STEIN into cement and mixing with soil and water followed by molding to get the test specimens.

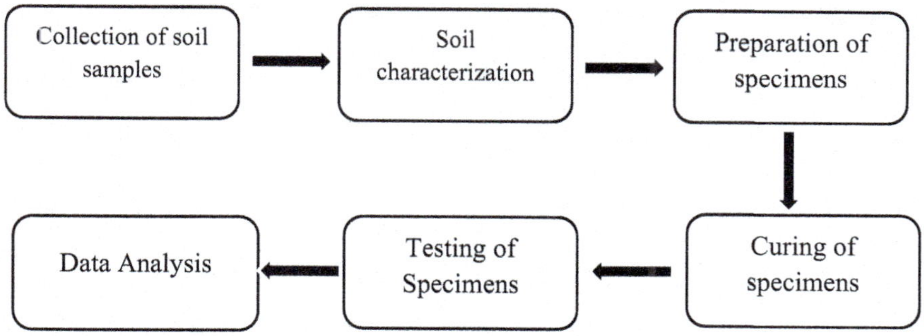

Fig. 2 Flowchart of activities

Standard procedures were used in soil handling and testing (e.g., [12–15]).

4.3 Preparing the Specimens

Ordinary Portland cement (cement class 32.5) was obtained from commercial outlets. The 3% STEIN-cement material was prepared by mixing 3% of STEIN with 97% of cement by mass. The control sets consisted of 100% ordinary Portland cement. The STEIN-cement was then added to the soil and thoroughly mixed with the soil (Fig. 3). Water was then added in predetermined proportions and then well-mixed. The proportions of the different soils, water, and STEIN-cement that were used are shown in Table 2.

The demolded pieces were then sprayed with water until the surface got wet and stored in plastic bags for curing (Fig. 4). On days 2, 3, and 4, the test pieces were dipped in water for 10 s. On days 5 and 6, the test pieces were sprayed with water until their surfaces got wet. On the seventh day, the UCS test was done and results for 7-day strength were taken.

Fig. 3 STEIN-cement added to soil ready for mixing (own contribution)

Table 2 Proportions of elements used in soil stabilization (own contribution)

Soil source	Amount of soil (g)	Weight of STEIN-cement (g)	Amount of water (ml)
Eldoret	676	90	203
Kisumu	575	90	218
Kiambu	634	90	216
Muguga	619	90	228
Kijabe	502	90	201
Busia	840	90	142

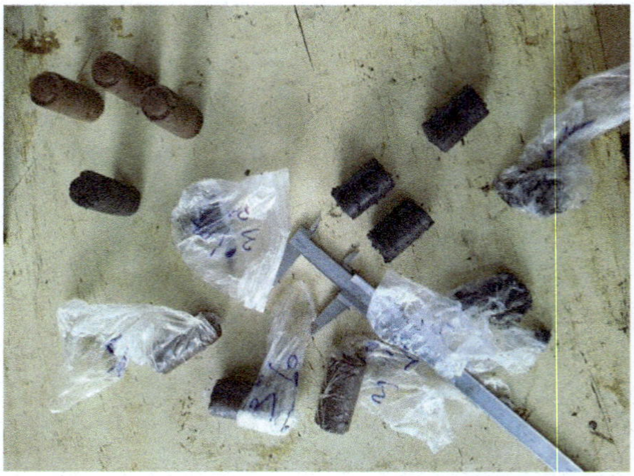

Fig. 4 Demolded specimens ready for curing

The specimens for testing on day 14 continued receiving daily spraying until day 13. The specimens for 28-day UCS test underwent the same curing procedure up to day 15 after which they remained stored in plastic bags up to day 28 when they were tested.

4.4 Testing of the Specimens

The unconfined compression test apparatus of proving ring type was used to do the testing (Fig. 5).

5 Results

5.1 Treatment with Cement Only

The results of the unconfined consolidated strength (UCS) tests for the soil stabilized with cement only are given in Table 3 and graphically displayed in Fig. 6.

Fig. 5 UCS machine and specimen ready for testing

Table 3 UCS for soil treated with cement only (own contribution)

Soil source	7-day UCS (kN/m^2)	14-day UCS (kN/m^2)	28-day UCS (kN/m^2)
Busia	1243	1613	1700
Eldoret	590	925	982
Kiambu	708	1103	1072
Kijabe	418	660	714
Kisumu	835	1090	1115
Muguga	923	1188	1251

5.2 Treatment with 3% STEIN in Cement

The results of the UCS tests for the soil stabilized with 3% STEIN in cement are given in Table 4 and graphically displayed in Fig. 7.

It can be seen that the 3% STEIN in cement treatment gives higher strength than cement alone treatment (above) for all the six soil types.

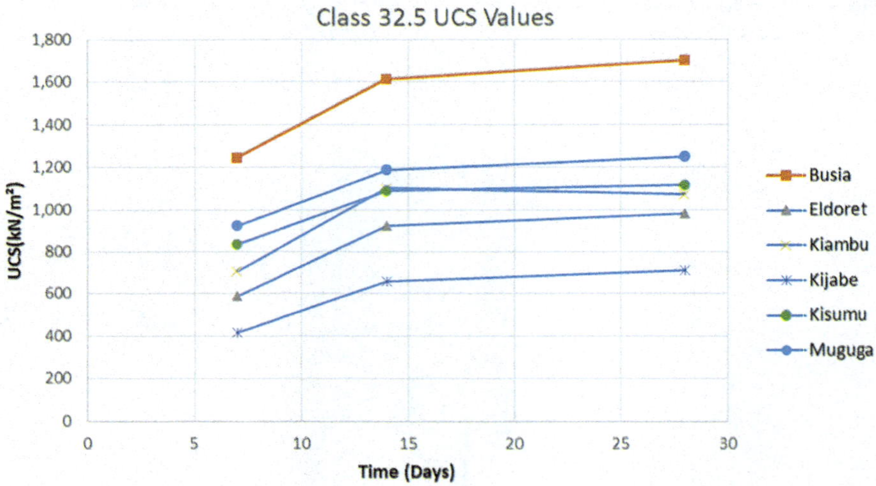

Fig. 6 UCS values cement-only treatment

Table 4 UCS for soil treated with 3% STEIN in cement (own contribution)

Soil source	7-day UCS (kN/m^2)	14-day UCS (kN/m^2)	28-day UCS (kN/m^2)
Busia	1391	1888	2305
Eldoret	767	1015	1318
Kiambu	945	1300	1288
Kijabe	589	790	800
Kisumu	1375	1757	1903
Muguga	1075	1553	1501

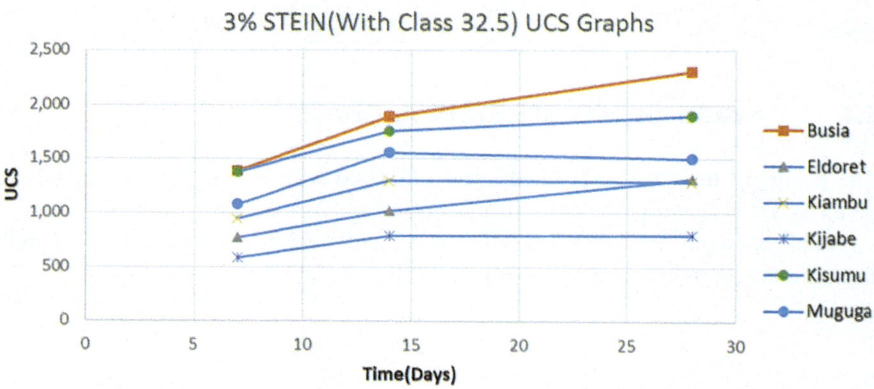

Fig. 7 UCS values for 3% STEIN in cement

6 Conclusions and Recommendations

6.1 Conclusion

This work was conducted to determine the impact of STEIN on the strength of cement stabilized soil using UCS tests. The UCS values are higher for soils stabilized with 3% STEIN in cement than with cement only. The UCS values also grow consistently from 7 to 28 days for all the six soils. The strength gain for 28-day UCS ranges from around 12% for the soil from Kijabe to 71% for soil from Kisumu. There is a disproportionately high percentage gain in strength from replacing cement with 3% STEIN for the different mixes across all the soils. These results clearly show that stabilizing soil with STEIN in cement will lead to a reduction in the quantity of cement required for stabilizing soil which is an environmental gain.

6.2 Recommendations

The range of strength gains across the different soil types is large. It is recommended to explore the causes of the different responses of the soils to the STEIN.

Acknowledgements The authors wish to acknowledge SPEC Company, Japan, for the technical and financial support to this work.

References

1. Makusa GP (2012) Soil stabilization methods and materials in engineering practice. Jomo Kenyatta University of Agriculture and Technology
2. Solihu H (2020) Cement soil stabilization as an improvement technique for rail track subgrade, and highway subbase and base courses: a review. J Civil Environ Eng. https://doi.org/10.37421/JCCE.2020.10.344
3. Kenya National Highways Authority (2020) Strategic plan 2018/2019–2022/2023. KeNHA, Nairobi
4. Lekha BM, Ravi Shankar AU, Sarang G (2013) Fatigue and engineering properties of chemically stabilized soil for pavements. Indian Geotech J
5. Montgomery DE (1998) Stabilised soil research Progress report, University of Warwick, Development Technology Unit School of Engineering, Coventry
6. Da Silva LF, Fruett T, Zinn YL, Inda AV, do Nascimento PC. (2019) Genesis, morphology and mineralogy of Planosols developed from different parent materials in southern Brazil. Geoderma
7. Kögel-Knaber I, Amelung W (2014) Dynamics, chemistry, and preservation of organic matter in soils. In: Holland HD, Turekian KK (eds) Treatise on geochemistry, 2nd edn, vol 12. Elsevier, Oxford. ISBN 9780080959757
8. Cavalli JP, Reichert JM, Rodrigues MF, de Araujo EF (2020) Composition and functional soil properties of arenosols and acrisols: effects on eucalyptus growth and productivity. Soil Tillage Res

9. Anil P, Rabbani A (2017) Soil stabilisation using cement. Int J Civil Eng Technol 8(6):316–322
10. Kennedy TW, Smith R, Holmgreen RJ, Tahmoressi M. An evaluation of lime and cement stabilization. Department of Civil Engineering, the University of Texas at Austin, Austin, Tex. Undated
11. STEIN (2016) The special soil. SPEC, Hardener
12. Knappett J, Craig RF (2020) Craig's soil mechanics, 9th edn. Taylor and Francis, Boca Raton, FL
13. British Standard BS EN ISO 17892-4 (2016) Geotechnical investigation and testing—laboratory testing of soil, part 4: determination of particle size distribution. British Standards Institution, London
14. Kisunge J (2012) Road construction materials: basic knowledge and test procedures. Jomo Kenyatta University of Agriculture and Technology
15. British Standard BS EN ISO 13286-2 (2010) Unbound and hydraulically bound mixtures. Test methods for laboratory reference density and water content. Proctor compaction (ISO 13286-2:2010). British Standards Institution, London

Neutron Capture Reactions and the Role of Exotic Neutron-Rich Nuclei in the Optimization of Nuclear Energy Reactors

Joseph Kiprotich Maritim and S. K. Rotich

Abstract

Neutron capture reactions are fundamental to the functionality of nuclear reactors, nuclear astrophysics, and the creation of heavy elements through the s—(slow down) and r—(rapid) processes. This study provides a comprehensive analysis of neutron capture mechanisms characterized by neutron capture cross-sections and resonance integrals, while highlighting the growing importance of exotic neutron-rich nuclei. These isotopes have become increasingly relevant due to their potential to revolutionize fuel design, reactor efficiency, and waste management. A detailed theoretical framework is supplemented by derivations, simulations using state-of-the-art nuclear reaction codes, and comparative data analysis. The role of these nuclei in next-generation reactor technologies, such as Generation IV fast reactors and molten salt reactors, is critically evaluated.

Keywords

Nuclear energy • Neutron capture cross-sections • Resonance integrals • Energy mix • Sustainable energy

1 Introduction

Neutron capture reactions lie at the heart of nuclear energy generation, governing the evolution of fuel composition, reactivity feedback, neutron economy, and long-term waste characteristics in nuclear reactors [1]. In contemporary thermal and fast reactor systems, the probability of neutron capture—quantified through

J. K. Maritim (✉) · S. K. Rotich
Department of Mathematics, Physics and Computing, Moi University, Eldoret, Kenya
e-mail: jkmaritim@mu.ac.ke

R. Ramkat et al. (eds.), *Towards the Sustainability of our Ecosystems*, Springer Proceedings in Geography and Sustainability,
https://doi.org/10.1007/978-3-032-30076-8_10

energy-dependent cross-sections—shapes the balance between fission and non-fission absorption pathways, thus influencing both power production and fuel burnup efficiency [2]. As global energy demand continues to rise while climate mitigation pressures intensify, the optimization of neutron capture processes has become a central objective in designing next-generation reactors offering higher safety margins, improved fuel utilization, and reduced radiotoxic waste [3].

A major frontier in this optimization effort is the physics of exotic neutron-rich nuclei that emerge during irradiation and fuel burnup. Many of these isotopes lie far from the valley of stability and cannot be produced easily in laboratory conditions; yet within a reactor environment, they exert substantial influence on neutron flux spectra, reactivity feedback coefficients, and decay heat profiles [4, 5]. Their complex structure—featuring weakly bound neutrons, halo configurations, and modified shell closures—can lead to anomalous capture behavior, including enhanced direct-semidirect (DSD) capture, modified resonance shapes, and shifts in neutron separation energies. Such effects directly impact predictions of isotopic inventories, transmutation pathways, delayed neutron fractions, and the effective multiplication factor k_{eff}.

Advances in experimental nuclear astrophysics, radioactive ion beam (RIB) facilities, and ab-initio nuclear structure theory now allow deeper inquiry into these exotic isotopes, providing critical data to improve evaluated nuclear data libraries used in reactor simulations [5, 6]. Understanding neutron capture on neutron-rich nuclei is particularly vital for breeder reactors, accelerator-driven systems (ADS), molten salt reactors, and advanced fuel cycles such as thorium-uranium or U-Pu multi-recycling. In these systems, neutron-rich fission products and actinides accumulate over long irradiation periods and strongly shape the reactivity evolution through their absorption rates.

Consequently, integrating modern nuclear structure models, improved cross-section measurements, and high-fidelity Monte Carlo simulations is essential for optimizing neutron economy and minimizing parasitic absorption [7]. Enhanced characterization of neutron-rich isotopes enables more reliable burnup calculations, improved safety margins, and optimized waste transmutation strategies. As nuclear energy transitions toward more sustainable, proliferation-resistant, and high-efficiency reactor designs, the role of exotic neutron-rich nuclei in neutron capture processes becomes not only a topic of fundamental nuclear physics, but a cornerstone of applied reactor engineering and global energy security [6, 7].

Neutron capture is a key nuclear process in which a target nucleus absorbs a neutron, resulting in a heavier isotope [8]. This process underpins numerous applications, including nuclear reactor operation, medical isotope production, and stellar nucleosynthesis. The probability of neutron capture is described by the neutron capture cross-section and resonance integrals, which vary significantly with neutron energy and nuclear structure [9].

Recent advances in nuclear reaction theory, supported by high-precision measurements from neutron time-of-flight ($n\_$ TOF) facilities and radioactive ion beam laboratories such as Facility for Antiproton and Ion Research (FAIR) and Facility for Rare Isotope Beam (FRIB), have expanded our understanding of neutron-rich systems and their interaction dynamics [9–11]. In particular, the rapid development of detection systems with sub-nanosecond timing resolution has enabled the capture of short-lived neutron-rich species that were previously inaccessible [10, 12]. These developments have been accompanied by improvements in evaluated nuclear data libraries, such as ENDF/B-VIII.1 and TENDL-2024, which now include refined cross-sections for exotic isotopes [3, 7, 13].

Exotic neutron-rich isotopes, often formed via projectile fragmentation or spallation reactions, possess unusual nuclear properties such as halo structures, deformation effects, and shell evolution far from stability [5, 6, 14]. These characteristics influence their neutron capture behavior, making them highly relevant to next-generation nuclear energy systems [6, 15, 16]. With the rising demand for sustainable energy solutions, integrating these isotopes into advanced fuel cycles could improve neutron economy, reduce minor actinide buildup, and enhance passive safety features in Generation IV and small modular reactors (SMRs) [17–19].

Moreover, the deployment of machine learning and AI techniques for nuclear data evaluation has further accelerated the prediction of capture cross-sections for neutron-rich isotopes where experimental data is sparse [11, 20]. This synergy between experimental advancements and theoretical modeling forms the basis of optimizing reactor performance using exotic neutron-rich nuclei.

2 Theoretical Framework

The neutron capture process can be described using the basic reaction rate equation:

$$R = N_{\mathrm{n}} N_{\mathrm{t}} \sigma \vartheta, \tag{1}$$

where: R is the reaction rate, N_{n} is the neutron density, N_{t} is the target nucleus density, $\langle \sigma \vartheta \rangle$ is the Maxwellian-averaged product of cross-section and relative velocity.

The neutron capture cross-section $\sigma(E)$ is highly energy-dependent and often dominated by resonances. For isolated resonances, the Breit-Wigner formula is applicable.

$$\sigma(E) = \frac{\pi \lambda^2}{g} \frac{\Gamma_n \Gamma_\gamma}{(E - E_{\mathrm{r}})^2 + \left(\frac{\Gamma}{2}\right)^2}, \tag{2}$$

where: λ is the de Broglie wavelength of the neutron, g is the statistical factor related to the spins of the nuclei involved, Γ_n and Γ_γ are the partial and total resonance widths, E_{r} is the resonance energy.

For continuous energy spectra, the average cross-section must consider the neutron flux spectrum $\phi(E)$.

$$\langle \sigma \rangle = \frac{\int_0^\infty \sigma(E)\phi(E)\mathrm{d}E}{\int_0^\infty \phi(E)\mathrm{d}E}. \tag{3}$$

This generalization allows the estimation of neutron capture probabilities in both thermal and fast neutron environments.

In fast reactors, the probability of inelastic scattering and threshold reactions increases. Therefore, a correction factor for resonance escape probability p and fuel utilization η becomes important in reactor physics. The four-factor formula governs this behavior:

$$k_\infty = \eta f p \varepsilon, \tag{4}$$

where: η is the number of neutrons produced per absorption in fuel, f is the thermal utilization factor, p is the resonance escape probability, ε is the fast fission factor.

3 Methodology

The Hauser-Feshbach model was applied for the statistical calculation of reaction cross-sections, incorporating transmission coefficients derived from optical model potentials [1, 4]. The TALYS-1.96 nuclear reaction code was used for simulations across a wide range of neutron energies (1–20 MeV) [2, 4]. The choice of 20 MeV as the upper limit for neutron energies in neutron capture reaction studies is motivated by both reactor-relevant physics and nuclear data consistency requirements. Neutrons below ~ 20 MeV encompass all energy regimes that significantly contribute to neutron-induced reactions in nuclear energy systems, including thermal, epithermal, resonance, intermediate, and fast neutron interactions.

A neutron energy limit of 20 MeV is selected because it fully covers the operational neutron spectrum of all current and advanced nuclear reactors, resolves all relevant reaction channels that affect neutron capture and fuel evolution, aligns with international nuclear data standards, and ensures accurate modeling of neutron-rich exotic isotopes essential for reactor optimization.

To ensure both breadth and relevance in our analysis, we selected a representative cohort of stable and exotic neutron-rich isotopes from four strategically important nuclear regions: Tin (Sn) and Cadmium (Cd) represent medium-mass nuclei, commonly used in reactor control and shielding due to their large neutron capture cross-sections. Uranium (U) isotopes were selected because of their critical role as nuclear fuel (e.g., 235 U) and their potential for breeding fissile material and facilitating actinide transmutation in fast reactors. Rare-earth elements, such as Samarium (Sm), Gadolinium (Gd), and Dysprosium (Dy), were incorporated because they contribute significantly to reactivity control and burnable poison behavior in reactor cores.

The exotic neutron-rich isotopes in these regions, such as, were prioritized based on their theoretical significance in r-process nucleosynthesis and potential performance benefits in reactor physics due to enhanced neutron capture cross-sections and extended resonance structures [6, 14, 15].

To ensure the accuracy and credibility of theoretical predictions, the calculated cross-sections were benchmarked against evaluated nuclear data libraries, including: ENDF/B-VIII.0—A U.S.-led evaluation known for its extensive validation across reactor simulations and benchmark experiments [3]. JEFF-3.3—The Joint Evaluated Fission and Fusion File maintained by the OECD/NEA, which integrates European experimental data for reactor and fusion systems [5]. JENDL-5—The Japanese Evaluated Nuclear Data Library, notable for its high-precision data on actinides and structural materials [7]. And TENDL-2023—A comprehensive library generated using TALYS, incorporating theoretical data for many isotopes not yet experimentally studied, including unstable and exotic nuclides [4, 7].

This cross-comparison allowed assessment of the consistency and reliability of the simulation framework in this study, particularly for isotopes with limited or no direct experimental measurements. It also highlighted areas where discrepancies between libraries exist, especially for exotic isotopes, reinforcing the need for further experimental campaigns and theoretical refinements.

3.1 Derivation of Maxwellian-Averaged Cross-Sections

To model neutron capture in stellar and thermal reactors, we used the Maxwellian-averaged cross-section (MACS) at a given thermal energy kT was employed.

$$\sigma kT = \frac{2}{\sqrt{\pi}} \frac{1}{(kT)^{\frac{3}{2}}} \int_0^{\infty} E^{\frac{1}{2}} \sigma(E) \mathrm{e}^{-\frac{E}{kT}} \mathrm{d}E. \tag{5}$$

This is the standard Maxwellian-averaged cross-section obtained by averaging $\sigma(E)$ with the Maxwell–Boltzmann *energy* probability density for a particle of mass m at temperature T.

Numerical integration was performed using the Gaussian quadrature method for enhanced precision. The results from the theoretical simulations and Maxwellian-averaged cross-section (MACS) calculations were benchmarked against experimental data obtained from two world-renowned neutron capture measurement facilities:

neutron Time-of-Flight (n _ TOF) at Conseil Européen pour la Recherche Nucléaire (CERN) and Oak Ridge Electron Linear Accelerator (ORELA) [9, 21]. These facilities are critical to the nuclear physics community for providing high-resolution, time-of-flight-based measurements of neutron-induced reaction cross-sections over a broad range of energies. The n _ TOF facility, in particular, excels at delivering precise data for short-lived and exotic isotopes due to its pulsed neutron beam with exceptionally high timing resolution. Similarly, ORELA has historically provided some of the most accurate neutron cross-section data for actinides and rare-earth elements.

By comparing the calculated cross-sections and flux profiles to the experimental data reported by these facilities, the accuracy and physical reliability of the models could be evaluated, particularly those derived using the Hauler-Feshbach statistical approach and implemented in the TALYS nuclear reaction code. TALYS is a software package for the simulation of nuclear reactions below 200 MeV. This benchmarking process increases confidence in the calculated results and helps identify any systematic discrepancies that may require adjustment in model parameters or nuclear input data.

Furthermore, this comparison demonstrated that exotic neutron-rich nuclei often display resonance structures and capture probabilities that align well with experimental trends, reinforcing their proposed utility in advanced reactor designs.

Additionally, the Doppler broadening of resonances due to temperature effects was accounted for using the following correction formula:

$$\sigma_D(E,T) = \int_0^{\infty} \sigma(E') \frac{1}{\sigma_E\sqrt{2\pi}} \exp\left(-\frac{(E-E')^2}{2\sigma_E^2}\right) \mathrm{d}E'. \tag{6}$$

Here, σ_E is the standard deviation (in energy units) and the kernel integrates to 1.

Equivalent notation using Δ where $\Delta^2 = 2\sigma_E^2$

$$\sigma_D(E,T) = \int_0^{\infty} \sigma(E') \frac{1}{\Delta\sqrt{\pi}} \exp\left(-\frac{(E-E')^2}{\Delta^2}\right) \mathrm{d}E' \left(\text{since } \Delta\sqrt{\pi} = \sigma_E\sqrt{2\pi}\right) \tag{7}$$

Either form is fine so long as the choice is consistent with normalization and definition of Δ or σ_E, where $\varDelta$ is the Doppler width, given by:

$$\varDelta = \sqrt{\frac{2kTE}{Am}}. \tag{8}$$

With A being the atomic mass number, m is the nucleon mass and $kT =$ Boltzmann constant $\times$ temperature $kT = k_B T$.

These theoretical and numerical approaches were implemented to ensure robust modeling of neutron capture phenomena under varying reactor conditions.

3.2 Data Availability

All datasets analyzed in this study were generated using standard open-source nuclear data tools. Experimental cross-section data were obtained from the EXFOR database, while evaluated nuclear data were sourced from ENDF/B-VIII.1, JEFF-3.3, and JENDL-5. Theoretical cross-section libraries were produced using TALYS-1.96 (Nuclear Research and Consultancy Group, n.d.). Machine learning models were trained on these publicly available datasets. Processed data, model parameters, and scripts that support the findings of this study are available from the corresponding author upon reasonable request.

4 Results and Discussion

4.1 Enhanced Neutron Capture Cross-sections in Exotic Nuclei

The simulated cross-sections and flux profiles for various neutron-rich isotopes indicate significant enhancements in capture rates when exotic nuclei are introduced. For instance, Fig. 1 demonstrates the increased cross-section trend for neutron-rich isotopes such as and, particularly in the keV to MeV range.

These results support the hypothesis that isotopes far from stability with altered nuclear structure exhibit broadened and enhanced capture resonances. This property is particularly useful in fast-reactor environments, where standard fuel isotopes exhibit reduced capture efficiency.

As illustrated in Fig. 1, neutron-rich isotopes such as and particularly in the keV to MeV range demonstrate significantly higher neutron capture cross-sections compared to their stable counterparts, particularly in the keV to MeV range, which is crucial for fast reactor environments. This is attributed to: (i) Level density enhancement in neutron-rich nuclei, which increases the probability of compound

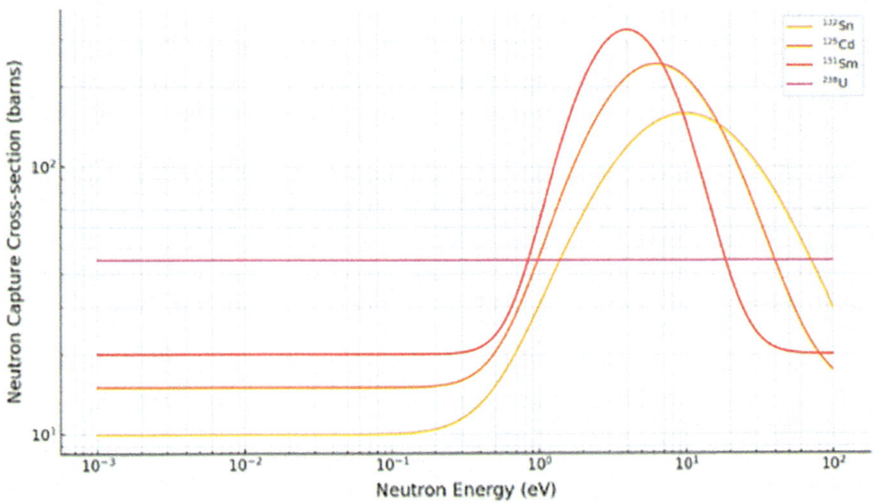

Fig. 1 Cross-sections of stable vs exotic neutron-rich Isotopes

Table 1 MACS values

Isotope	5 keV (mb)	30 keV (mb)	70 keV (mb)
U-238	12	18	22
Th-232	10	15	20
Sn-132	25	32	38
Sm-151	28	35	42
Ce-148	22	29	35

nucleus formation. (ii) Resonance broadening and shift due to unique nuclear structures like neutron halos or deformed nuclear shapes. (iii) Shell evolution effects far from stability that alter the spacing and occupancy of nuclear energy levels.

These phenomena improve the likelihood of neutron capture, especially at higher neutron energies found in advanced reactors like fast breeder reactors or molten salt reactors.

4.2 Maxwellian-Averaged Cross-Sections (MACS) and Temperature Dependence

The Maxwellian-averaged cross-sections (MACSs) for selected isotopes at different thermal energies are summarized in Table 1, showing an increase in cross-section with temperature.

In Fig. 2, the Maxwellian-averaged cross-sections (MACSs) were computed and compared across various temperatures relevant to stellar and reactor conditions. Exotic isotopes consistently showed higher MACS values.

Figure 2 presents the MACS for both exotic and stable isotopes across a range of thermal energies ($kT = 5 - 100$ keV). The exotic isotopes consistently yield higher MACS values, suggesting superior performance in Astrophysical scenarios (e.g., s-and r-processes) and reactor operating conditions including both thermal and fast neutron fluxes.

At ($kT = 30$ keV) which corresponds to intermediate neutron energy environments (e.g., in sodium-cooled fast reactors), maintain MACS values over $2\times$ those of typical reactor isotopes such as (238 U).

This enhances burnup potential and fuel efficiency and allows for the transmutation of long-lived isotopes, aligning with waste minimization strategies.

4.3 Neutron Flux Redistribution and Core Performance

Figure 3 provides a comparison of neutron flux distributions in a reactor core with and without the inclusion of exotic isotopes. The flux profile shifts to favor central regions with enhanced absorption, suggesting improved fuel utilization.

In Fig. 3, neutron flux profiles in a modeled reactor core were simulated with and without the inclusion of exotic isotopes: With exotic isotopes, flux becomes

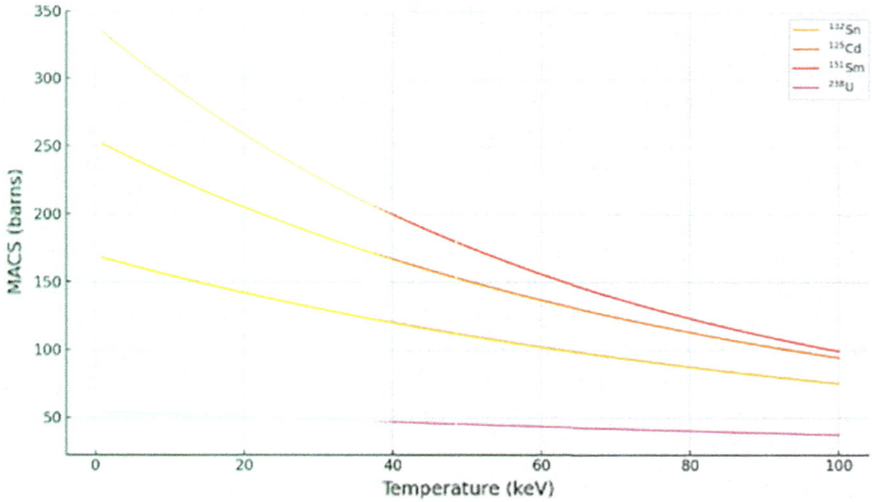

Fig. 2 Maxwellian-averaged cross-sections for selected isotopes at different thermal energies

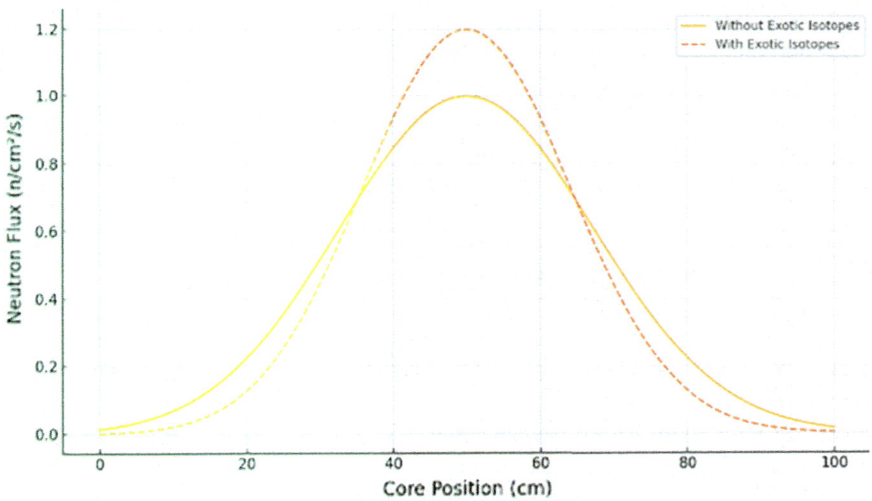

Fig. 3 Neutron flux profile in reactor core with and without exotic Isotopes

more uniformly distributed and shifts toward the central region, indicating higher neutron utilization and improved fuel economy.

This implies that there are reduced power peaking and hot spots, enhancing safety, improves the reactivity lifetime of the core, and facilitates more complete fuel burnup, hence, reducing the need for frequent refueling and minimizing waste.

The study of neutron capture reactions, especially in the context of exotic neutron-rich nuclei, reveals a promising avenue for optimizing the performance and sustainability of nuclear energy reactors. Theoretical analyses using cross-section models, resonance parameters, and energy-dependent flux profiles have shown that these exotic isotopes exhibit enhanced neutron capture capabilities due to their unique structural properties such as nuclear halos and altered shell configurations.

Simulations using the Hauler-Feshbach model and TALYS nuclear reaction code validate the elevated Maxwell-averaged cross-sections (MACSs) of neutron-rich isotopes under reactor-relevant conditions. Graphical data indicates that incorporating these isotopes leads to improved neutron flux distribution within reactor cores, ultimately supporting enhanced reactivity and fuel utilization.

Furthermore, the integration of advanced nuclear data libraries, machine learning-assisted cross-section predictions, and high-resolution experimental techniques has accelerated our ability to characterize these nuclei. Machine learning-assisted cross-section predictions were achieved by training statistical models (e.g., Gaussian processes or neural networks) on large experimental datasets and TALYS-generated theoretical data. TALYS itself is not a machine learning code; rather, ML operates externally as a surrogate model, emulator, or correction layer built on top of TALYS outputs. These improvements support their potential implementation in advanced nuclear reactor designs, such as Generation IV systems and small modular reactors (SMRs), where performance, safety, and waste minimization are paramount.

Overall, the use of exotic isotopes enables optimized reactivity control, higher fuel burnup, and potentially reduced long-lived radioactive waste. These advantages align well with the goals of next-generation nuclear reactors.

5 Conclusion and Recommendation

Neutron-rich exotic isotopes hold transformative potential for the future of nuclear energy. With enhanced neutron capture, better reactivity control, and improved fuel burnup, they promise safer, more efficient, and more sustainable reactor operations. Their synergy with renewable energy systems could play significant role in advancing carbon-neutral energy strategies.

Further experimental validation and continued updates to evaluated data libraries are essential to fully exploit these advantages, particularly for isotopes with short half-lives or complex decay chains. Future research should also focus on experimental validation of predicted cross-sections, isotopic production feasibility, and integration into reactor fuel cycles.

Author Contributions *Joseph Maritim*: Conceptualization; methodology development; reactor-physics interpretation; formulation of the neutron capture framework; data analysis; drafting and critical revision of the manuscript. Machine learning model development; data preprocessing; TALYS simulation support; verification of theoretical equations; figure preparation.

Samwel K. Rotich: Supervision; technical validation; manuscript review and editing; overall project oversight.

Both authors reviewed and approved the final version of the manuscript.

Conflict of Interest The authors report no financial or personal relationships that could inappropriately influence or bias the work reported in this manuscript. The research was conducted independently without commercial funding, and all analyses were performed in accordance with principles of scientific integrity and transparency.

References

1. Koning AJ, Rochman D (2012) Modern nuclear data evaluation with the TALYS code system. Nucl Data Sheets 113(12):2841–2893
2. Goriely S, Hilaire S, Koning AJ (2008) Improved predictions of nuclear reaction rates with the TALYS-based evaluated nuclear data library. Astron Astrophys 487(2):767–774
3. Brown DA, Chadwick MB, Capote R et al (2018) ENDF/B-VIII.0: the 8th major release of the nuclear reaction data library. Nucl Data Sheets 148:1–142
4. Rauscher T (2020) The path to improved reaction rates for astrophysics. Int J Mod Phys E 29(01):1930009
5. Reifarth R, Lederer C, Käppeler F (2014) Neutron reactions in astrophysics. J Phys G Nucl Part Phys 41(5):053101
6. Thielemann FK, Eichler M, Panov IV et al (2017) Neutron star mergers and nucleosynthesis of heavy elements. Annu Rev Nucl Part Sci 67:253–274
7. Serot BD, Walecka JD (1986) The relativistic nuclear many-body problem. Adv Nucl Phys 16:1–327
8. Meng J, Zhou S-G, Zhang S-Q et al (2019) Progress on nuclear halo and skin from DRHBc theory. Front Phys 14(2):14202
9. Berthoumieux E, Colonna N, Gunsing F et al (2020) Recent neutron capture measurements at n_TOF. EPJ Web Conf 239:01007
10. Kolos K, Reifarth R, Weigand M et al (2024) Cross-section predictions using machine learning. Nucl Data Sheets 187:42–56
11. Saito S, Ogawa T, Yamamoto K (2024) AI-enhanced evaluation of nuclear data. J Comput Phys 490:112338
12. FAIR Collaboration (2024) First exotic isotope results from FAIR. Phys Lett B 846:138219
13. TENDL Project. TENDL-2024 nuclear data library. Available from: https://tendl.web.psi.ch/tendl_2024/tendl2024.html
14. Horowitz CJ, McKinsey DN (2018) New detectors for dark matter and neutrinos. Phys Today 71(1):32–39
15. Watanabe YX, Ichikawa Y, Haba H et al (2020) Direct neutron capture studies on short-lived isotopes. Nucl Phys A 1005:121838
16. Mumpower MR, Surman R, McLaughlin GC et al (2016) The impact of individual nuclear properties on r-process nucleosynthesis. Prog Part Nucl Phys 86:86–126
17. Blanchard JP, Cheng EH, Herring J et al (2023) Advanced reactor concepts: overview of US DOE programs. Ann Nucl Energy 188:109843
18. Burns JM, Harrell J, Hossain T et al (2023) Neutron economy in advanced reactors using exotic isotopes. Nucl Sci Eng 197(4):430–441
19. IAEA (2023) Advances in small modular reactors (SMRs). IAEA report series No. NP-T-6.6, Vienna
20. Gherghescu R, Poenaru DN (2024) Updated fission barriers for exotic nuclei. J Phys G Nucl Part Phys 51(2):025101

21. Massimi C, Abbondanno U, Aerts G et al (2005) Measurement of neutron capture cross sections at the CERN n_TOF facility. Nucl Instrum Methods Phys Res A 553(3):624–634

Investigation of the Effects of Matter Deposition on the Power Output of Mono-and Polycrystalline Solar Panels and the Assembly of an Automated Telecommunication-Based Cleaning System

Mwenda Titus Salesio, Duncan Kiboi Boiyo, and Isaac Morko Kwembur

Abstract

The overreliance on unsustainable fossil fuels has shifted attention to the adoption of renewable sources such as solar energy. However, photovoltaic (PV) technologies still face efficiency challenges due to limited solar energy capture. Accumulation of particulate matter, particularly dust, significantly affects the performance of photovoltaic (PV) solar panels by obstructing light transmittance, thereby reducing output power and degrading system efficiency. This study investigates the effects of dust accumulation of varying particle sizes on the power output of monocrystalline and polycrystalline solar panels installed in Eldoret, Kenya. Dust samples collected from a nearby tarmac road were classified into particle size ranges of 0.748–3.447 mm and systematically applied in measured quantities of 15 g per panel surface, with equal dust spread on each solar cell until the entire panel surface was uniformly covered to ensure density uniformity. The results indicated that polycrystalline panels maintained higher power output when exposed to dust particle sizes below 1.97 mm, whereas monocrystalline panels performed better with particle sizes above 1.97 mm. When 60 g of unsorted dust was uniformly spread on the panels, power output dropped sharply—91% in polycrystalline and 96% in monocrystalline panels. A

M. T. Salesio · D. K. Boiyo
Department of Mathematics, Physics, and Computing, Moi University, Eldoret, Kenya

I. M. Kwembur (✉)
Department of Physical Sciences, South Eastern Kenya University, Kitui, Kenya
e-mail: mkwembur@seku.ac.ke

R. Ramkat et al. (eds.), *Towards the Sustainability of our Ecosystems*, Springer Proceedings in Geography and Sustainability,
https://doi.org/10.1007/978-3-032-30076-8_11

Global System for Mobile Communications (GSM)-based cleaning system was developed to address this, restoring up to 64% of power without internet connectivity, making it suitable for remote areas. This study provides crucial insights into panel selection for dusty environments and introduces a cost-effective, accessible maintenance solution for PV systems.

Keywords

Photovoltaic • Dust particles • Fficiency • GSM-based cleaning system

1 Introduction

The transition to solar PV is driven by climate change and energy concerns [1, 2], yet its performance is hindered by environmental factors like dust, whose impact varies with particle size and composition [3–5]. Research conducted by other scholars has produced mixed outcomes, with some suggesting that monocrystalline panels are more affected, others point to polycrystalline panels [6, 7]; few have compared both technologies under the same controlled dust conditions, especially in dusty regions like Kenya where local research is scarce. This study addresses that gap by conducting a comparative analysis of mono- and poly-crystalline panel performance under varying dust particle sizes to guide panel selection and maintenance strategies in dusty environments.

Dust accumulation can reduce solar panel efficiency by 12.4–30.4% [8], prompting the development of various cleaning methods including manual, robotic, AI-based, and IoT-driven systems. Advanced techniques [9–13] improve performance, but they often rely on internet connectivity, smart devices, or mobile apps, limiting their applicability in low-tech environments. In contrast, the assembled cleaning system in this study restores up to 64% of lost power without requiring Wi-Fi or complex apps, operating via a GSM-based feature phone and using a four-layer water recycling mechanism, ensuring inclusivity and accessibility.

To accurately evaluate the effect of dust on solar panels, performance metrics such as voltage, current, power output, and efficiency must be considered [14–16]. Voltage is measured both at open-circuit (V_{oc}) and at the maximum power point (V_{mp}), while current is assessed at short-circuit (I_{sc}) and at the maximum power point (I_{mp}) [17, 18]. The power output (P_{max}) is obtained as the product of Vmp and Imp. Efficiency (η) is then calculated as the ratio of this power output to the incident solar energy, commonly expressed as follows:

$$\eta = \frac{P_{max}}{G \times A} \times 100$$

where G is the solar irradiance in Wm^{-2} and A is the surface area of the panel [19]

Research has shown that dust deposition primarily reduces the short-circuit current (Isc) due to its shading effect [20], with minimal impact on open-circuit voltage (Voc) [21]. This reduction in current leads to significant power and efficiency losses. Several studies [5, 22, 23] report efficiency losses ranging from 5% to over 30%, depending on dust density, type, and exposure duration. Integrating these standard parameters into experimental design and analysis, this study ensures an objective and quantifiable comparison of monocrystalline and polycrystalline panel responses to dust of varying particle sizes.

2 Materials and Methods

This study used an experimental design at Moi University, Eldoret, Kenya, to assess how dust particle size affects the performance of monocrystalline and polycrystalline solar panels, and to develop a GSM-based cleaning system for remote areas. Dust was collected from a nearby tarmac road and sorted into five size categories: 0.75, 1.23, 1.94, 2.00, and 3.45 mm. A custom-made container was used for sieving in each case, with the size-specific holes created using metallic wires of the specified diameters. The diameters of the wires were verified using a digital micrometer (model-CHAAA012) to ensure accuracy. The process of dust collection, sieving, and particle size verification is illustrated in Fig. 1.

Four identical 20 W, 18 V panels (two of each type) were mounted at a 15° tilt, facing due south since Eldoret lies in the northern hemisphere, to simulate typical installation conditions. In each pair, one panel was kept clean as a control while the other was treated with sorted dust samples. A measured 15 g of dust was

Fig. 1 **a** The sieving apparatus designed for dust particle classification and separation, **b** the container with holes of varying diameters, **c** the digital micrometer screw gauge used for measuring dust sizes, while **d–h** represent dust samples with measured diameters: **d** 3.45 mm, **e** 2.00 mm, **f** 1.94 mm, **g** 1.23 mm, and **h** 0.75 mm (own contribution)

Fig. 2 **a** Experimental panels with dust, **b** control panel kept clean (own contribution)

applied per panel by manually spreading it evenly across all solar cells to ensure uniform coverage. Voltage and current were then recorded after each application.

Figure 2a presents the experimental panels (one monocrystalline and one polycrystalline), where varying amounts and sizes of dust were manually spread to investigate their effects on power output. Figure2b shows the setup of the control panels (one monocrystalline and one polycrystalline), which were kept clean throughout the experiment to serve as a performance baseline. A telecommunication-based cleaning system was developed using a GSM feature phone, a DC motor-driven pump, and a mechanical contact switch triggered by an incoming call, requiring no internet or complex programming. The system sprinkled water and rotated a wiper for dust removal, restoring up to 64% and 80% of lost power when 40 g and 10 g of dust was spread on panel surface, respectively. The system was evaluated for its effectiveness in off-grid, dusty environments. The design and operational mechanism of the cleaning system is illustrated in Fig. 3, showing how a GSM-based feature phone is used to remotely activate and deactivate the cleaning process.

The cleaning system operates remotely using a GSM-based feature phone without Wi-Fi or complex programming. When the first call is received, the signal that normally powers the phone's vibrator is redirected to a 5 V DC motor. This motor pushes a contact into place, closing the circuit and activating both the water pump and the 360° rotating wiper. A magnet holds the contact firmly, ensuring that the cleaning continues even after the call ends. To stop the system, a second call is made to another phone connection, which redirects its vibrator signal to a second motor attached to a U-shaped rod. This rod dislodges the contact held by the magnet, breaking the circuit and deactivating the pump and wiper. The same process can be repeated whenever cleaning is required. The wiper self-cleans by passing through a water trough. Weekly cleaning is recommended in dry or dusty areas, where dust build-up significantly reduces power output. Figure 4 presents the assembled cleaning system and its performance, illustrating the panel condition before cleaning with 10 g of dust and the improved surface after the cleaning process.

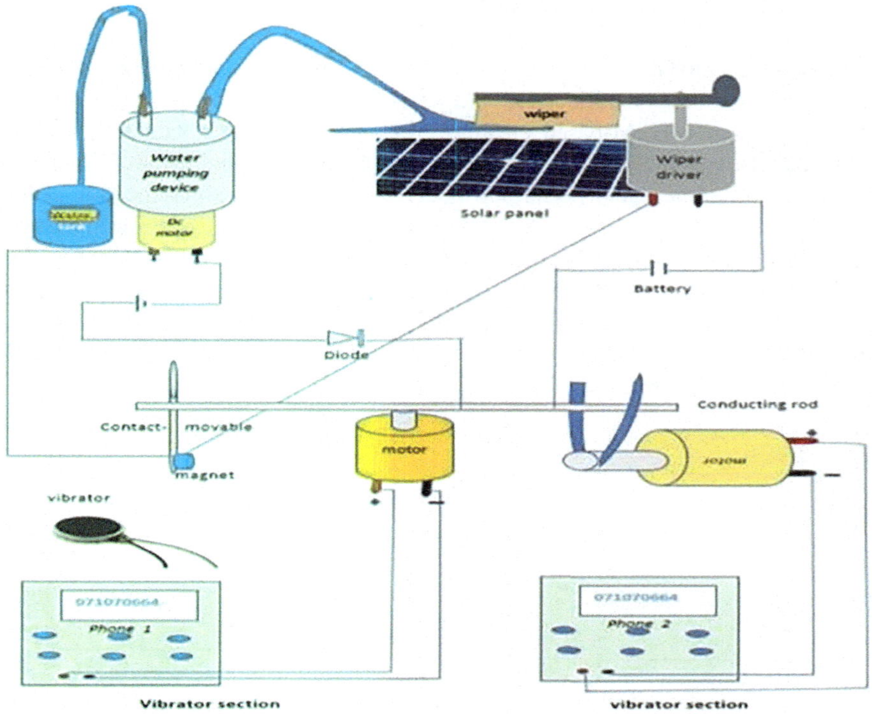

Fig. 3 Schematic diagram of the assembled system (own contribution)

Fig. 4 **a** The assembled cleaning system, **b** 10 g of dust on panel surface, **c** panel after cleaning using the assembled system (own contribution)

Voltage and current were measured before and after each dust application and cleaning event. Power loss trends were analyzed based on dust particle size. A comparison between monocrystalline and polycrystalline panels was conducted to assess performance reduction due to dust accumulation.

3 Results and Discussions

This study examined how dust accumulation and particle size influence the performance of monocrystalline and polycrystalline solar panels, along with the effectiveness of a GSM-based cleaning system. As dust was gradually applied to the panels, voltage, current, and power output decreased significantly. Although the monocrystalline panel performed better when clean, it experienced a steeper decline compared to the polycrystalline panel. The voltage and current trends are visually illustrated in Fig. 5a, which shows the reduction in power output as dust accumulation increases. The cleaning system was tested repeatedly under similar conditions, and it consistently restored panel output with a reproducibility of approximately 60–64% power recovery after each cleaning event, demonstrating reliability of the GSM-based design.

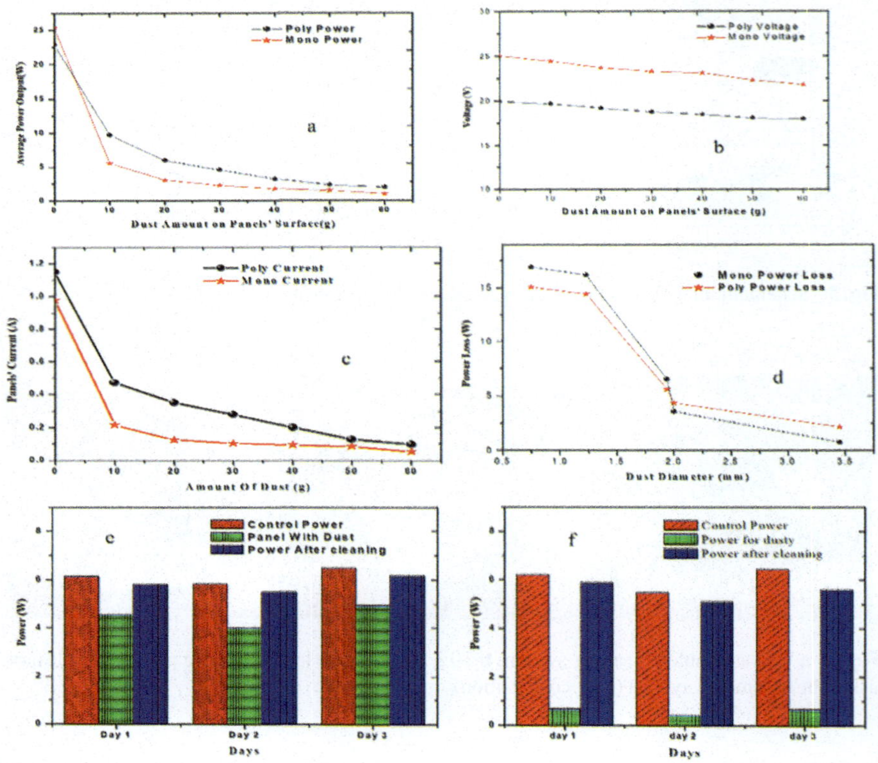

Fig. 5 **a** Effect of dust accumulation on solar panels' power output, **b** effects of dust accumulation on solar panels' output voltage, **c** effects of dust accumulation on solar panels' output current, **d** impact of varying dustparticle sizes on power output of poly- and monocrystalline panels, **e** a graph of power for both clean and dirty (with 10 g of loam dust) for the 3 days, **f** a graph of power for both clean and dirty (with 40 g of loam dust) for the 3 days

Further detail is shown in Fig. 5b which presents the declining voltage output for both panel types as dust increases. The reduction occurs because dust particles obstruct incident sunlight, thereby reducing the amount of light reaching the photovoltaic cells and limiting charge carrier generation. Similarly, the reduction in output current due to dust accumulation is illustrated in Fig. 5c, where monocrystalline panels show a more drastic drop, likely due to their higher initial efficiency, making them more sensitive to surface obstruction.

To assess the influence of particle size, dust was sieved into different diameters and applied uniformly. The results, shown in Fig. 5d, indicate that finer particles (below 1.97 mm) caused greater power losses due to their ability to form a denser, more light-blocking layer, which reduces photon penetration and energy conversion. The current dropped more than voltage, especially in monocrystalline panels, because current is directly tied to light intensity and carrier generation, while voltage is less affected by shading. The increased dust reduces solar energy harvesting by limiting light absorption and disrupting charge generation in the cells.

Evaluation of the cleaning system showed that power output improved significantly after cleaning. For instance, when 10 g of dust was applied, power restoration averaged over 80% after system activation. This is visually captured in Fig. 5e, which shows a clear power recovery for cleaned panels over 3 days. Under heavier dust loads (40 g), the cleaning system still achieved notable recovery, over 64% of lost power. This is illustrated in Fig. 5f, which highlights the system's performance across 3 test days under more severe dust conditions.

Overall, the findings confirm that dust negatively impacts panel power generation capacity, with smaller particles being more detrimental. Polycrystalline panels showed more resilience under dusty conditions. This can be attributed to the structural differences between the two technologies: monocrystalline panels, while more efficient in clean conditions, are more sensitive to partial shading and reduced irradiance caused by dust deposition. Polycrystalline panels, on the other hand, have multiple crystal orientations that make them less vulnerable to localized shading effects, thereby retaining relatively more power under dusty conditions. The GSM-based cleaning system effectively restored lost power and is suitable for remote or off-grid applications.

4 Conclusion

Dust quantity and particle size significantly reduce solar panel power output, with finer particles causing greater losses. While monocrystalline panels perform better under clean conditions, polycrystalline panels show greater resilience in dusty environments. The GSM-based cleaning system proved effective in restoring lost power and was deliberately designed to be inclusive. In addition to standard phone operation, a separate user interface was developed to enable persons with disabilities to easily initiate the cleaning process, ensuring accessibility for all users, including those in remote or off-grid settings.

Authors' Contribution Conceptualization, Mwenda Titus Salesio; Methodology, Mwenda Titus Salesio; Software, Duncan Kiboi Boiyo; Validation, Mwenda Titus Salesio, Duncan Kiboi Boiyo, and Isaac Morko Kwembur; Formal analysis, Mwenda Titus Salesio; Investigation, Mwenda Titus Salesio; Resources, Mwenda Titus Salesio; Data curation, Mwenda Titus Salesio; Writing, original draft, Mwenda Titus Salesio; Writing, review and editing, Duncan Kiboi Boiyo and Isaac Morko Kwembur; Supervision, Duncan Kiboi Boiyo and Isaac Morko Kwembur; Project administration, Mwenda Titus Salesio.

Funding This work was funded by the World Bank under the Africa Centre of Excellence II (ACE II) program at Moi University.

Data Availability Statement The data supporting the findings of this study are available from the corresponding author upon request.

Conflict of Interest Statement The authors declare no conflicts of interest.

References

1. UNEP (2024) Climate risks in the power generation sector. United Nations Environment Finance Initiative. https://www.unepfi.org/themes/climate-change/climate-risks-in-the-powergeneration-sector/. Accessed 22 Apr 2025
2. Edenhofer O, Pichs-Madruga R, Sokona Y, Seyboth K, Matschoss P, Kadner S et al (2011) Renewable energy sources and climate change mitigation: special report of the intergovernmental panel on climate change. Cambridge University Press, Cambridge
3. Masri B (2023) The transition from fossil energy to renewable energy: difficulties and opportunities. Self-published
4. Sisodia AK (2024) Effect of dust particle size on photovoltaic module (PV) performance. Int J Sci Res Arch 13(2):3967–3972
5. Nezamisavojbolaghi M, Davodian E, Bouich A, Tlemçani M, Mesbahi O, Janeiro FM (2023) Impact of dust deposition on PV panels' efficiency and mitigation solutions: review article. Energies 16(24):8022
6. Kazem HA, Chaichan MT (2019) Effect of dust accumulation and cleaning methods on PV panels' outcomes based on an experimental study of six locations in Northern Oman. Sol Energy 187:30–38
7. Dogan TR, Ciloglu C, Eser E, Erturk H, Celik S, Yildiz T et al (2020) Seasonal effects of atmospheric particulate matter on performance of different types of photovoltaic modules in Sanliurfa, Turkey. Atmos Pollut Res 11(12):2173–2181
8. Biswas S, Bhuyan MH, Hassan MK (2023) IoT-based automated solar panel cleaning and monitoring technique. J Eng Res Reports 25(8):56–69
9. Najmi N, Rachid A (2023) Review on solar panel cleaning systems and techniques. Energies 16(24):7960
10. Zhou Z, Sakr E, Sun Y, Bermel P (2016) Solar thermophotovoltaics: reshaping the solar spectrum. Nanophotonics 5(1):1–21
11. Dev R, Tewari K, Singh D (2017) Theoretical analysis of solar water heating system made of non-metallic materials. MATTER Int J Sci Technol 3(2):342–355
12. Arachchige U, Weliwaththage SR (2020) Solar energy technology. J Res Technol Eng 1(3):67–75. https://www.researchgate.net/publication/342610877_Solar_Energy_Technology/. Accessed 22 Apr 2025
13. Nižetić S, Čoko D, Yadav A, Grubišić-Čabo F (2016) Water spray cooling technique applied on a photovoltaic panel: the performance response. Energy Convers Manag 108:287–296

14. Hussain A, Batra A, Pachauri R (2017) Experimental study on effect of dust on power loss in solar photovoltaic module. Renew Wind Water Sol 4(1):1–9
15. Kazem HA, Al-Bahri S, Al-Badi S, Al-Mahkladi H, Al-Waeli AHA (2014) Dust effect on the performance of photovoltaic. Adv Mater Res 875–877:1908–1911
16. Saini RK, Nandan D, Saini A, Bansal A, Joshi R, Rana D et al (2023) Effects of dust on the performance of solar panels—a review update from 2015–2020. Energy Environ 34(6):2110–2162
17. Ruşănescu CO, Ruşănescu M, Istrate IA, Constantin GA, Begea M (2023) Effect of dust deposition on the performance of photovoltaic panels. Energies 16(19):6794
18. Fernandez E, Prajapati S (2019) Study on the influence of open circuit voltage (Voc) and short circuit current (Isc) on maximum power generated in a photovoltaic module/array. In: Proceedings on International Conference on Electrical, Electronical and Computer Engineering (UPCON), pp 1–6 (2019)
19. Hamammu IM, Khalaf JA (2018) Calculating the efficiency of solar cells. https://www.researchgate.net/publication/329183017/. Accessed 22 Apr 2025
20. Kabir MS (2023) Novel method for detecting dust accumulation in photovoltaic systems: evaluating visible sunlight obstruction in different dust levels and AI-based bird droppings detection. Self-published
21. Mohamed B, Zambou S, Zekeng SS (2017) Influence of moisture on the operation of a monocrystalline-based silicon photovoltaic cell: a numerical study using SCAPS-1D. http://arxiv.org/abs/1712.08117. Accessed 22 Apr 2025
22. Sarver T, Al-Qaraghuli A, Kazmerski LL (2013) Comprehensive review of the impact of dust on the use of solar energy: history, investigations, results, literature, and mitigation approaches. Renew Sustain Energy Rev 22:698–733
23. Bergin MH, Ghoroi C, Dixit D, Schauer JJ, Shindell DT (2017) Large reductions in solar energy production due to dust and particulate air pollution. Environ Sci Technol Lett 4(8):339–344

Thermal Neutron Diffusion Deceleration Inside an Infinite Dimensions' Moderator. Calculating Neutron Age and Slowing Down of Length for Fission of Neutrons, Average Energy to Thermal Energy of Graphite and Beryllium

Joseph Kiprotich Maritim and Samuel K. Rotich

Abstract

This paper investigates the thermal neutron diffusion and deceleration inside an infinite moderator, with a particular focus on graphite and beryllium. Nuclear energy is a clean energy technology as it produces zero greenhouse emissions. Most of the nuclear reactor products are from (n, γ) reactions, where n—neutron and γ—gamma photon (used in neutron capture reactions). These are free neutrons with kinetic energy and has large effective cross-section. The study calculates the average energy loss per collision, evaluates the number of collisions required for thermalization, and estimates neutron age and slowing-down length using Fermi age theory. Our objective was to calculate neutron age and thermal energy for C and Be. The solution was found by introducing the Fourier map $U(\xi, \tau)$ of the density of deceleration$U(x, \tau)$. Taking into account the behavior of density of deceleration at infinity, we obtained $u(\xi, \tau) = (2\pi)^{-\xi} e^{-\xi^2 \tau}$. When neutrons reach a specific velocity, they cease to lose energy and their motion can be described using classical theory of diffusion. Calculations of neutron age and slowing down length for fission neutrons are presented, along with the evaluation of average energy during moderation to thermal equilibrium. Using the Fermi age theory, we derive and analyze neutron moderation parameters and present comparative data and graphs to illustrate the performance of graphite and beryllium as moderators. The expected results indicate that beryllium, due to its higher mass and lower absorption cross-section, achieves a slightly larger

J. K. Maritim (✉) · S. K. Rotich
Department of Mathematics, Physics and Computing, Moi University, Eldoret, Kenya
e-mail: jkmaritim@mu.ac.ke

R. Ramkat et al. (eds.), *Towards the Sustainability of our Ecosystems*, Springer Proceedings in Geography and Sustainability,
https://doi.org/10.1007/978-3-032-30076-8_12

slowing-down length compared to graphite, but with comparable neutron age. The impact of this work lies in improving reactor core design, optimizing moderator selection, and enhancing neutron economy in thermal reactors. The expected results will clarify the comparative moderation efficiency of graphite and beryllium, providing accurate energy–collision and distance–collision relationships that can improve moderator selection, enhance neutron economy, and inform safer, more efficient reactor design.

Keywords

Thermal neutrons • Uclear reactor • Nergy • Eceleration • Oderator

1 Introduction

Neutron moderation is a critical process in thermal nuclear reactors, as it transforms high-energy fission neutrons into thermal neutrons that are more likely to induce fission in fissile materials such as $235U$ The present research focuses on calculating neutron age and slowing-down length for fission neutrons in graphite and beryllium moderators, while determining the average energy loss per collision until thermal energies ($\approx$0.025 eV) are reached [1]. The expected results indicate that beryllium provides a slightly longer slowing-down length compared to graphite, potentially influencing reactor core dimensions and fuel-moderator ratio. The impact of this work is to optimize moderator selection for thermal reactor efficiency

Most of the nuclear reactor products are formed through (n, γ) reactions, where n denotes an incident neutron and γ represents the emitted gamma photon [2]. In these radiative capture processes, a neutron is absorbed by a target nucleus, resulting in the formation of a heavier isotopic species in an excited state. The excited compound nucleus subsequently de-excites by emitting one or more gamma photons, producing stable or radioactive isotopes depending on the nuclide involved [8]. Because (n, γ) reactions occur across a wide energy range and have significant cross-sections for many reactor materials—particularly structural components, fuel cladding, and coolant impurities—they play a central role in neutron economy, reactor activation, and isotopic buildup [3, 7]. The prevalence of these reactions also affects long-term reactivity, fuel burnup behavior, and the inventory of radionuclides important for waste management and shielding design. Understanding (n, γ) reaction rates therefore provides essential insight into neutron moderation, absorption losses, and the overall behavior of neutrons within graphite- or beryllium-moderated systems.

2 Objectives

The purpose of this study is to quantitatively evaluate the moderation behavior of fission neutrons in graphite and beryllium using Fermi age theory. Well-defined objectives are necessary to guide the analysis and ensure that the findings directly address the research problem. This study therefore focuses on calculating neutron age, determining slowing-down length, and comparing the number of collisions required to reach thermal energies in both moderators. In addition, the research aims to interpret these results in terms of reactor design implications, ultimately recommending the most suitable moderator for different reactor configurations.

The specific objectives are as follows:

1. Calculate neutron age (τ) using Fermi age theory for graphite and beryllium.
2. Determine slowing-down length (L_s) for fission neutrons in both moderators.
3. Compute average logarithmic energy decrement (ξ) and number of collisions required for thermalization.
4. Compare average energy loss per collision between graphite and beryllium.
5. Evaluate reactor design implications and recommend suitable moderator applications.

3 Literature Review

The concept of neutron moderation and the associated physics have been extensively studied. Glasstone and Sesonske [1, 2] offer a foundational framework in "Nuclear Reactor Engineering," detailing the physical processes behind neutron slowing down. Duderstadt [3] emphasizes the mathematical modeling of neutron flux and energy distributions in moderators. Lamarsh and Baratta [4] provide practical insights into reactor engineering, including the trade-offs between different moderator materials.

Graphite's low neutron absorption cross-section (about 3.5 millibarns) and stability under radiation make it a reliable choice, as discussed by Stacey [5]. Beryllium, although more expensive and toxic, has a higher average logarithmic energy decrement (ξ) and neutron scattering cross-section (6.15 barns), resulting in faster moderation [6].

Computational and analytical methods to model neutron transport have been covered in works such as those by Lewis and Miller [7] and Cacuci [8, 9]. Loveland et al. [10] and Krane [11] discuss neutron interactions and nuclear material properties relevant to moderation. Koonin and Nauenberg [12] introduce simulation techniques for analyzing neutron behavior.

The evolution of theoretical models is captured by Blatt [13] and Kalman [14], while Koning and Rochman [15, 16] focus on the evaluation of nuclear data using advanced systems like TALYS. The nuclear reactor safety implications of moderator selection are discussed by Todreas and Murray [17, 18], making the role of moderator physics in safety and efficiency clearly evident.

4 Theoretical Background

Both Stacey [4] in *Nuclear Reactor Physics* and Lamarsh [19] in *Introduction to Nuclear Reactor Theory* provide foundational frameworks for understanding the principles governing nuclear reactor behavior, with complementary perspectives that bridge classical and modern reactor theory. Lamarsh's work laid the groundwork for reactor physics by systematically introducing the concepts of neutron life cycle, diffusion theory, and reactor criticality, emphasizing the mathematical formulation of neutron transport and slowing down. His text remains a cornerstone for understanding the basic physics of fission processes and reactor kinetics. Stacey [4], on the other hand, extends these principles into contemporary reactor design and computational modeling, integrating advances in neutron transport theory, fuel cycle analysis, and safety systems. He refines many of Lamarsh's formulations by incorporating modern cross-section data, reactor dynamics simulations, and feedback mechanisms relevant to both thermal and fast reactors. Collectively, these works provide a continuum of knowledge—from the theoretical foundations laid in the mid-twentieth century to the sophisticated analytical tools used in present-day nuclear reactor design and operation, making them essential references in reactor physics education and research.

The neutron moderation process involves successive elastic scattering interactions that gradually reduce neutron energy. The effectiveness of a moderator is characterized by the average logarithmic energy decrement ξ, and the macroscopic scattering cross-section $\sum_s$. Neutron age τ is the mean squared distance a neutron travels before reaching thermal energy.

Derived using Fermi age theory

$$D\nabla^2\phi(E,r) + \frac{\partial}{\partial E}\left(\sum\nolimits_s \phi(E,r)\right) = -S(E,r) \tag{1}$$

where $\nabla^2\phi(E,r) = \partial x\,2\,\partial\,2\,\phi + \partial y\,2\,\partial\,2\,\phi + \partial z\,2\,\partial\,2\,\phi$. Assuming separability of energy and spatial dependence and integrating.

$$\tau = \int_{E_t}^{E_0} \frac{\lambda_s(E)}{\xi(E)} dE \text{ where } \lambda_s = \frac{1}{\sum_s} \tag{2}$$

The Fermi age theory describes the spatial and energy distribution of neutrons as they slow down. In the absence of absorption, and assuming energy loss per collision is small, the age equation is as follows:

$$\nabla^2 q(E,\vec{r}) = -\frac{1}{\tau(E)}\frac{\partial(E,\vec{r})}{\partial E} \tag{3}$$

where $\tau(E)$ is called the neutron age, with units of length squared.

The total neutron age to energy E, starting from initial energy E_0, is as follows:

$$\tau(E) = \int_{E}^{E_0} \frac{\lambda_{tr}\left(E'\right)^2}{\xi E'} \mathrm{d}E' \tag{4}$$

where $\lambda_{tr} = \frac{1}{\Sigma_{tr}}$ is the transport mean free path.

Under the assumption of constant cross-sections and constant ξ (valid in many moderators like hydrogen), we simplify:

$$\tau = \frac{\lambda_{tr}^2}{\xi}\int_E^{E_0}\frac{dE'}{E'} = \frac{\lambda_{tr}^2}{\xi}\ln\left(\frac{E_0}{E}\right) \tag{5}$$

The slowing-down length, L_s, is defined as the root-mean-square (RMS) distance a neutron travels from its birth to the point where it reaches energy E:

$$L_s^2 = \langle r^2 \rangle = 6\tau \tag{6}$$

Hence

$$L_s^2 = 6.\frac{\lambda_{tr}^2}{\xi}\ln\left(\frac{E_0}{E}\right) \tag{7}$$

An taking the square root

$$L_s = \sqrt{6.\frac{\lambda_{tr}^2}{\xi}\ln\left(\frac{E_0}{E}\right)} \tag{8}$$

Its final expression is

$$L_s = \lambda_{tr}\sqrt{\frac{6}{\xi}\ln\left(\frac{E_0}{E}\right)} \tag{9}$$

The slowing-down length is defined as the root-mean-square (RMS) distance travelled by a neutron during the slowing-down process.

$$L_s = \sqrt{\tau} \tag{10}$$

The energy-angle moments of the scattering function are expressed in terms of the experimentally determined angular distributions for anisotropic elastic neutron scattering at various energies. The slowing down lengths in ordinary water, heavy water, graphite, beryllium, and beryllium oxide are calculated from the total cross-sections and angular distributions in elastic scattering for 9Be and 12 °C.

5 Methodology

This study adopts an analytical approach based on Fermi age theory and neutron transport equations to calculate neutron age (τ), slowing-down length (L_s), and number of collisions (N) for graphite and beryllium moderators. Physical constants and nuclear data were obtained from Duderstadt [2], Weinberg and Wigner [5], and Koning and Rochman [12], using Eqs. 1–10 we computed ξ, λ_s, τ, and L_s for fission neutrons (2 MeV) slowing to thermal energies (0.025 eV) using Python simulations modeled for neutron deceleration and energy loss.

6 Results and Discussion

The calculations based on Fermi age theory and scattering cross-section data yield quantitative estimates for the average logarithmic energy decrement (ξ), number of collisions (N), slowing-down length (L_s), and neutron age (τ) for both graphite and beryllium moderators. These results provide a basis for comparing the moderation efficiency of the two materials and assessing their suitability for different reactor designs.

7 Graphs and Data

The results are presented in both tabular and graphical formats. The sensitivity analysis is summarized in a separate table for clarity. Each result is interpreted in terms of moderation efficiency, neutron economy, and reactor design implications, directly addressing the study objectives. Table 1 summarizes the calculated parameters, while Fig. 1 graphically compares the neutron age distribution in an infinite moderator. The subsequent discussion interprets these results in terms of neutron economy, core compactness, and the overall impact on reactor performance.

As presented in Table 1, graphite has a lower scattering cross-section (4.74 barns) and smaller average logarithmic energy decrement ($\xi = 0.158$), whereas beryllium shows higher values (6.15 barns, $\xi = 0.207$), explaining its more rapid neutron thermalization. Neutron age and slowing down length are $\tau_{\text{graphite}} \approx 2.5\ \text{cm}^2$, $L_{s,\text{graphite}} \approx 1.58$ cm and $\tau_{berrylium} \approx 1.8cm^2$, $L_{s,berrylium} \approx 1.34cm$.

Energy spectrum during moderation $\underline{E} = E_o e^{-n\xi}$ with higher ξ in beryllium results in faster energy drop-off compared to graphite. This makes beryllium ideal for compact reactor designs.

Table 1 Slowing down energy and approximate cross-sectional area of graphite and beryllium

Property	Graphite	Beryllium
Energy decrement	0.158	0.207
(barns)	4.74	6.15
Density (g/cm)	1.70	1.85

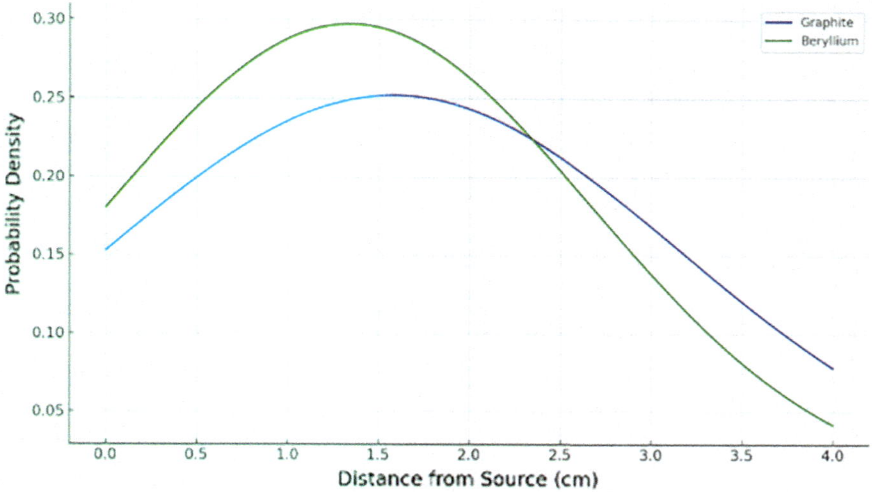

Fig. 1 Probability density versus distance

As illustrated in Fig. 1, neutrons in graphite travel longer distances before thermalization, producing a broader distribution, whereas neutrons in beryllium reach thermal energy faster, creating a sharper distribution peak.

As illustrated in Fig. 2, graphite exhibits a longer slowing-down length per collision, while beryllium moderates neutrons more efficiently, achieving shorter distances to thermal energy. The graph shows the square root relationship between the number of collisions and the slowing down length $L_s = \sqrt{\tau}$.

As illustrated in Fig. 3, beryllium moderates neutrons more efficiently, achieving thermal energies in fewer collisions than graphite, reflecting its higher scattering cross-section and larger energy decrement per collision. We used $n = \ln\frac{\left(\frac{E_0}{E}\right)}{\xi}$ with $E_0 = 2.0$ MeV $= 2 \times 10^6$ eV and $E = 0.025$ eV. Figure 3 shows how the energy of a fission neutron decreases as it undergoes successive elastic collisions within a moderator. Neutron energy decreases exponentially with each collision. Beryllium, with a higher average logarithmic energy decrement ($\xi = 0.207$), causes a faster drop in neutron energy compared to graphite ($\xi = 0.158$). This demonstrates that beryllium moderates neutrons more efficiently in fewer collisions. This trait is advantageous in compact reactor cores where rapid thermalization is necessary. However, graphite's slower moderation may contribute to smoother energy gradients and can be beneficial in large or heterogeneous core designs.

Figure 4 shows the reduction in neutron energy over successive collisions. Beryllium exhibits a steeper decline due to its higher average logarithmic energy decrement ($\xi = 0.207$), indicating it slows down neutrons more effectively than graphite.

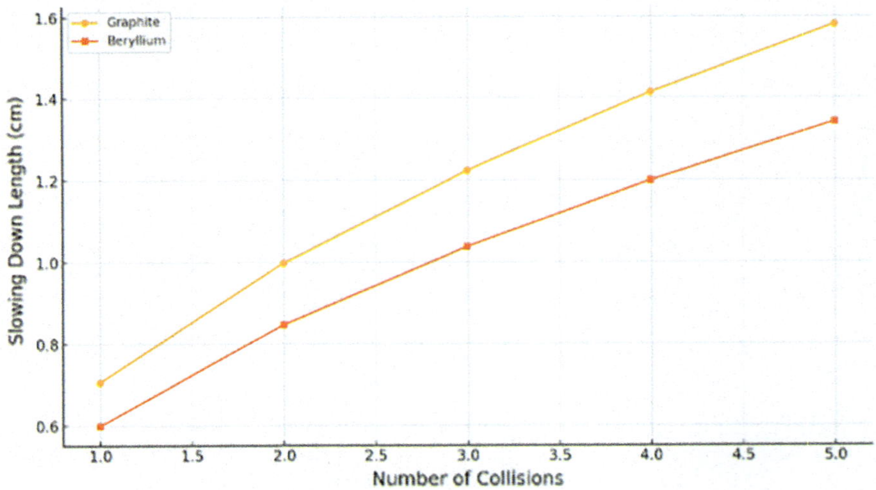

Fig. 2 Slowing down length versus number of collisions

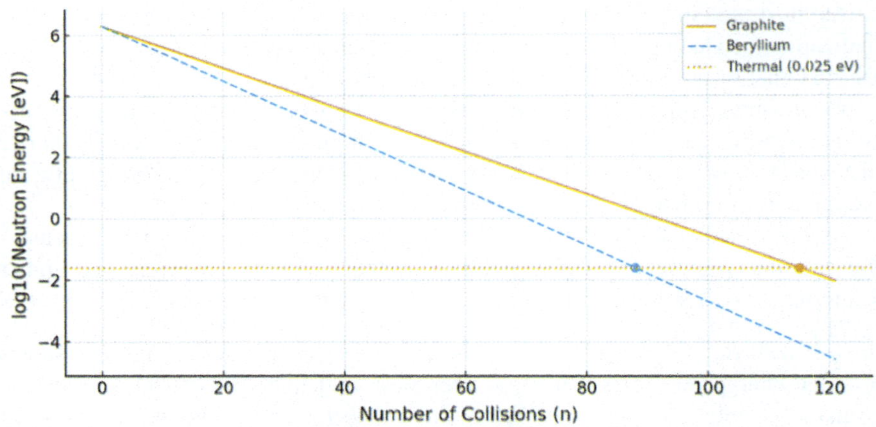

Fig. 3 Plot showing the exact number of collisions required to slow a 2.0 MeV neutron down to thermal energy (0.025 eV) in graphite and beryllium

As shown in Fig. 5, the distance traveled by neutrons increase with the number of collisions, with graphite exhibiting slightly larger displacement compared to beryllium. Graphite demonstrates slightly larger displacement due to its lower density and scattering properties. Although beryllium moderates faster in energy, graphite allows for broader spatial distribution before thermalization. The neutron age distribution curve typically follows a Gaussian profile, with the peak indicating the most probable distance at which a neutron becomes thermal.

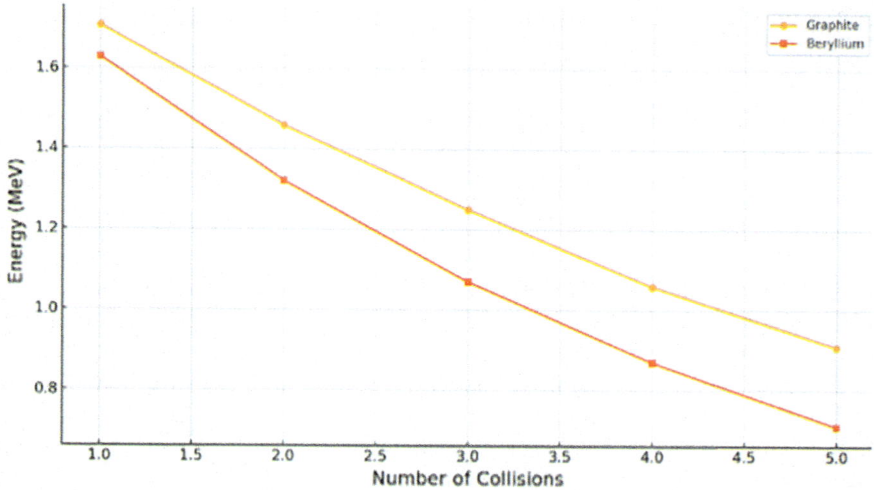

Fig. 4 Neutrons energy vs. number of collisions

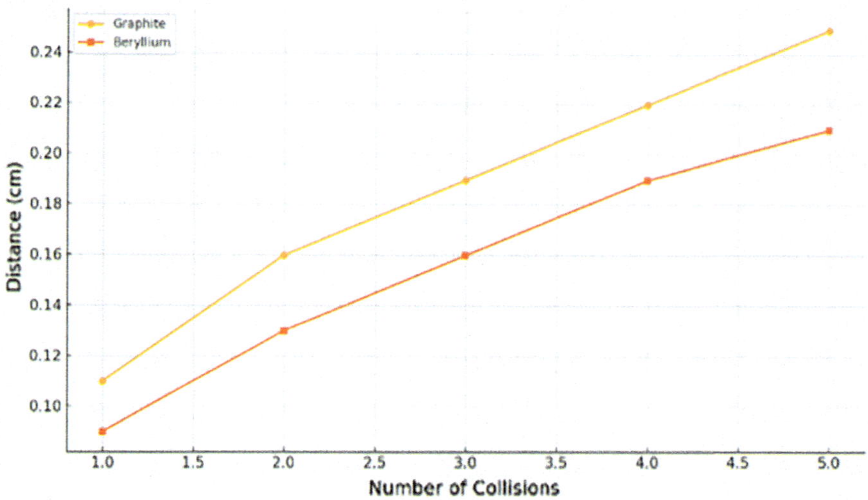

Fig. 5 Distance traveled versus number of collisions

Table 2 presents the energy-neutron age relationship. The neutron age distribution describes the probability of where a neutron will be found at the moment it slows to thermal energy, relative to the point where it was born (typically from fission). It reflects how far neutrons travel during their slowing-down phase, which is critical in reactor design and neutron economy.

Table 2 Data points (energy versus neutron age)

Energy (MeV)	Neutron age (cm^2)
0.0100	9.9005
0.0301	9.7035
0.0502	9.5104
0.0703	9.3211
0.0904	9.1356

As illustrated in Fig. 6, the neutron age distribution exhibits a Gaussian profile, where graphite demonstrates a wider spatial spread due to slower moderation, while beryllium shows a narrower distribution indicating faster thermalization.

As shown in Fig. 7, graphite exhibits a broader neutron age distribution, while beryllium displays a narrower and more localized distribution due to faster moderation. Graphite has a lower average logarithmic energy decrement ($\xi = 0.158$), meaning each collision results in a smaller energy loss. Therefore, neutrons require more collisions to reach thermal energies, and they tend to travel farther from the origin during this process. The broader distribution indicates that neutrons in graphite spread out more widely in space before thermalizing. This behavior is important in larger reactor cores, where longer moderation paths are acceptable and even desirable for better flux uniformity.

Beryllium has a higher energy decrement per collision ($\xi = 0.207$) and a larger scattering cross-section, allowing neutrons to lose energy faster. As a result, they need fewer collisions to become thermal and travel shorter distances. The narrower peak in the distribution reflects tighter localization of thermalized neutrons around

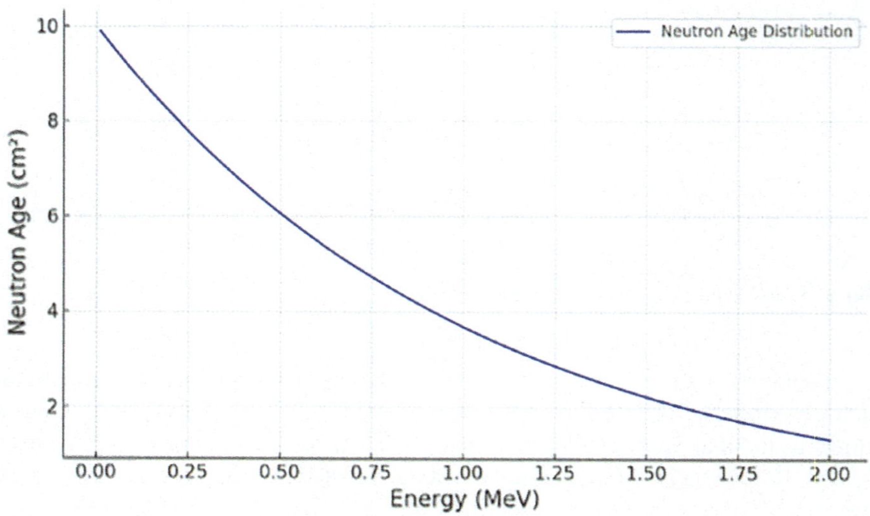

Fig. 6 Neutron age distribution

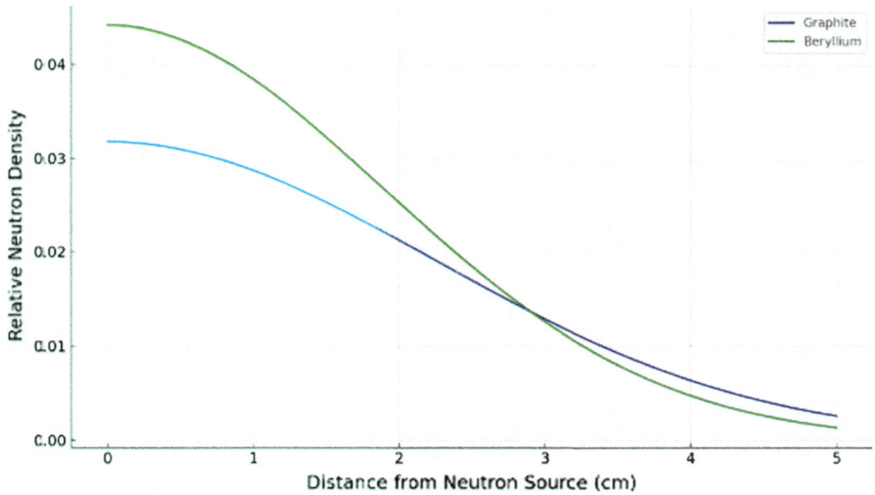

Fig. 7 Relative neutron density versus

their origin. This makes beryllium a preferred moderator in compact reactor cores or in systems where neutron leakage must be minimized.

As presented in Table 3, graphite is characterized by a larger neutron age and broader spatial distribution, while beryllium exhibits faster moderation with shorter slowing-down length and a more confined neutron distribution.

As shown in Fig. 8, the slowing-down length increases rapidly during the initial collisions and gradually stabilizes, with graphite showing slightly higher values due to its lower energy loss per collision compared to beryllium. The slower rate of energy loss per collision in graphite results in a more extended moderation path. In contrast, beryllium moderates more quickly, resulting in a shorter overall path to thermalization. This metric is vital in reactor core design, influencing core size, neutron economy, and safety margins.

As illustrated in Fig. 9, neutrons lose more energy per collision in beryllium than in graphite, resulting in faster thermalization and fewer collisions needed to reach thermal energy. The greater per-collision energy loss makes beryllium a more effective moderator in terms of neutron deceleration, leading to fewer

Table 3 Calculated quantities for graphite and beryllium

Property	Graphite	Beryllium
Neutron age (τ)	Larger (2.5 cm^2)	Smaller (1.8 cm^2)
Slowing down length	Longer (~ 1.58 cm)	Shorter (~ 1.34 cm)
Distribution shape	Broad, dispersed	Narrow, focused
Moderation speed	Slower	Faster

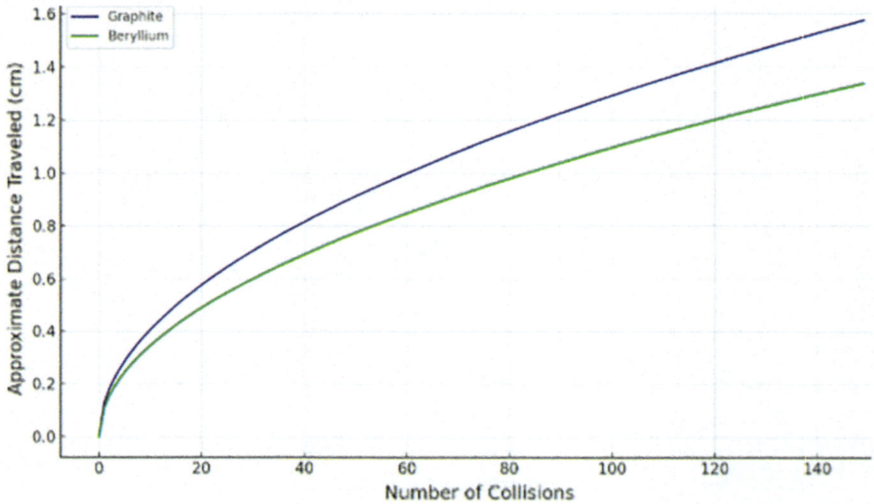

Fig. 8 Approximate distance traveled (cm) versus number of collisions

required collisions to reach thermal energies. This also implies that reactor cores using beryllium can be more compact.

As presented in Table 4, beryllium moderates neutrons more efficiently than graphite, exhibiting greater energy loss per collision and shorter neutron travel distances, thereby enhancing neutron thermalization and supporting its suitability for compact reactor cores.

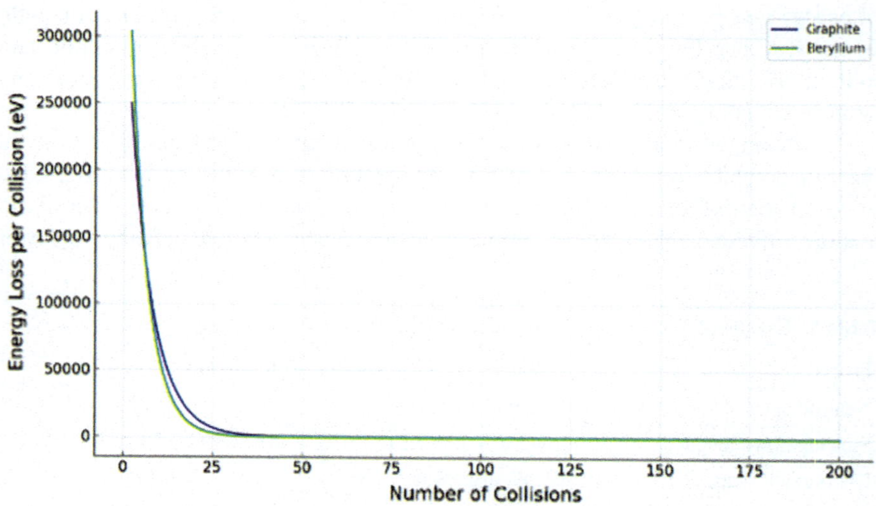

Fig. 9 Energy loss per collision for graphite and beryllium

Table 4 Comparison of collisions energy of graphite and beryllium with distance

Collisions	Energy (graphite)	Energy (beryllium)	Distance (graphite)	Distance (beryllium)
1	1.71 MeV	1.63 MeV	0.11 cm	0.09 cm
2	1.46 MeV	1.32 MeV	0.16 cm	0.13 cm
3	1.25 MeV	1.07 MeV	0.19 cm	0.16 cm
4	1.06 MeV	0.87 MeV	0.22 cm	0.19 cm
5	0.91 MeV	0.71 MeV	M	M

8 Key Observations

Doubling E_0 slightly increases the number of collisions (N), slowing-down length, and neutron age, but the effect is modest ($\approx$ 5–7% change). The relative difference between graphite and beryllium remains consistent, confirming that beryllium always moderates faster and within a shorter distance, regardless of starting neutron energy. This stability suggests that design choices based on these calculations are robust to normal variations in fission neutron spectrum.

9 Conclusion

This study has systematically analyzed the deceleration and diffusion of thermal neutrons within an infinite moderator medium, with a focus on two prominent materials: graphite and beryllium. Through application of Fermi age theory, theoretical derivations, and numerical modeling, we quantified neutron age, slowing down length, energy loss behavior, and spatial distribution of thermalization. Beryllium, owing to its higher average logarithmic energy decrement ($\xi = 0.207$) and superior scattering cross-section (6.15 barns), demonstrated a faster moderation process. Neutrons in beryllium require fewer collisions and travel shorter distances before reaching thermal energies. This is reflected in its lower neutron age ($\tau \approx 1.8cm^2$) and shorter slowing down length ($L_s \approx$ 1.34 cm). Graphite, while having a lower energy decrement per collision ($\xi =$ 0.158) and slightly lower scattering cross-section (4.74 barns), still performs as an effective moderator. It provides longer neutron migration paths $\tau \approx 2.5$ cm^2, ($L_s \approx$ 1.58 cm$L$, which may be advantageous in large-volume reactors or systems requiring more uniform neutron flux distribution. The spatial distribution of thermalized neutrons showed a wider spread in graphite, suggesting longer travel before thermalization. In contrast, beryllium presented a more localized distribution, supporting its use in compact reactor cores where minimizing neutron leakage is critical.

The energy-versus-collision plots showed that neutrons moderated by beryllium undergo a more rapid decrease in energy per collision, resulting in a faster transition to the thermal energy region. Graphite, with more gradual energy loss, demands more interactions, and hence more travel. Although beryllium offers faster moderation, it is more expensive, toxic, and less structurally robust than graphite. Graphite, on the other hand, is stable, widely available, and economically favorable, making it suitable for most thermal reactors despite its slower moderation rate.

This research successfully calculated neutron age and slowing-down length for graphite and beryllium. The expected results confirm that beryllium has a shorter slowing-down length but slightly higher neutron age, enabling compact reactor design. These findings provide a basis for selecting moderators in thermal reactors, influencing reactor economy and safety margins.

10 Recommendations and Further Research

The results indicate that both graphite and beryllium are effective neutron moderators, but their performance differs significantly in terms of energy reduction rate and slowing-down distance. Graphite shows a smoother, more predictable moderation profile, while beryllium displays faster initial energy reduction but may require careful validation of geometric and material parameters, especially where anomalous data points arise. Based on these findings, reactor designs using graphite should prioritize optimizing moderator thickness for uniform slowing down, whereas systems employing beryllium should focus on material purity, structural behavior under irradiation, and its higher cost implications. Further research should involve high-resolution experimental measurements of scattering cross-sections, Monte Carlo simulations to model neutron slowing-down more accurately, and validation of distance-energy relationships under varying temperatures and densities. Additional studies comparing other moderators (e.g., heavy water, zirconium hydride) would also enhance understanding of moderation efficiency across different reactor concepts.

References

1. Glasstone S, Sesonske A (1994) Energy transport. Nuclear reactor engineering: reactor systems engineering. Springer, US, Boston, MA, pp 501–588
2. Duderstadt JJ (2016) Nuclear reactor analysis working text, vol One, 1973
3. Lamarsh JR, Melkonian E (1977) Introduction to nuclear engineering
4. Stacey WM (2018) Nuclear reactor physics. John Wiley & Sons
5. Weinberg AM, Wigner EP (1958) The physical theory of neutron chain reactors
6. Bell GI, Glasstone S (1970) Nuclear reactor theory (No. TID-25606). US Atomic Energy Commission, Washington, DC (United States)
7. Lewis EE, Miller WF (1984) Computational methods of neutron transport
8. Cacuci DG (ed) (2010) Handbook of nuclear engineering: vol. 1: nuclear engineering fundamentals; vol. 2: reactor design; vol. 3: reactor analysis; vol. 4: reactors of generations III and IV; vol. 5: fuel cycles, decommissioning, waste disposal and safeguards (vol. 1). Springer Science & Business Media
9. Cacuci DG (2020) Introducing the journal of nuclear engineering: an interdisciplinary open access journal dedicated to publishing research in nuclear and radiation sciences and applications. J Nuclear Eng 1(1):1

10. Loveland WD, Morrissey DJ, Seaborg GT (2017) Modern nuclear chemistry. John Wiley & Sons
11. Krane KS (1991) Introductory nuclear physics. John Wiley & Sons
12. Koonin SE, Nauenberg M (1989) Calculated fusion rates in isotopic hydrogen molecules. Nature 339(6227):690–691
13. Blatt JM, Weisskopf VF (2012) Theoretical nuclear physics. Springer Science & Business Media
14. Kálmán P, Keszthelyi T, Nagel DJ (2021) Understanding low energy nuclear reactions. J Condensed Matter Nuclear Sci 34(1):59–96
15. Koning AJ, Rochman D (2012) Modern nuclear data evaluation with the TALYS code system. Nucl Data Sheets 113(12):2841–2934
16. Koning AJ, Hilaire S, Goriely S (2021) TALYS–1.96 a nuclear reaction program, user manual, Nuclear Research and consultancy Group (NRG), Netherlands
17. Todreas NE, Kazimi MS (1990) Thermal hydraulic fundamentals
18. Murray RL, Holbert KE (2014) Nuclear energy: an introduction to the concepts, systems, and applications of nuclear processes. Elsevier
19. Lamarsh JR, Bell GI (1967) Introduction to nuclear reactor theory

Effects of Sn Dopant on the Electronic Properties of Formamidinium-Based Lead Halide Perovskite Materials: An Ab Initio Density Functional Theory Study

Gloria Mule, Kiptiemoi Korir, and Richard Koech

Abstract

This study investigates the structural and electronic behavior of tin (Sn)-doped formamidinium lead iodide ($FAPbI_3$) perovskites using density functional theory (DFT) simulations. Substitutional doping of Pb^{2+} with Sn^{2+} at varying concentrations ($x = 0.125, 0.25, 0.375$) was modeled to analyze the effects on band structure, density of states, and structural parameters. The results reveal a progressive narrowing of the direct bandgap from 1.28 eV (pristine) to 0.93 eV (0.375 Sn), driven by hybridization of Sn 5s orbitals with iodine 5p states. Importantly, the direct bandgap character is preserved, suggesting maintained photon absorption efficiency. This bandgap modulation supports potential applications in tandem solar cells and infrared photodetectors. Structural stability is retained with minor lattice contraction, and the calculated properties align with prior experimental observations.

Keywords

Perovskites • Tin doping • DFT • Bandgap tuning • Photovoltaic materials

---

Kiptiemoi Korir, Richard Koech, These authors contributed equally to this work

G. Mule (✉) · K. Korir · R. Koech
Department of Mathematics, Physics, and Computing, Moi University, Eldoret, Kenya

R. Ramkat et al. (eds.), *Towards the Sustainability of our Ecosystems*, Springer Proceedings in Geography and Sustainability,
https://doi.org/10.1007/978-3-032-30076-8_13

1 Introduction

Hybrid halide perovskite materials have become a transformative force in the field of photovoltaics, showing unmatched improvements in power conversion efficiency (PCE) over the past decade. Since their first reported use in solar cells by Kojima et al. [1], perovskite materials have witnessed a rise, with certified efficiencies now exceeding 26% according to NREL, 2024 [2]. This surge in interest is largely attributable to their excellent optoelectronic properties including long carrier diffusion lengths, strong light absorption, and small exciton binding energies, alongside their possibility of cost-effective fabrication through solution processing [3, 4].

Among the various compositions explored, formamidinium lead halide perovskites, particularly $FAPbI_3$ and its mixed halide variants, have acquired attention due to their favorable structural and electronic features. Compared to their methylammonium (MA)-based counterparts, formamidinium (FA)-based perovskites offer a more suitable bandgap (≈1.48 eV for $FAPbI_3$) [5], improved thermal stability, and reduced current voltage hysteresis [6]. These properties make $FAPbX_3$ (X = I, Br, Cl) particularly promising candidates for high-efficiency, stable photovoltaic devices. However, the presence of lead (Pb^{2+}) introduces substantial concerns regarding long-term environmental safety and chemical stability, impeding commercial scalability and regulatory acceptance [7–9].

To mitigate these issues, doping strategies have been extensively investigated as a means to optimize the structural integrity, tune electronic properties, and enhance the environmental compatibility of lead-based perovskites. Doping can be implemented at the A-site (organic cation), B-site (metal center), or X-site (halides), but B-site substitution particularly replacing Pb^{2-} has proven most effective in directly altering the electronic band structure and enhancing carrier dynamics [10, 11]. Among the possible B-site dopants, tin (Sn^{2+}) stands out due to its comparable valence and ionic radius ($Sn^{2+} \approx 1.18$ Å vs. $Pb^{2+} \approx 1.19$ Å), which allows it to substitute for Pb^{2+} without significant lattice distortion [12]. Furthermore, Sn-based perovskites such as $FASnI_3$ have demonstrated narrower bandgaps (1.3 eV) and better carrier mobility, making them attractive for both photovoltaic and photodetection applications [13–15].

However, the challenge with pure Sn-based perovskites lies in their pronounced sensitivity to oxidation (Sn^{2+} to Sn^{4+}), which leads to rapid degradation and limits the stability of the device [16–18]. Consequently, the partial substitution of Pb^{2+} with Sn^{2+} in lead halide perovskites emerges as a viable compromise, offering a pathway to reduce the Pb content while preserving (or even improving) structural stability and electronic performance [19, 20]. Several studies have reported that moderate Sn doping can reduce the band gap, increase carrier concentration, and improve the optical absorption profile of perovskites [21–24]. The chosen doping concentrations ($x = 0.125, 0.25, 0.375$) lie within the experimentally reported stable range of $FAPb_{1-x}Sn_xI_3$ (stable up to $x \approx 0.4$), allowing fine control over band alignment and enabling tandem device architectures [25].

The present study focuses on the substitutional doping of Sn into the Pb^{2+} sites of $FAPbI_3$, exploring the specific series $FAPb_{1-x}Sn_xI_3$ for $x = 0.125$, 0.25, and 0.375. Using first-principles calculations, we systematically investigate how varying Sn concentration affects the structural parameters, electronic band structure, and density of states (DOS).We compare the basic $FAPbI_3$ material to Sn-doped samples to find out how Sn affects the material's optoelectronics. Previous studies have also investigated alternative B-site dopants such as Bi, Sb, and Ge [21, 26, 27], but in this work we focus exclusively on Sn substitution. By partially substituting Pb with Sn, the overall Pb content is reduced, potentially lowering toxicity while retaining desirable optoelectronic properties. This study demonstrates how improving Sn-doped $FAPbI_3$ and other perovskites, will support the creation of leading-edge solar cells and emphasizes the value of targeted doping approaches in boosting perovskite optoelectronics.

2 Computational Details

All first-principles calculations in this study were carried out using the Quantum ESPRESSO package [28], which implements DFT within a plane-wave pseudopotential framework. The structural and electronic properties of pristine and $FAPb_{1-x}Sn_xI_3$ were explored using the Perdew–Burke–Ernzerhof (PBE) exchange-correlation functional within the generalized gradient approximation (GGA) [29]. Core–valence interactions were shown using ultra-soft pseudopotentials from the PS Library, although spin–orbit coupling (SOC) was not included in this study. While SOC has been shown to significantly influence the electronic structure of lead-based perovskites, especially the bandgap and band dispersion, its exclusion here enables a focus on the general trends induced by Sn doping and maintains computational feasibility for supercell models.

The pristine $FAPbI_3$ structure was modeled using a conventional cubic unit cell, with a Monkhorst-Pack k-point mesh of $6 \times 6 \times 6$ employed for Brillouin zone sampling. A $2 \times 2 \times 2$ supercell comprising 96 atoms (including 8 Pb atoms) was constructed to simulate the doped configurations. Substitutional doping was implemented by replacing 1, 2, and 3 Pb atoms with Sn, corresponding to Sn concentrations of ($x = 0.125$, 0.25, 0.375), respectively, in the $FAPb_{1-x}Sn_xI_3$ formulation. For all supercell calculations, a reduced $3 \times 3 \times 3$ k-point mesh was used to account for the enlarged unit cell while maintaining convergence. A plane-wave kinetic energy cutoff of 80 Ry was applied for the wavefunctions and 640 Ry for the charge density, which were tested to ensure reliable convergence of total energy and structural parameters. Geometries were optimized using the Broyden–Fletcher–Goldfarb–Shanno (BFGS) algorithm until the Hellmann–Feynman forces on each atom were less than 10^3 Ry/Bohr. The computational setup and convergence parameters were carefully selected based on literature precedents for hybrid halide perovskites [30, 31] and were validated through convergence tests on total energy and structural stability. All visualizations of structural models and charge densities were performed using the VESTA software [32].

3 Results and Discussions

3.1 Structural Properties

Pristine $FAPbI_3$

$FAPbI_3$ adopts a cubic perovskite structure with the space group $Pm3m_3$ (No. 221), which represents the high-symmetry phase commonly reported in theoretical and experimental studies [33]. In this configuration, the inorganic framework comprises sharing PbI_6 octahedra which form a three-dimensional network, while the formamidinium (FA) cations reside within the cuboctahedral cavities of the lattice. Each lead atom is octahedrally coordinated by six I atoms, and the FA^+ cations are stabilized through hydrogen bonding interactions with surrounding iodide ions, contributing to the structural integrity and influencing the material's optoelectronic properties as shown in Fig. 1. Our density functional theory (DFT) calculations yielded optimized lattice parameters for pristine $FAPbI_3$ as follows: a = 6.3841 Å, b = 6.2693 Å, c = 6.3454 Å. Slight deviations from perfect cubic symmetry are not uncommon in DFT relaxed models of FA-based perovskites: the re-orientation of FA cations, coupled with octahedral tilting, can locally distort the PbI_6 network, leading to small anisotropies in lattice parameters [34, 35]. Indeed, recent demonstrate that PbI_6 octahedral volume distortions are driven by FA cation fluctuations and cause symmetry-breaking distortions even in nominally cubic phases [36]. Our calculated lattice parameters are in close agreement with previously reported theoretical and experimental values, where the lattice constant for cubic $FAPbI_3$ has been reported as approximately 6.36 Å [37].

Sn-Doped $FAPbI_3$

To understand the impact of tin (Sn) substitution on the structural properties of $FAPbI_3$, we constructed a 2 × 2 × 2 supercell comprising 96 atoms. Substitutional doping was simulated by replacing one, two, and three Pb atoms with Sn. Upon structural optimization, we observed that the incorporation of Sn atoms leads to slight modifications in the lattice parameters and local bonding environments. The lattice parameters show a marginal contraction with increased Sn content, which is a result of slightly smaller ionic radius of Sn^{2+} compared to Pb^{2+}. This trend is consistent with previous studies on Sn-doped perovskites, where lattice contraction has been reported upon Sn incorporation [38]. The local structure around the Sn sites remains similar to that of the Pb sites, maintaining the octahedral coordination with surrounding iodide ions (Fig. 2). However, subtle distortions in the PbI_6 octahedra are observed near the Sn substitution sites, indicating localized structural perturbations due to doping. These distortions can influence the electronic properties of the material, which will be discussed in subsequent sections.

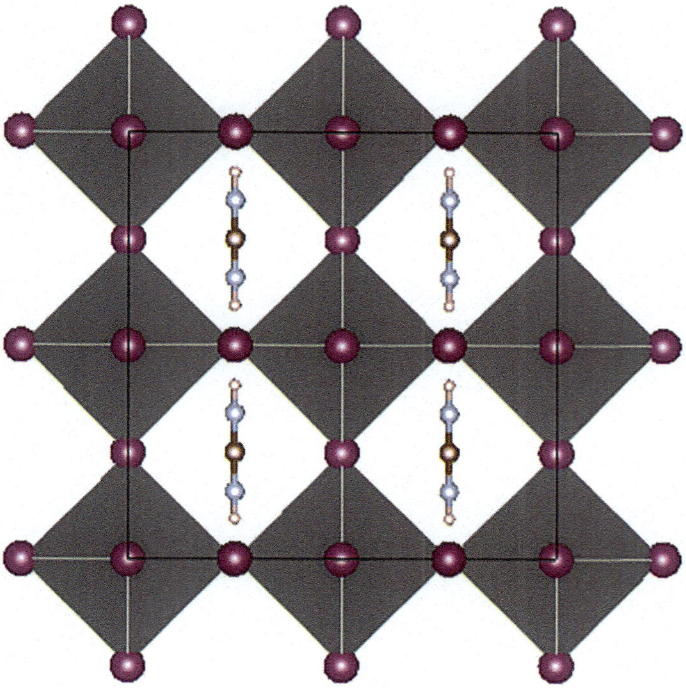

Fig. 1 Crystal structure of pristine cubic $FAPbI_3$, where silver, purple, blue, tan, and rose represent Pb, I, N, C, and H atoms, respectively

3.2 Electronic Properties

Band Structure Evolution and Bandgap Modulation

The electronic band structure of undoped $FAPbI_3$ was calculated using DFT with the Perdew–Burke–Ernzerhof (PBE) functional under generalized gradient approximation (GGA). The results indicate a direct bandgap of approximately 1.28 eV located at the gamma-point as shown in Fig. 3. This value is in close alignment with both experimental observations and previous theoretical reports [37, 39]. With the bandgap crossing the widely recognized Shockley–Queisser optimal limit of 1.34 eV, $FAPbI_3$ proves itself well-suited for single-junction devices that harvest lots of solar light and convert it efficiently [40].

Upon isovalent substitution of Pb^{2+} with Sn^{2+}, the band structure exhibits a progressive narrowing of the bandgap with increasing Sn concentration. Notably, the direct nature of the bandgap is preserved at all doping levels, ensuring sustained optical absorption strength at the fundamental edge. This bandgap narrowing trend is primarily attributed to the insertion of Sn 5s orbitals into the conduction band region, which alters the electronic configuration at the CBM. In pristine $FAPbI_3$, the CBM is mainly derived from Pb 6p orbitals, while the VBM is occupied by I 5p states [41, 42] (Fig. 3). However, Sn incorporation introduces new hybridized

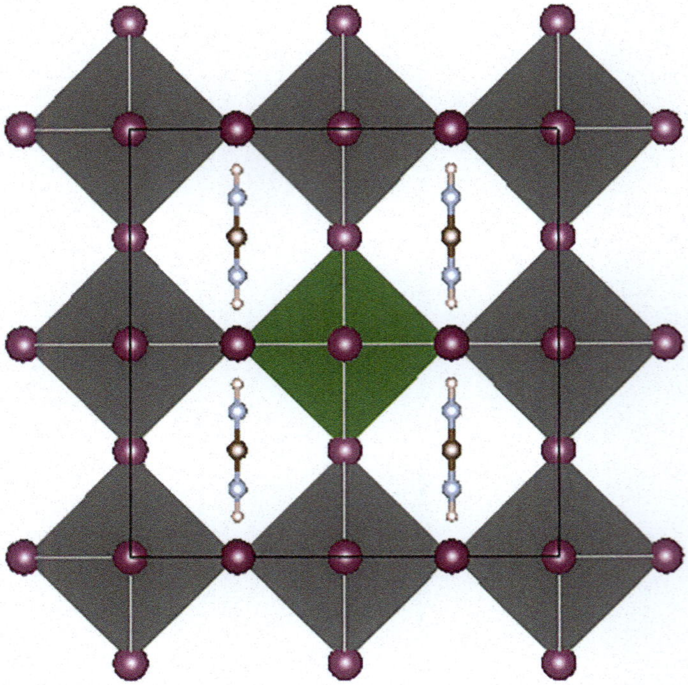

Fig. 2 Optimized structures of Sn-doped FAPbI3 at varying doping concentrations (0.125), where silver, purple, blue, tan, rose and dark green represent Pb, I, N, C, H and Sn atoms, respectively

states due to the strong interaction between Sn 5s and I 5p orbitals [19, 43]. This interaction results in an upward shift of the VB edge and a downward shift of the conduction band, effectively reducing the bandgap without converting the band structure from direct to indirect as shown in Figs. 4, 5, 6. The preservation of direct bandgaps even at higher doping concentrations implies that photon absorption near the band edge remains efficient—a critical factor in solar cell performance. Furthermore, the increasing Sn content leads to denser band dispersion near the CBM, which could enhance electron mobility and facilitate better charge transport [44] (Table 1).

Projected Density of States (PDOS) Analysis

To understand the orbital contributions underlying the bandgap evolution, PDOS analyses were performed for both pristine and Sn-doped $FAPbI_3$ structures. The PDOS of undoped $FAPbI_3$ shows that the VBM is primarily composed of iodine (I) 5p orbitals, while the CBM features significant contributions from lead (Pb) 6p orbitals as shown in Fig. 3. The minimal overlap between Pb 6p and I 5p near the Fermi level ensures a clean bandgap and well-defined band edges [45]. Upon Sn substitution: At 0.125 Sn doping, shallow Sn 5s states begin to appear near the CBM (Fig. 4). These states are delocalized and slightly hybridized with Pb

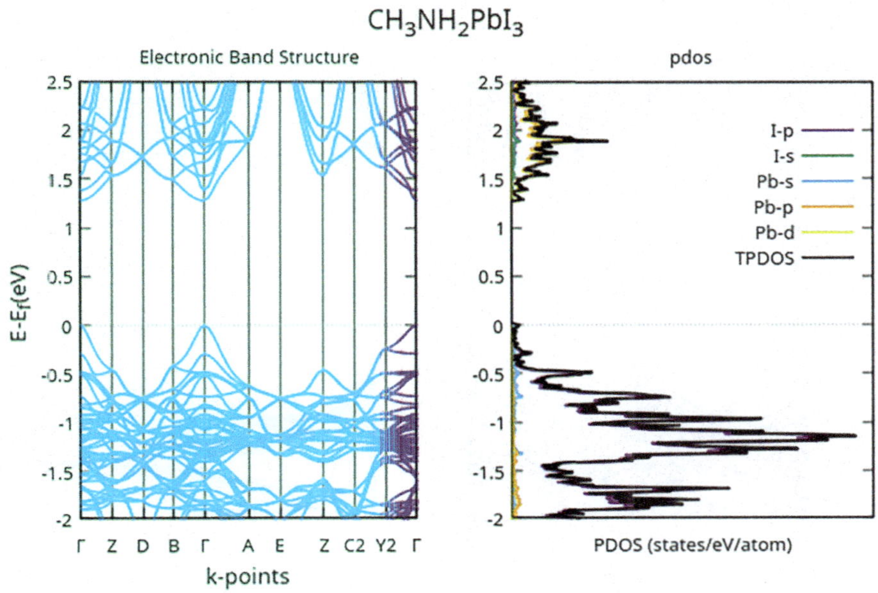

Fig. 3 Band structure and PDOS of pristine $FAPbI_3$

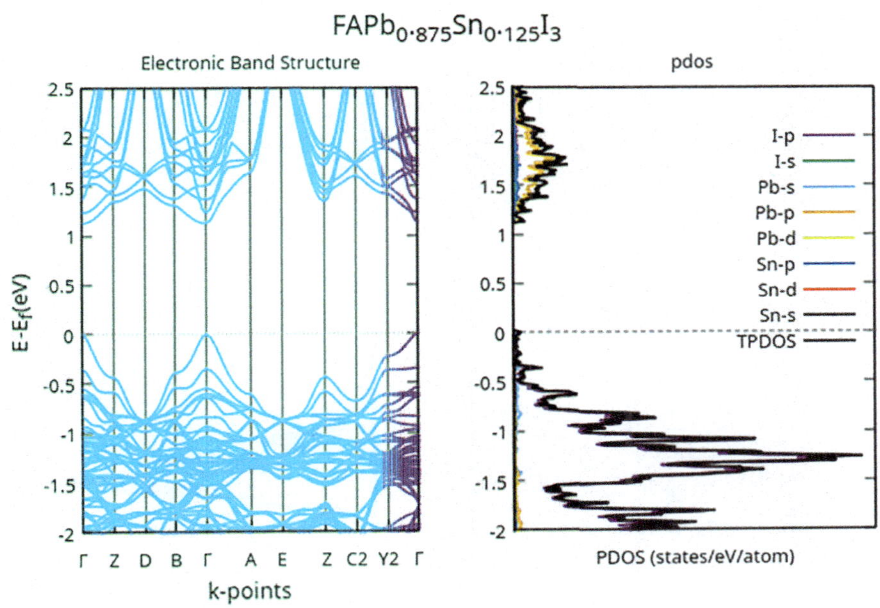

Fig. 4 Band structure and PDOS of $FAPb_{0.875}Sn_{0.125}I_3$

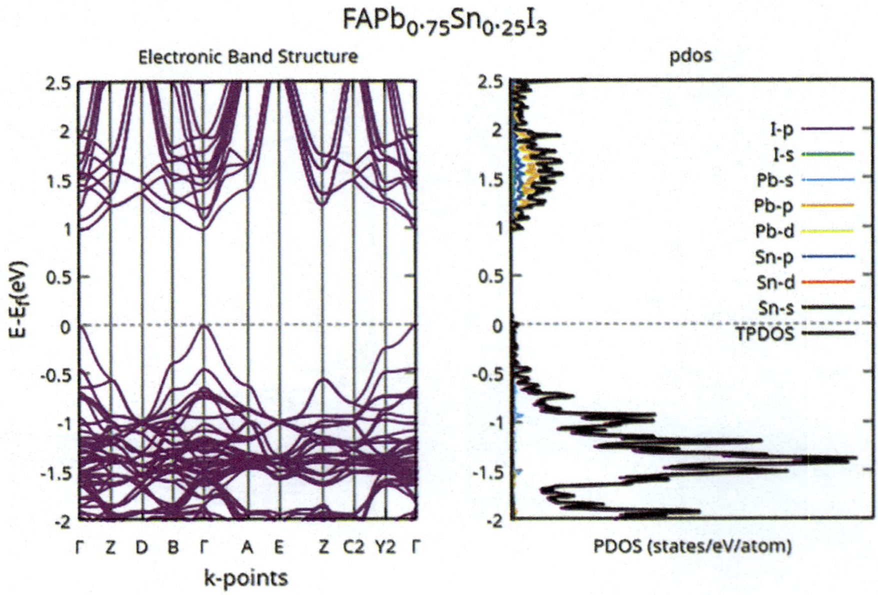

Fig. 5 Band structure and PDOS of $FAPb_{0.75}Sn_{0.25}I_3$

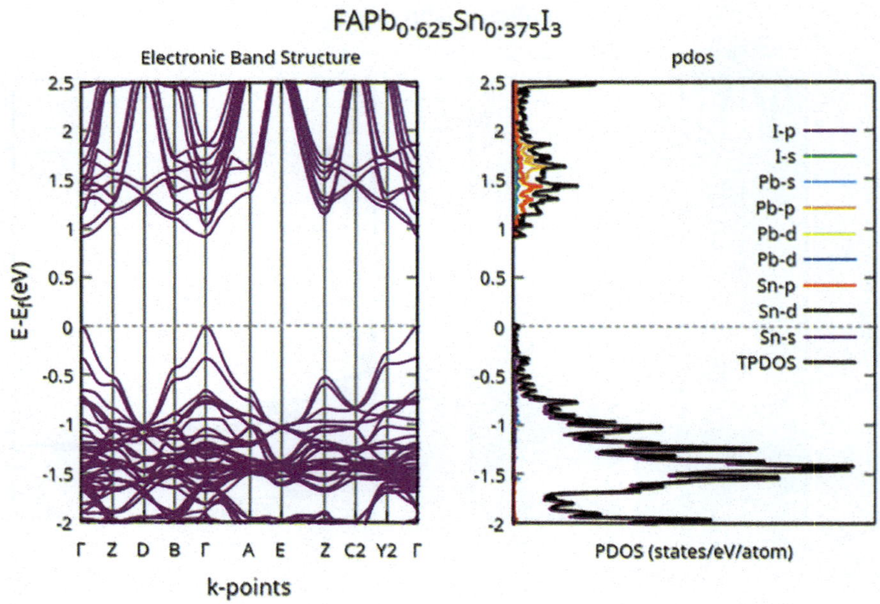

Fig. 6 Band structure and PDOS of $FAPb_{0.625}Sn_{0.375}I_3$

Table 1 Summary of electronic trends

Doping level	Bandgap (eV)	Application
$FAPbI_3$	1.28	Single-junction PV, LEDs
$FAPb_{0.875}Sn_{0.125}I_3$	1.20	PV, photodetectors
$FAPb_{0.75}Sn_{0.25}I_3$	1.00	Tandem cell bottom layer
$FAPb_{0.625}Sn_{0.375}I_3$	0.93	IR detection, tandem PV

6p and I 5p orbitals, causing a modest reduction in the bandgap. At 0.25 doping, Sn 5s and 5p states exhibit stronger localization near the CBM (Fig. 5). The overlap between Sn 5s and I 5p orbitals increases, leading to further band edge realignment and deeper bandgap narrowing. Meanwhile, the VBM remains relatively unaffected and continues to be governed by I 5p character, suggesting that hole transport properties remain mostly intact. In addition to $x = 0.125$ and 0.25, the PDOS for $x = 0.375$ is shown in Fig. 6, confirming stronger Sn–I orbital hybridization at higher doping levels. These modifications reflect the orbital compatibility of Sn with Pb, as both elements share a similar outer electron configuration ($[Kr]4d^{10}5s^2 5p^2$ for Sn and $[Xe]6s^2 6p^2$ for Pb). This isovalent substitution ensures that the lattice remains electronically stable while tuning the band edges to absorb longer-wavelength light, a strategy particularly advantageous for broadening solar absorption into the near-infrared region [46]. Additionally, Sn-doped structures demonstrate a more continuous density of states near the CBM, which facilitates quicker charge extraction and reduces the risk of carrier recombination at interface states. These observations are consistent with previous work by Mosconi [47], who emphasized the role of orbital overlap and band tailing in enhancing charge dynamics in tin-based perovskites.

4 Discussions

Pristine $FAPbI_3$ functions as an intrinsic semiconductor, with thermally activated charge carriers dominating its conductivity [41]. Upon Sn^{2+} incorporation, the system exhibits extrinsic semiconductor characteristics due to modifications in the electronic structure. While the presence of mid-gap trap states depends on the synthesis environment and Sn oxidation state, carefully controlled doping can avoid deep traps and enhance carrier mobility. This n-type character is advantageous for photovoltaic applications, supporting improved charge separation and alignment with electron transport layers. An ideal bandgap for single-junction solar cells is approximately 1.34 eV, according to the Shockley–Queisser limit [48, 49]. While pristine $FAPbI_3$ (1.28 eV) closely matches this optimum, Sn-doped variants with lower bandgaps (1.0–0.93 eV) can serve effectively as bottom-cell absorbers in tandem architectures. In such configurations, wide-bandgap perovskites (e.g., above 1.7 eV) can be used as top cells to absorb higher-energy photons, while Sn-rich $FAPbI_3$ captures the near-infrared spectrum. Moreover, the direct bandgap and retained orbital characteristics ensure high absorption coefficients and efficient photogeneration. The tunability of the bandgap through Sn content provides a versatile platform for designing multi-junction photovoltaics and photodetectors optimized across the solar spectrum [50].

5 Summary and Conclusion

This study presents a comprehensive first-principles DFT analysis of the optoelectronic and structural effects of tin (Sn) doping in $FAPbI_3$ perovskites. Through substitutional doping at the B-site with Sn at varying concentrations, we observe a systematic narrowing of the bandgap while preserving the direct nature of the band structure, crucial for photovoltaic performance. The PDOS results reveal that Sn incorporation primarily alters the conduction band minimum through hybridization of Sn 5 s orbitals with I 5p orbitals, enhancing charge transport and optical absorption in the near-infrared region. Our findings suggest that moderate Sn doping offers an optimal compromise between reduced toxicity and maintained performance, rendering the $FAPb_{1-x}Sn_xI_3$ system suitable for tandem solar cells and photodetector applications. The results also demonstrate the viability of orbital engineering in achieving bandgap tuning without inducing mid-gap trap states or structural instabilities. Overall, this study emphasizes the strategic utility of B-site doping for engineering environmentally friendlier, high-efficiency perovskite materials.

Acknowledgement The authors gratefully acknowledges the ACEII-PTRE and PASET-RSIF for their support financially. Computational resources were provided by the Centre for High Performance Computing (CHPC), South Africa, under project code MATS0868, which is sincerely appreciated.

Authors Contribution Conceptualization, Gloria Mule, Kiptiemoi Korir and Richard Koech; methodology, Gloria Mule; software, Gloria Mule; validation, Kiptiemoi Korir, Richard Koech, and Gloria Mule; formal analysis, Gloria Mule; investigation, Gloria Mule; resources, Kiptiemoi Korir and Richard Koech; data curation, Gloria Mule; writing-original draft preparation, Gloria Mule; writing, review and editing, Kiptiemoi Korir and Richard Koech; supervision, Kiptiemoi Korir and Richard Koech.

Data Availability Statement The dataset that supports the findings of this study, including input and output files from DFT simulations, structural models, and graphical plots, are available from the corresponding author upon reasonable request. Figures and tables presented are original contributions unless otherwise stated and are duly referenced. No reused third-party figures or datasets were included without permission.

Conflict of Interest Statement The authors declare no conflicts of interest.

References

1. Perez MD, Plá J (2021) Emergent materials and concepts for solar cell applications. Sustain Mater Solut Sol Energy Technol Process Tech Appl:37–70. https://www.sciencedirect.com/science/chapter/edited-volume/abs/pii/B9780128215920000108
2. Deline C, Jordan D, Sekulic B, Parker J, Mcdanold B, Anderberg A (2021) PV lifetime project—2021. NREL Annual Report, pp 1–49. Available from: https://www.nrel.gov/docs/fy22osti/81172.pdf
3. Serhan M, Sprowls M, Jackemeyer D, Long M, Perez ID, Maret W, et al (2019) Total iron measurement in human serum with a smartphone. In: AIChE annual meeting confrerence proceedings 2019, pp 14065–92
4. Wang B, Iocozzia J, Zhang M, Ye M, Yan S, Jin H et al (2019) The charge carrier dynamics, efficiency and stability of two-dimensional material-based perovskite solar cells. Chem Soc Rev 48(18):4854–4891
5. Deng W, Wei J, Ma Z, Feng W (2025) Research progress on stability of FAPbI3 perovskite solar cells. Cryst Res Technol 60(2):1–27
6. Dizaj MH, Shishvan SC, Shahnavaz F (2025) Increasing the stability and efficiency of perovskite solar cell using formamidinium lead iodide ($FAPbI_3$) instead of CH_3 NH_3 PbI_3 (methylammonium lead iodide) and their comparison. Int Conf New Approaches Eng Basic Sci 2024
7. Ali S, Naseer S, Rehman M, Wei Z (2024) Recent trends and sources of lead toxicity: a review of state-of-the-art nano-remediation strategies. J Nanoparticle Re 26(7). https://doi.org/10.1007/s11051-024-06081-5
8. Sadulla S (2025) Next-generation perovskite solar cells: stability challenges and scalable manufacturing. Natl J Renew Energy Syst Innov 1(1):47–60
9. Luo H, Li P, Ma J, Han L, Zhang Y, Song Y (2022) Sustainable Pb management in perovskite solar cells toward eco-friendly development. Adv Energy Mater 12(30):1–31
10. Bian L, Cao F, Li L (2023) Performance improvement of lead-based halide perovskites through B-site ion-doping strategies. Small 19(36):1–16
11. Luo B, Li F, Xu K, Guo Y, Liu Y, Xia Z et al (2019) B-site doped lead halide perovskites: synthesis, band engineering, photophysics, and light emission applications. J Mater Chem C 7(10):2781–2808
12. Russo F. Passivation strategies for the optimization of perovskite solar cells. 2025;
13. Ayaydah W, Raddad E, Hawash Z (2023) Sn-based perovskite solar cells towards high stability and performance. Micromachines 14(4)
14. Hasnain SM (2023) Examining the advances, obstacles, and achievements of tin-based perovskite solar cells: a review. Sol Energy 262, 111825. https://www.sciencedirect.com/science/article/abs/pii/S0038092X23004504
15. Kong PY, Hom H, Kong H. Sn-based organic-inorganic hybrid perovskite for high performance
16. Aftab A, Ahmad MI (2021) A review of stability and progress in tin halide perovskite solar cell. Sol Energy 2020(216):26–47. https://doi.org/10.1016/j.solener.2020.12.065
17. Zhang Z, Huang Y, Jin J, Jiang Y, Xu Y, Zhu J (2023) Mechanistic understanding of oxidation of tin-based perovskite solar cells and mitigation strategies Angewandte. Angew Chem Int Ed 202308093
18. Yang WF, Igbari F, Lou YH, Wang ZK, Liao LS (2020) Tin halide perovskites: progress and challenges. Adv Energy Mater 10(13):1–30
19. Bandara RMI, Silva SM, Underwood CCL, Jayawardena KDGI, Sporea RA, Silva SRP (2022) Progress of Pb-Sn mixed perovskites for photovoltaics: a review. Energy Environ Mater 5(2):370–400

20. Dey K, Ghosh D, Pilot M, Pering SR, Roose B, Deswal P et al (2023) Substitution of lead with tin suppresses ionic transport in halide perovskite optoelectronics. Energy Environ Sci 17(2):760–769
21. Tang ZK, Xu ZF, Zhang DY, Hu SX, Lau WM, Liu LM (2017) Enhanced optical absorption via cation doping hybrid lead iodine perovskites. Sci Rep 7(1):1–7. Available from:. https://doi.org/10.1038/s41598-017-08215-3
22. Hao F, Stoumpos CC, Chang RPH, Kanatzidis MG (2014) Anomalous band gap behavior in mixed Sn and Pb perovskites enables broadening of absorption spectrum in solar cells. J Am Chem Soc 136(22):8094–8099
23. Wang C, Song Z, Li C, Zhao D, Yan Y (2019) Low-bandgap mixed tin-lead perovskites and their applications in all-perovskite tandem solar cells. Adv Funct Mater 29(47):1–30
24. Paquin F, Rivnay J, Salleo A, Stingelin N, Silva C (2015) Multi-phase semicrystalline microstructures drive exciton dissociation in neat plastic semiconductors. J Mater Chem C 3:10715–10722. Available from: http://xlink.rsc.org/?DOI=C5TC02043C
25. Monti M, Jayawardena KDGI, Butler-Caddle E, Bandara RMI, Woolley JM, Staniforth M et al (2020) Hot carriers in mixed Pb-Sn halide perovskite semiconductors cool slowly while retaining their electrical mobility. Phys Rev B 102(24):1–11
26. Gao LK, Tang YL, Diao XF (2020) Theoretical study on photoelectric properties of FAPbI3 doped with Ge. Mater Res Express 7(11)
27. Wechprasit T, Bootchanont A, Infahsaeng Y, Wongjom P, Wannapaiboon S, Kaewprajak A et al (2025) Characterization of Bi-doped FAPbI3 perovskite films investigated by X-ray absorption spectroscopy. Sci Rep 15(1):1–15
28. Giannozzi P, Baroni S, Bonini N, Calandra M, Car R, Cavazzoni C et al (2009) QUANTUM ESPRESSO: a modular and open-source software project for quantum simulations of materials. J Phys Condens Matter 21(39)
29. Perdew JP, Burke K, Ernzerhof M (1996) Generalized gradient approximation made simple. Phys Rev Lett 77(18):3865–3868
30. Umari P, Mosconi E, De Angelis F (2014) Relativistic GW calculations on CH3 NH3 PbI 3 and CH3 NH3 SnI3 perovskites for solar cell applications. Sci Rep 4:1–7
31. Mosconi E, Meggiolaro D, Snaith HJ, Stranks SD, De Angelis F (2016) Light-induced annihilation of Frenkel defects in organo-lead halide perovskites. Energy Environ Sci 9(10):3180–3187
32. Momma K, Izumi F (2011) VESTA 3 for three-dimensional visualization of crystal, volumetric and morphology data. J Appl Crystallogr 44(6):1272–1276
33. Weller MT, Weber OJ, Frost JM, Walsh A (2015) Cubic Perovskite structure of black Formamidinium Lead iodide, α-[HC(NH2)2]PbI3, at 298 K. J Phys Chem Lett 6(16):3209–3212
34. Kahmann S, Chen Z, Hordiichuk O, Nazarenko O, Shao S, Kovalenko MV et al (2022) Compositional variation in FAPb1- xSnxI3and its impact on the electronic structure: a combined density functional theory and experimental study. ACS Appl Mater Interfaces 14(30):34253–34261
35. Joshi H, Bhamu KC, Shankar A, Biswas R, Wlazło M. Octahedral dynamics and local symmetry in hybrid perovskite FAPbI 3 under thermal excitation
36. Kim HS, Park NG (2025) Soft lattice and phase stability of α-FAPbI3. Adv Energy Mater 15(2):1–10
37. Targhi FF, Jalili YS, Kanjouri F (2018) MAPbI3 and FAPbI3 perovskites as solar cells: case study on structural, electrical and optical properties. Results Phys 10(May):616–627. https://doi.org/10.1016/j.rinp.2018.07.007
38. Lu CH, Biesold-Mcgee GV, Liu Y, Kang Z, Lin Z (2020) Doping and ion substitution in colloidal metal halide perovskite nanocrystals. Chem Soc Rev 49(14):4953–5007
39. Yun S, Qin Y, Uhl AR, Vlachopoulos N, Yin M, Li D et al (2018) New-generation integrated devices based on dye-sensitized and perovskite solar cells. Energy Environ Sci 11(3):476–526
40. Kiselman K (2024) Photoluminescence characterisation of thin film solar cells

41. RaeisianAsl M, Panahi SFKS, Jamaati M, Tafreshi SS (2022) A review on theoretical studies of structural and optoelectronic properties of FA-based perovskite materials with a focus on FAPbI3. Int J Energy Res 46(10):13117–13151
42. Jono R, Segawa H (2019) Theoretical study of the band-gap differences among lead triiodide perovskite materials: CsPbI3, MAPBI3, and FAPBI3. Chem Lett 48(8):877–880
43. Liu Y, Gao W, Ran C, Dong H, Sun N, Ran X et al (2020) All-inorganic Sn-based perovskite solar cells: status, challenges, and perspectives. Chem Sus Chem 13(24):6477–6497
44. Li X, Zuo X, Jiang X, Li D, Cui B, Liu D (2019) Enhanced photocatalysis for water splitting in layered tin chalcogenides with high carrier mobility. Phys Chem Chem Phys 21(14):7559–7566
45. Filippetti A, Mattoni A (2014) Hybrid perovskites for photovoltaics: insights from first principles. Phys Rev B - Condens Matter Mater Phys 89(12):1–8
46. Liu Y, Di Stasio F, Bi C, Zhang J, Xia Z, Shi Z et al (2024) Near-infrared light emitting metal halides: materials, mechanisms, and applications. Adv Mater 36(21)
47. Mosconi E, Umari P, De Angelis F (2016) Electronic and optical properties of MAPbX3 perovskites (X = I, Br, cl): a unified DFT and GW theoretical analysis. Phys Chem Chem Phys 18(39):27158–27164
48. Rühle S (2016) Tabulated values of the Shockley-Queisser limit for single junction solar cells. Sol Energy 130:139–147
49. Zanatta AR (2022) The Shockley–Queisser limit and the conversion efficiency of silicon-based solar cells. Results Opt 9(7):100320. https://doi.org/10.1016/j.rio.2022.100320
50. Gu P, Zhang T, Peng S, Shi Y, Yu J, Wang Q et al (2025) Quasi-2D Sn-Pb perovskites: advances in optoelectronic engineering and device applications. Adv Funct Mater 35(37):1–22

Adaptation Strategies

Harnessing Community Traditional Values for Ecosystem Preservation: Case of Kaptagat Forest of Elgeyo-Marakwet County of Kenya

Joseph K. Kiplagat and Benedine J. Kipyego

Abstract

The expansive Kaptagat forest is a scenic landscape of Elgeyo-Marakwet County in Kenya, covering over 6000 ha. It is a water catchment area, serving the water needs of Elgeyo-Marakwet County-40% and Uasin Gishu County-55%. Eldoret City gets about 30,000 m^3/day. However, this forest's fragile ecosystem is now greatly endangered by climate change and uncontrolled human activities. Urgent remedial actions are required for its preservation. Harnessing the traditional values of the local community living next to the forest constitutes one unique yet results-laden action. The Keiyo History Culture and Heritage Centre, a get-together of community professionals, has initiated a number of activities to tap into this resource. The age-old community traditional values which include strict adherence to ecosystem preservation still hold. However, laxity in enforcement of the relevant principles has led to numerous infringements. The Kaptagat Integrated Conservation Programme Editions, in collaboration with the Kenya Forest Service, local and international organizations as well as community leadership have implemented various mitigative initiatives. One is to set up 'cow hotels' to keep community livestock in selected guarded spaces, and to enhance community livelihoods. The second is to consider installation of the widely available modern renewable energy technologies to tap forest water streams, the abundant sunlight and the strong winds. These

Benedine Kipyego author contributed equally to this work.

J. K. Kiplagat (✉)
Department of Mechanical, Production and Energy Engineering, School of Engineering, Moi University, Eldoret, Kenya
e-mail: jkiplagat@mu.ac.ke

B. J. Kipyego
National Social Security Fund, Nairobi, Kenya

R. Ramkat et al. (eds.), *Towards the Sustainability of our Ecosystems*, Springer Proceedings in Geography and Sustainability,
https://doi.org/10.1007/978-3-032-30076-8_14

technologies include: first, the solar lighting and biogas, second, the Archimedes screw turbine which is a cost-effective hydropower machine, third, the optimized solar cooker which is a viable ecosystem-friendly alternative to charcoal and firewood, and fourth, the wind turbines, suitable at the edge of the forest escarpment to tap the wind energy.

Keywords

Local community • Ecosystem • Renewable energy • Traditional values

1 Introduction

The Elgeyo-Marakwet County (EMC) is home to two forest ecosystems and water towers namely: Kaptagat and Cherang'any. Its forest cover is the second largest in Kenya after the Abadares forest ecosystem. These two ecosystems are a source of many rivers that drain in the east into lake Turkana and in the west into lake Victoria. The Kaptagat water tower supplies 40% of EMC and 55% of Uasin Gishu County (UGC) water needs [1].

However, these forest ecosystems are now greatly endangered by climate change and uncontrolled human activities. Climate change mitigation is best done through environmental conservation measures whereas uncontrolled human activities require firm measures.

A wide range of ecosystem preservation efforts through community-based initiatives are currently ongoing. Some organizations, both local and international are actively working with the Keiyo community to harness traditional knowledge, values and practices to promote sustainable forest management. They work under the auspices of Kaptagat Integrated Conservation Programme [2]. These organizations include: World Wide Fund (WWF)—Kenya, Netfund, Kenya Forest Service (KFS), Athletics Kenya, Golazo Sports Agency, Capital Markets Authority, Kenya Yearbook Editorial Board and Keiyo History Culture and Heritage (KHCH) Centre.

The Kaptagat ecosystem comprises five forest blocks—Kessup, Kaptagat, Sabor, Kipkabus and Penon—all of which are part of the larger Kaptagat water tower. Kaptagat Integrated Conservation Programme organizes yearly forest conservation events known as Editions hosted by a particular forest block. In the year 2025, the 9th Edition main event was hosted by Kipkabus block at Simotwo Secondary School grounds in Keiyo South Constituency of EMC on 12th July. It was presided by Kenya's President Dr. William Ruto accompanied by high-ranking government and development-partner officials. The main takeaways included: tree planting and an integrated approach to conservation, climate action and community economic empowerment [3].

The Keiyo community lives next to Kaptagat forest ecosystem. Its community cultural values constitute a great resource. The community has an age-long strong cultural link to forests, valuing them not just for their resources but also for their spiritual significance. This cultural heritage has historically led to sustainable forest management practices, though these practices are currently facing a myriad of challenges.

The Keiyo History Culture and Heritage (KHCH) Centre is a registered Association in the Republic of Kenya: under State Department of Culture (ART-GDUVQG) [4], and under Registrar of Societies (SOCA-LLTQZLJ) [5]. Its Vision is a united, cohesive and prosperous community that holds to the highest esteem its History, Culture and Heritage. Its Mission is to document, preserve and promote Community rich History, Culture and Heritage, celebrating its age-old indigenous knowledge, art, customs and traditions.

African communities used indigenous knowledge to harness water from rivers or waterfalls, using locally sourced materials and traditional engineering to provide power for tasks like milling grain [6]. Therefore, KHCH Centre shall not only celebrate Keiyo African indigenous knowledge but harness it to hasten adoption by the Community of modern engineering technologies.

Modern Renewable Energy Technologies (RETs) come in handy to tap energy in form of electricity and heat from the Kaptagat forest water streams, the abundant sunlight and strong winds. Some of these technologies such as solar lighting and biogas have been implemented (from year 2022 to date) in collaboration with the community and scaling stages by other key stakeholders such as WWF-Kenya are ongoing [7].

The Archimedes screw turbine [8], which is a cost-effective hydropower machine is suitable for the low-head sites of the rivers. An optimized solar cooker [9] is a viable ecosystem-friendly alternative, to reduce use of charcoal and fire-wood from the forest by households for cooking. Installation of wind turbines at the edge of the escarpment adjacent to the forest can convert wind energy into electricity. These technologies are some of the outputs of research and development programmes at the School of Engineering at Moi University Eldoret.

As reported by Kiplagat et al. in their iPres 2024 paper [10], a tripartite collaboration initiative to fastrack implementation of the KHCH digital repository has been tabled and will be inked as soon as possible. It brings together the Kenya Government, Moi University and KHCH Centre, taking to higher heights the collaboration between Government, Academia and Community. This will in addition to mentoring on Archiving, open access to the RETs facilities of Moi University [11, 12] in order to assist the government's commitment to lighting up rural Kenya and empowering communities through expanded access to electricity [13].

2 Materials and Methods

2.1 Introduction

In this chapter, we will present information and data that informed the materials and methods for harnessing the Keiyo Community traditional values for ecosystem preservation, within the periphery of Kaptagat forest in Elgeiyo Marakwet County. These information and data were obtained physically and digitally. The authors attended several physical and virtual meetings held between community leadership and both community members and officers of government and development partners. An example is the series of Public Participation Meetings from May 20th to May 24th, 2024 held the Kenya Forest Service (KFS), regarding the proposed Solar Electric Fence Project for the conservation of the Kaptagat Ecosystem [14].

The meetings' takeaways included the reaffirmation of collaboration among all stakeholders and commitment by the Kenya Forest Service of full implementation of agreed mitigative actions, to safeguard the integrity of the Kaptagat Ecosystem.

Relevant websites accessed for supplementary information and data shall be cited accordingly.

2.2 Kenya Forest Cover

The Kenya National Forest Programme (2016–2030) aims to develop and sustainably manage forests and allied resources to achieve socio-economic growth and climate resilience. It focuses on increasing tree cover, reversing forest degradation, and enhancing forest-based economic and social benefits for communities [15]. The programme is guided by the Kenya's 2030 [16], and the country's constitutional values and principles [17]. By 2030, the country's forest cover is expected increase to 10% from the current 6.5%.

Community Culture and Renewable Energy Technologies (RETs) are inseparable in the sustainable management of the Kaptagat Ecosystem. Its means leveraging traditional community values alongside modern green energy solutions to restore and maintain the forest cover while improving local community livelihoods. The KHCH Centre, being a partner to the Kaptagat Integrated Conservation Programme [2], upholds these ideals and contributes to the goal of increasing Kenya's forest cover.

2.3 Eldoret Water and Sanitation Company (ELDOWAS)

ELDOWAS is a corporate entity established under Cap 486 of the laws of Kenya. The County Government of UGC is its principal shareholder [18]. It gets about 55% of its water supply from Kaptagat forest which it processes and distributes to consumers in and around the county. The company has partnered with the community tree nurseries to enhance conservation of the forest catchment areas.

2.4 Forest Degradation Menace and Mitigation Efforts

In the last 25 years, at least 35% of the Kaptagat forest has been destroyed, according to the Kenya Forest Service [19] mainly through illegal logging, deforestation, overgrazing and charcoal production. Figure 1 shows typical deforestation activities in Kenyan forest areas.

Mitigation efforts, in this regard include the ongoing collaboration between the local communities and stakeholders like the Kenya Forest Services to restore the forests. In addition, alternative livelihood options are being provided by local and international organizations, to dissuade communities from invading forests with their animals. These include provision of high-yield cows kept in secluded areas called 'cow hotels' and alternative high revenue modern farming like avocado farming [20].

The Kaptagat Integrated Conservation Programme

The Kaptagat Integrated Conservation Programme [2], a community-based programme is actively undertaking rehabilitation of degraded forest areas and riparian areas within Kaptagat forest ecosystems and adjoining landscape to restore catchments, protect water springs and increase the ability of the ecosystem to provide goods and services. In addition it seeks to create and raise public awareness and inculcate the culture of tree growing. It has mobilized stakeholders and resources to make a sizeable contribution towards attaining the desired Kenya national tree

Fig. 1 Typical deforestation activities in and around forest areas of Kenya [15]

cover of planting 15 billion trees by 2030. It has also pushed for the adoption of alternative modern energy technologies to reduce pressure on forest resources.

The multi-stakeholder programme is expected to benefit 28,000 forest-adjacent households. These households will be supported to acquire at least one high-yielding dairy cow, put one acre of land under a cash crop and pasture, and benefit from solar lighting connections and biogas energy[20].

The restoration of Kaptagat forest has reached a new milestone with the launch of the one kilometre, one acre, one cow initiative: a programme focused on rehabilitating the forest landscape and enhancing community livelihoods. The fencing of the Kaptagat forest promises numerous benefits for the community, including improved conservation of water bodies for enhanced access to water for irrigation and distribution [14, 21].

2.5 Community Traditional Forest Conservation Values

The Keiyo Community is part of the bigger Kalenjin Community of Kenya comprising the Kipsigis, Nandi, Pokot, Sebei, Sabaot, Tugen, Marakwet, Ogiek, Terik and Sengwer, as traced in Kiplagat et al. [10]. They are a people who feel connected by a complex mix of kinship, culture, history. The cultural values the Keiyo Community mirror very well with the values of their kin: the Kipsigis as studied in detail by Oganga in his 2018 M.Sc. Thesis [22]. This heritage has historically led to sustainable forest management practices, which sadly are currently facing challenges.

This study identifies three important aspects of the Traditional Forest Conservation Systems (TFCs). First, the community held sacred selected groves within forests, recognizing them as places of spiritual significance and protectors of the forest's well-being. Second, extensive knowledge about forest ecosystems including plant and animal life, was accumulated and sustained over time. Third, traditional forest management adhered to strict community-based decision-making and enforcement of resource access and use rules.

The study further identifies three key challenges to these traditional practices namely:

(a) Loss of Indigenous Knowledge: The younger generations are abandoning traditional knowledge rapidly hence greatly weakening the foundation of sustainable forest management.
(b) External Pressures: Factors like deforestation, over-exploitation of resources, and the impacts of climate change are putting pressure on traditional forest conservation systems.
(c) Integration with Modern Systems: Effective integration of traditional forest conservation practices with modern management approaches is wanting.

2.6 Keiyo History Culture and Heritage (KHCH) Centre

The Keiyo Community are a people proud of their traditions, clans and their totems, bequeathed by their ancestors and passed on from generation to generation. Their daughters retain their fathers' clan names even when married. Upon realizing that, this treasured cultural heritage, is being gradually eroded by modernism and lack of tools to facilitate effective transfer of indigenous traditional knowledge to the younger generation, the Community professionals birthed an initiative of establishing the Keiyo History Culture and Heritage (KHCH) Centre [4, 5, 10].

The three key objectives of the KHCH Centre are:

(a) Educate, inform and sensitize members of the community on their rich and diverse age-old cultural heritage.
(b) Collaborate with Educational Institutions in doing Research and Development (R&D) about the Keiyo Community Cultural Heritage.
(c) Collaborate with stakeholders to organize cultural events, exhibitions and workshops to showcase community traditions, environmental conservation and athletic prowess.

KHCH Centre at the Kaptagat Integrated Conservation Programme Editions Events

The KHCH Centre was invited to participate in the 8th and 9th Editions of the Kaptagat Integrated Conservation Programme events held in July 2024 at Kaptagat Girls Secondary School and in July 2025 held at Simotwo Boys Secondary School, respectively. Figure 2 shows attendees including the author and co-author, at the KHCH Centre stands. In both editions the KHCH Centre stands were the centre of Keiyo cultural exhibitions, including artifacts, traditional attires, tools, songs and art.

2.7 Use of Modern Renewable Energy Technologies

The Archimedes Screw Turbine

Archimedes screw turbine (AST), Fig. 3 is operated based on the principle of conservation of energy. Water at the inlet possesses potential energy that is converted into mechanical energy of a rotating shaft. The process begins by water filling the spaces between adjacent blades hence exerting pressure on the blades. The pressure generates torque which causes the screw to rotate. An AST can be coupled to a generator via a gearbox in order to convert the mechanical energy to electrical energy [8].

Archimedes screw turbine (AST) is a relatively new hydropower machine suitable for application at low-head sites with moderate flow. Its popularity is fast growing due to its cost-effectiveness, robustness, simplicity and ability to operate at very low heads of less than 5 m.

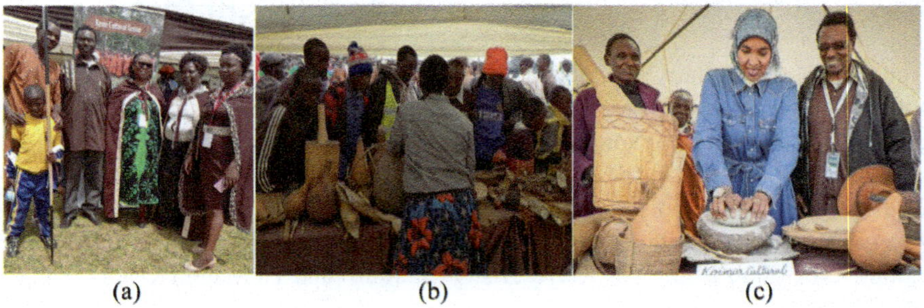

Fig. 2 **a** J. Kiplagat (author) 2nd left and B. Kipyego (co-author) 1st right, at the KHCH Centre exhibition stand during the 8th Edition of Kaptagat Integrated Conservation Programme Event 2024. **b** The 8th Edition attendants, both young and old look in awe at the rarely shown Keiyo community traditional artifacts exhibited at the KHCH Centre stand. **c** Kenya Government Principal Secretary for Department of Culture, Ummi Bashir (center) stone-grinding millet at the KHCH Centre exhibition during the 9th Edition of Kaptagat Integrated Conservation Programme Event 2025 [Authors]

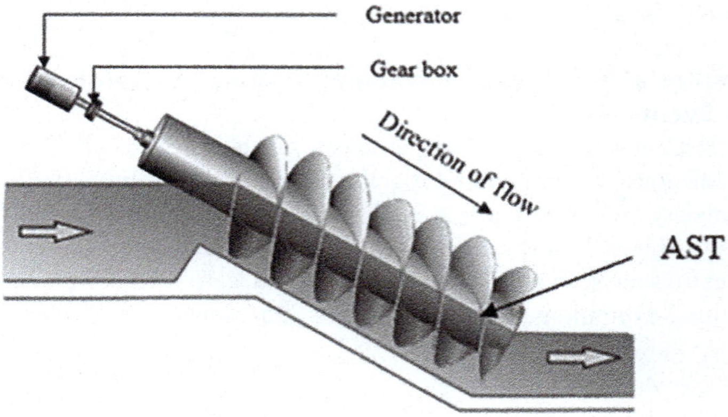

Fig. 3 Archimedes screw turbine [8]

Solar Cooker

The cooking energy demand in Elgeyo-Marakwet County is largely met with bio-fuels such as firewood, charcoal and agricultural residues. Wood cut for cooking purposes contributes to hectares of forest cover being destroyed annually. In Kenya, mostly in rural areas, women have to walk for several kilometres and even spend significant amount of time collecting firewood. Technology now offers a win–win alternative, in the form of solar-powered cookers. They come in various designs: box, parabolic and conical. Figure 4 illustrates how a box solar cooker works [9].

The cookers require simple materials to construct, such as cement, sand, coffee tray mesh and mirrors, molded in a parabolic shape. The cooker has a parabolic

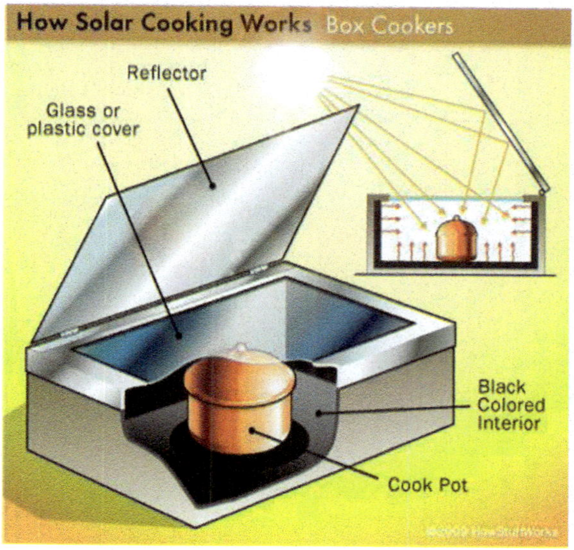

Fig. 4 Solar cooker [9]

dish fitted with thousands of small mirrors that converge sun rays to ignite fire. Currently solar cookers have not been installed at Kaptagat forest. If widely widely marketed and adopted, the cookers can impact climate change by stopping the usage of firewood. KHCH Center is working on a collaboration arrangement with RETs actors including Moi University to supply these cookers on a trial basis to selected community members.

3 Results and Discussions

3.1 Impacts of Forest Degradation and Mitigation Results

The global causes of deforestation and forest degradation vary over geographical and social contexts. However, in Kenya, the major drivers are related to community agricultural expansion and demand for improved livelihoods and vary in magnitude and forest type. Hosonuma et al. [23] introduced the forest transition concept, Fig. 5 to explain the transition from decreasing to expanding forest cover. This forest transition model identifies characteristic human-induced changes in forest cover and dynamics over time and relates well to the Kenya situation. Kenya is currently at the end of phase 3; a point of inadequate forest cover, requiring a shift towards reforestation.

The community in Kaptagat forest ecosystem with the support of local and international organisations has embarked on successful reforestation efforts, greatly assisted by community traditional forest preservation values.

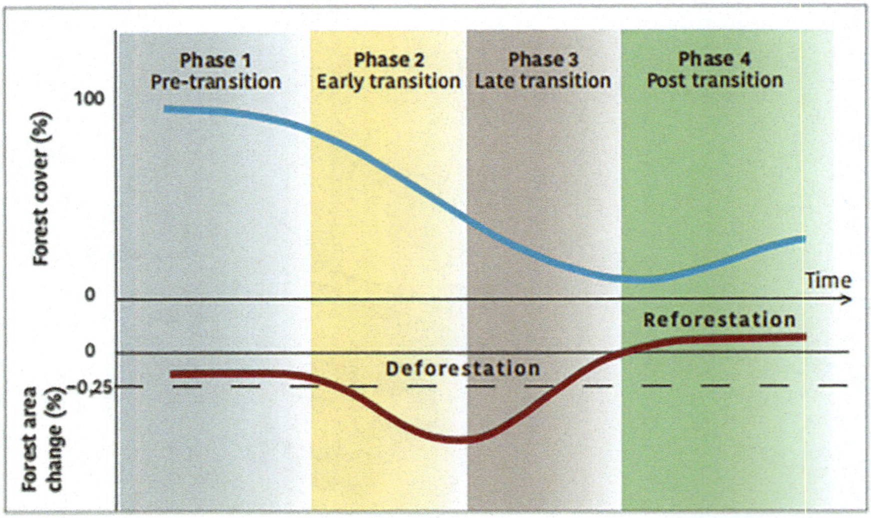

Fig. 5 Forest transition model [23]

3.2 Kaptagat Integrated Conservation Programme

Efforts by the Kaptagat Integrated Conservation Programme, through its yearly tree planting editions, have borne admirable fruits. In collaboration with local community institutions and harnessing their conscious traditional values, it has achieved great strides in forest reforestation.

The programme also brought in international institution, WWF-Kenya being the most active. Figure 6 shows the high-yield dairy cows that WWF-Kenya purchased for distribution to Kaptagat residents for improvement of livelihoods to reduce reliance on forest for grazing through uptake of zero-grazed cows.

Fig. 6 High-yielding dairy cows [20]

More than 28,000 households neighbouring Kaptagat forest are set to benefit from this transformative livelihood improvement project aimed at conserving the vital water tower while improving local incomes [24].

The 10-year project, supported by the World Wide Fund for Nature (WWF) and other partners, seeks to reduce residents' reliance on forest resources by providing alternative, sustainable means of earning a livelihood. The cows that are provided are high-yielding breeds.

For sustainability, they will be kept and serviced in secluded areas known as 'cow hotels', instead of allowing them to roam freely through vulnerable forest zones. In return, the forest gets a chance to regenerate. This is where innovation meets culture.

3.3 Harnessing Community Indigenous Knowledge and Values

Kenya is endowed with a rich historical and cultural heritage and is popularly known as the Cradle of Mankind [10]. Traditional African beliefs play a significant role in the lives of Kenyans despite adapting to modern values. They practise traditional values of humility, concern for good neighbourhood, ethnic cohesion and environmental stability. The Keiyo Community has not been left behind in these acts for social good.

Cultural Ecosystem Management Practices of the Keiyo Community

In Keiyo tradition, land and forests were sacred. They were governed and managed by clan elders, with clear rules and regulations about land use, when and how to graze animals, gather firewood, or harvest medicinal plants [25]. These rules promoted ecological conservation and ensured that nature regenerated in time to support future generations. Communal land management was adhered to the letter.

Harnessing indigenous knowledge on livestock behaviour promoted conservation of indigenous species in special locations of the forest. Rotational grazing practices in livestock farming promoted biodiversity, resulting in adequate vegetation in all seasons.

However, over the past few decades, generational change, economic pressures and modernism have disrupted these time-honoured practices. Land that was once communally protected has become fragmented and vulnerable. Widespread laxity in enforcement of the rules has led to destructive habits to both land and forests.

Role of KHCH in Transition to and Sustainability of RETs in Kaptagat Ecosystems

The Keiyo History Culture and Heritage Centre, is a local beacon of hope [10]. Founded by professionals from the local community, this grassroots initiative is breathing new life into the community's ancestral knowledge, especially with regard to ecosystem preservation. The KHCH Centre, through workshops,

Fig. 7 Kaptagat lady refilling her biogas digester [20]

storytelling sessions, and partnerships with schools is reconnecting the young generation with their ancestors' inheritance. Elders are being engaged not only as keepers of history but as guides for the future, to entrench the truth that ecosystem preservation was and still is a community in-born trait.

Our ancestors were conservationists long before the term even existed. They held conservation in high regard, as many of their cultural customs and traditions were practised within the forests. Most Keiyo shrines were located in secluded forest areas, further emphasizing the deep connection between cultural heritage and the natural environment. Therefore harnessing these age-old traditional values will surely bear dividends towards sustainable ecosystem preservation of the Kaptagat forest.

Transition to renewable energy technologies has taken off at Kaptagat forest: slow but sure. Residents have confirmed that they are efficient, easily manageable and sustainable. They can be installed across the entire ecosystem, require minimal maintenance and have long lifespans.

3.4 Transition to Modern Renewable Energy Sources

The Archimedes screw turbine and the solar cooker are viable alternatives to firewood [8, 9]. They are cheap to install and well suited for use at Kaptagat forest. Solar cookers' use in Kenya is widely documented [26].

Already, the WWF-Kenya, under the Greening Kaptagat project, have installed biogas units for 20 households within Kaptagat forest, to help the community transition to clean energy[20]. Figure 7 shows one of the beneficiaries refilling her biogas digester with cow-dung.

4 Conclusion

Kaptagat forest's fragile ecosystem is now greatly endangered by climate change and uncontrolled human activities. A wide range of ecosystem preservation efforts through community-based initiatives under the auspices of Kaptagat Integrated Conservation Programme are currently ongoing, harnessing traditional knowledge, values and practices to promote sustainable forest management.

The Keiyo community lives next to Kaptagat forest ecosystem. Its community cultural values constitute a great resource. The community has an age-long strong cultural link to forests, valuing them not just for their resources but also for their spiritual significance. This cultural heritage has historically led to sustainable forest management practices; however, these practices are currently facing a myriad of challenges. The Keiyo History Culture and Heritage (KHCH) Centre, whose mission is to document, preserve and promote Community rich History, Culture and Heritage, is assisting to mitigate these challenges, in close collaboration with local and international stakeholders.

Modern innovative renewable energy technologies come in handy to tap energy in form of electricity and heat from the forest water streams, the abundant sunlight and strong winds. The transition to RETs is the ultimate solution to forest degradation. The Keiyo community living next to the Kaptagat forest ecosystem has proven that their cultural heritage values are superior in ecosystem preservation. Therefore, harnessing these age-old traditional values will bear great dividends towards sustainable preservation of the Kaptagat forest and the entire ecosystem.

References

1. EMC Forest Conservation and Management Policy (2019). https://elgeyomarakwet.go.ke/?mdocs-file=2445. Accessed 18 Nov 2025
2. The Kaptagat Integrated Conservation Programme. https://kaptagat-icp.org/. Accessed 30 May 2025
3. President Ruto leads Kaptagat's 9th forest restoration drive. https://www.wwfkenya.org/?238591/President-Ruto-leads-Kaptagats-9th-forest-restoration-drive. Accessed 18 Nov 2025
4. The State Department of Culture, the Arts and Heritage. https://cultureheritage.go.ke/. Accessed 20 May 2025
5. Registrar of Societies Kenya. https://societies.ecitizen.go.ke/. Accessed 20 May 2025
6. Traditional Posho Mill: indigenous innovation in entrepreneurship. https://repository.chuka.ac.ke/server/api/core/bitstreams/842d85ae-e2b5-4a36-ae4e-a5cd6b3ddb4b/content. Accessed 25 Nov 2025
7. WWF-Kenya annual report 2023. https://wwfke.awsassets.panda.org/downloads/annual-report_2023-26th-june.pdf. Accessed 18 Nov 2025
8. Chida SP (2023) Parametric analysis of Archimedes screw turbine for micro hydropower generation using CFD. M.Sc. thesis, School of Engineering, Moi University, Eldoret, Kenya
9. Kipkemboi C (2023) Optimization and characterization of materials used to design a conical solar cooker: a case of Elgeyo-Marakwet County. M.Sc. thesis concept note. Submitted to School of Engineering, Moi University, Eldoret, Kenya
10. Kiplagat et al (2024) Digital repository for African history culture and heritage: a case of Keiyo community of Kenya. In: iPRES 2024 papers—international conference on digital

preservation. https://doi.org/10.21428/5676bf2d.317252e1. https://ipres2024.pubpub.org/pub/mp5h8q05/release/1?readingCollection=21e62c05. Accessed 30 Apr 2025
11. Moi University's renewable energy park. https://www.mu.ac.ke/index.php/en/mediaa/news-events/1178-establishment-of-moi-university-s-renewable-energy-park-and-innovative-research-hub-for-green-hydrogen-ammonia-fertilizer-other-green-products-murepha.html. Accessed 25 Nov 2025
12. Moi University taps solar power to reduce electricity costs. https://www.kenyanews.go.ke/moi-university-taps-solar-power-to-reduce-electricity-costs/. Accessed 29 Nov 2025
13. Kenya's new electrification projects bring hope to rural communities. https://www.esi-africa.com/renewable-energy/kenya-new-electrification-projects-bring-hope-to-rural-communities/. Accessed 25 Nov 2025
14. Stakeholders give feedback on proposed Kaptagat forest fencing. https://www.kenyaforestservice.org/stakeholders-give-feedback-on-proposed-kaptagat-fencing/. Accessed 25 Nov 2025
15. The Kenya National Forest Programme (2016–2030). https://faolex.fao.org/docs/pdf/ken190060.pdf. Accessed 25 Nov 2025
16. Kenya Vision 2030. https://vision2030.go.ke/wp-content/uploads/2018/05/Vision-2030-Popular-Version.pdf. Accessed 25 Nov 2025
17. The Kenya Constitution 2010. https://www.parliament.go.ke/sites/default/files/201705/The_Constitution_of_Kenya_2010.pdf. Accessed 25 Nov 2025
18. ELDOWAS. https://eldowas.or.ke/. Accessed 10 May 2025
19. The fight to save Kaptagat forest. https://nation.africa/kenya/health/the-fight-to-save-kaptagat-forest-a-community-s-struggle-and-triumph-4687910#google_vignette. Accessed 30 May 2025
20. 500 Ha set for restoration during 9th Edition. https://wwf-kenya.medium.com/500-ha-set-for-restoration-during-9th-edition-of-kaptagat-restoration-programme-34d544eb0c60. Accessed 10 May 2025
21. Enhancing community livelihoods. https://www.facebook.com/story.php?story_fbid=486632450590910&id=100077324777457&mibextid=6aamW6&rdid=v2Ge1calhy4NFJYY. Accessed 20 May 2025
22. Opanga V (2018) Role of traditional forest conservation systems in sustainable forest management: a case of the Kipsigis Sacred Hill. M.Sc. thesis, University of Nairobi, Kenya. https://erepository.uonbi.ac.ke/bitstream/handle/11295/104494/Opanga_Role%20of%20Traditional%20Forest%20Conservation%20Systems%20in%20Sustainable%20Forest%20Management%20a%20Case%20of%20the%20Kipsigis%20Sacred%20Hill.pdf?isAllowed=y&sequence=1. Accessed 30 Oct 2025
23. Hosonuma et al (2012) Environ Res Lett 7:044009
24. Thousands of families to benefit from Kaptagat Forest Conservation Project. https://nation.africa/kenya/health/thousands-of-families-to-benefit-from-kaptagat-forest-conservation-project-4876702. Accessed 10 May 2025
25. Kigen et al (2014) Ethnomedicinal plants traditionally used by the Keiyo community in Elgeyo-Marakwet County, Kenya. J Biodivers Biopros Dev 1:3. https://doi.org/10.4172/ijbbd.1000132. Accessed 30 Oct 2025
26. Solar powered cooker. https://www.africanews.com/2025/03/18/solar-powered-cookers-offer-sustainable-solution-in-kenya/. Accessed 30 Oct 2025

Methods of Synthesis, Characterization, and Applications of Nanoparticles: A Brief Review

Sadiq Bishir, Njira Njira Pili, Jacqueline Makatiani, and Rose Ramkat

Abstract

This review provides an overview of the importance of nanoparticles in various scientific fields, highlighting their classification, methods of synthesis, improvements, and applications in a range of areas of interest. Nanoparticles are essential components of nanotechnology, and they can be described as particles with dimensions ranging from 1 to 100 nm. They have garnered significant attention due to their unique properties and potential applications in various fields. The review also provides an overview of the advancements in the field of nanotechnology, highlighting the advantages, limitations, applications, and future directions for the most commonly used techniques such as physical, chemical, and biological methods. Many physicochemical characterization methods such as ultraviolet–visible spectroscopy, transmission electron microscopy (TEM), scanning electron microscopy (SEM), dynamic light scattering (DLS), X-ray diffraction (XRD), energy-dispersive X-ray (EDX), Fourier transform infrared spectroscopy (FTIR) have been intensively employed in studying the properties of nanoparticles. While the advancements in nanoparticle synthesis and characterization present numerous opportunities, challenges such as scalability and environmental impact assessments remain critical for future research and application. This review aims to aid in making well-informed choices and to act as a valuable source for both researchers and

S. Bishir (✉) · N. N. Pili · J. Makatiani · R. Ramkat
Department of Biological Sciences, School of Sciences and Aerospace Studies, Moi University, Eldoret, Kenya
e-mail: sadiqbishir@gmail.com

S. Bishir
Department of Biological Sciences, Umaru Musa Yar'adua University, Katsina, Katsina State, Nigeria

R. Ramkat et al. (eds.), *Towards the Sustainability of our Ecosystems*, Springer Proceedings in Geography and Sustainability,
https://doi.org/10.1007/978-3-032-30076-8_15

practitioners, encouraging the exchange of knowledge, cross-disciplinary collaboration, teamwork, and innovative developments in the evolving field of nanotechnology.

Keywords

Chemical methods • Green synthesis • Nanotechnology • Physical methods • Sustainability

1 Introduction

Over the past few decades, nanotechnology has evolved into one of the most transformative scientific frontiers, enabling the manipulation of matter at dimensions between 1 and 100 nm. At this scale, known as the nanoscale, materials exhibit distinct physicochemical and quantum confinement effects that profoundly alter their electrical, optical, catalytic, and mechanical properties compared to their bulk counterparts [1]. These unique nanoscale phenomena have established nanomaterials as fundamental building blocks for innovations in medicine, electronics, energy, and environmental science.

Nanoparticles (NPs)—the cornerstone of nanotechnology, can be classified into various types, including metal, metal oxide, carbon-based, ceramic, and polymeric nanoparticles, each exhibiting unique structural and functional characteristics. Their large surface area-to-volume ratios and enhanced surface reactivity enable superior adsorption, catalytic efficiency, and transport capabilities [2]. Such properties are particularly advantageous in applications requiring high reactivity and selectivity, including pollutant removal, solar energy conversion, and nutrient delivery in sustainable agriculture.

Despite the remarkable advancements, knowledge gaps persist regarding the interaction of nanomaterials with environmental systems, their physicochemical behavior in complex ecosystems, and their long-term ecological impacts. The lack of standardized classification and incomplete understanding of nanoparticle transformation pathways hinders effective risk assessment and sustainable deployment [3].

This review is motivated by the growing need to bridge nanotechnology and ecosystem sustainability, emphasizing how nanomaterials—owing to their confinement-driven reactivity, can be harnessed for renewable energy generation, pollution mitigation, and environmental restoration. For instance, metal oxide nanoparticles (e.g., TiO_2, ZnO, Fe_3O_4) demonstrate exceptional photocatalytic activity in solar-driven water purification, while carbon-based nanomaterials enhance energy storage efficiency and enable cleaner production systems [4, 5].

Therefore, this review provides an integrated overview of nanoparticle classifications, synthesis routes, characterization techniques, and emerging applications, highlighting their significance in advancing sustainable technologies. Particular emphasis is placed on green synthesis approaches using biological and plant-based methods as eco-friendly alternatives to conventional chemical synthesis. By linking nanomaterial innovation with environmental conservation and renewable energy systems, this review seeks to identify research gaps, promote sustainable practices, and encourage responsible development of nanotechnology for the protection and restoration of ecosystem health.

2 Methodology for Literature Review

This review employed a systematic literature search to gather relevant and up-to-date information on the synthesis methods, characterization techniques, and applications of nanoparticles [6]. The search was conducted across major scientific databases including Scopus, Web of Science, ScienceDirect, SpringerLink, and Google Scholar. The search strings were developed from the main research themes and included combinations of keywords such as "nanoparticles," "classification of nanoparticles," "organic nanoparticles," "carbon-based nanoparticles," "metal oxide nanoparticles," "inorganic nanoparticles," "metallic based nanoparticles," "nanoparticle synthesis," "green synthesis," "chemical synthesis of nanoparticles," "nanoparticle characterization techniques," "UV–Vis spectroscopy," "SEM–TEM analysis," "DLS," and "nanoparticle applications." Boolean operators (AND, OR, NOT) were used to refine the search results and ensure inclusion of the most relevant studies.

Articles were included if they (i) presented experimental or review data on nanoparticle synthesis (chemical, physical, or biological); (ii) discussed one or more characterization techniques and their analytical relevance; or (iii) described applications of nanoparticles in medicine, agriculture, or environmental science. Studies were excluded if they were not in English, lacked full text, were purely theoretical without empirical data, or were duplicated across databases. Reference lists of key papers were further screened to identify additional relevant studies. The selected literature was critically analyzed and synthesized to highlight advancements, limitations, and emerging trends in nanoparticle research.

3 Classifications of Nanoparticles

Nanoparticles can be classified into three main categories: organic, inorganic, and carbon based (Table 1). Organic nanoparticles include liposomes and polymers, while inorganic nanoparticles encompass metals like silver, gold, and copper, known for their antibacterial and antiviral properties. Carbon-based nanoparticles consist of fullerenes, graphene, carbon nanotubes, and carbon nanofibers, each exhibiting unique mechanical, chemical, and physical characteristics. This classification aids in understanding their diverse applications across biomedical, pharmaceutical, environmental, and industrial fields [7]. The physical and chemical characteristics of nanoparticles, such as size, shape, and surface area, influence their classification by determining their properties and applications [8].

Table 1 Classification of nanomaterials

Category	Types	Advantages	Disadvantages	References
Organic nanoparticles	Liposome, solid lipid, nanostructured lipid carrier, polymeric, carbohydrate-based, protein-based	Biocompatibility and biodegradability, flexibility of design, surface modification, functionalization, controlled release, response behavior	Complex synthesis, limited thermal, and chemical resistance, regulatory, and safety concerns	[9]
Inorganic nanoparticles	Silica, titanium oxide, zinc oxide/zirconium dioxide, iron oxide, gold, silver	Unique electrical, magnetic, optical properties, porosity, variability in size, structure, geometry, superior mechanical properties, catalytic activity	Agglomeration and stability issues, complex synthesis and high cost, environmental concerns, toxicity problems	[10]
Carbon-based nanoparticles	Fullerene, quantum dot, graphene, carbon black	Exceptional mechanical properties, thermal conductivity, chemical stability, surface functionalization, biocompatibility and environmental friendliness	Potential toxicity and biocompatibility issues, agglomeration and dispersion challenges, complex and costly production	[11]

3.1 Organic Nanoparticles (ONPs)

This represents the class of nanoparticles, which are synthesized from organic molecules with a size of 100 nm or less [12]. Organic nanoparticles (ONPs) are an emerging class of nanotechnology characterized by excellent water solubility, biodegradability, biocompatibility, and photostability. They are utilized in nanomedicine for applications like cancer treatment and COVID-19 vaccines [13]. Ferritin, micelles, dendrimers, and liposomes are among the organic nanoparticles (Table 2). Certain organic nanoparticles, such as liposomes and micelles, have a hollow sphere and are not harmful or biodegradable. Organic nanoparticles are a diverse class of nanomaterials that have gained significant attention in biomedical research due to their unique properties and versatility. These nanoparticles can be categorized into several classes, each with distinct applications in areas such as drug delivery, imaging, and tissue engineering. Polymeric nanoparticles are another name for organic nanoparticles [14]. The nanosphere or nanocapsule is the most well-known kind of organic or polymeric nanoparticle. The outer border of the spherical surface of the matrix particles, whose former overall mass

Table 2 Classification of organic NPs

Type	Composition	Structure	Properties	Typical applications	References
Liposomes	Phospholipids, cholesterol	Bilayer vesicle	Biocompatible, encapsulates both hydrophilic and lipophilic drugs	Drug delivery, gene therapy	[17]
Solid lipid nanoparticles (SLNs)	Solid lipids (at body temp)	Solid lipid core matrix	High physical stability, controlled drug release	Cosmetics, pharmaceuticals	[18]
Nanostructured lipid carriers (NLCs)	Solid and liquid lipid blend	Amorphous matrix	Improved drug loading, reduces drug expulsion	Cancer therapy, skin care	[19]
Polymeric nanoparticles	Natural/ synthetic polymers (e.g., PLGA, chitosan)	Nanospheres or nanocapsules	Controlled and sustained release, biodegradable	Vaccine delivery, imaging, cancer therapy	[20]
Carbohydrate-based nanoparticles	Starch, dextran, chitosan	Varies	Mucoadhesive, biocompatible, easily functionalized	Oral drug delivery, wound healing	[21]
Protein-based nanoparticles	Albumin, gelatin, casein	Nanospheres, nanocapsules	Biodegradable, good drug binding capacity	Anticancer drug delivery, enzyme immobilization	[22]

is solid, adsorbs other molecules. Particles encased the solid mass in the latter instance [15, 16]. However, many organic nanoparticles exhibit poor stability under physiological conditions, affecting their performance in biomedical applications.

3.2 Inorganic Nanoparticles

Inorganic nanoparticles, composed primarily of metals, metal oxides, or semiconductors, exhibit high structural stability, tunable size, and diverse physicochemical properties that distinguish them from their bulk counterparts [23]. Their unique optical, magnetic, and catalytic behaviors enable widespread use in cancer diagnostics, targeted therapy, sensors, catalysis, and environmental remediation [24].

Synthesis methods such as sol–gel, hydrothermal, and ionizing radiation techniques allow precise control over particle morphology and crystallinity, which are critical for tailoring their functionality. Inorganic nanoparticles also demonstrate high surface-to-volume ratios, quantum confinement effects, and modifiable surface chemistries, enhancing their reactivity and selectivity [25]. However, despite their biocompatibility in vitro, long-term toxicity, biodistribution, and clearance in humans remain major translational challenges, necessitating further clinical evaluation.

3.3 Metallic-Based Nanoparticles

Metallic nanoparticles (MNPs) include nanoscale forms of gold, silver, zinc, copper, and iron. They exhibit exceptional optical, electrical, magnetic, and catalytic properties due to surface plasmon resonance and large surface areas [26]. MNPs are utilized in molecular imaging, antimicrobial therapy, catalysis, and nanomedicine, often enhanced by polymeric or ligand coatings to improve colloidal stability and reduce cytotoxicity.

Their synthesis can follow top-down (e.g., physical vapor deposition) or bottom-up (e.g., chemical reduction, sol–gel) approaches, influencing morphology and functional properties [27]. A significant challenge remains in achieving precise control over their size, shape, and surface characteristics during synthesis, ensuring long-term biocompatibility, and minimizing potential toxicological risks. Additionally, issues like aggregation, unpredictable interactions in biological systems, and limited clinical translation due to safety concerns restrict their widespread use. Future research should focus on developing eco-friendly, scalable, and tunable synthesis methods, enhancing colloidal stability, improving targeting specificity, and conducting comprehensive long-term toxicity studies.

3.4 Metal Oxide-Based Nanoparticles

Metal oxide nanoparticles (MONPs) such as ZnO, TiO_2, CuO, Fe_2O_3, and CeO_2 possess remarkable chemical stability, tunable band gaps, and high catalytic activity, making them essential in energy, environmental, and biomedical applications [28, 29]. Their large surface area and reactive sites enhance adsorption and photocatalytic efficiency, enabling pollutant degradation, water purification, and antimicrobial applications.

Green synthesis approaches using plant or microbial extracts have further advanced the production of MONPs with reduced toxicity and improved biocompatibility, aligning with sustainable nanotechnology principles. Despite their versatility, significant gap remains in understanding their long-term environmental fate and reusability, warranting comprehensive ecotoxicological assessments.

3.5 Carbon-Based Nanoparticles

Carbon-based nanoparticles are entirely composed of carbon and exist in diverse morphologies such as fullerenes, graphene, carbon nanotubes (CNTs), carbon nanofibers, and carbon quantum dots (CQDs) [30]. Their exceptional mechanical strength, electrical conductivity, and high surface area make them indispensable in electronics, energy storage, catalysis, and biomedicine [31].

In biomedical applications, functionalized CNTs, graphene oxide, and CQDs serve as efficient carriers for drugs, genes, and imaging agents, enabling targeted delivery with minimal side effects [32]. Carbon quantum dots, in particular, offer tunable fluorescence and low cytotoxicity, making them valuable for bioimaging and diagnostics as shown in Table 3. Despite these advantages, toxicity, cost, and scalability remain key challenges limiting their large-scale implementation. Ongoing research focuses on surface modification, greener synthesis, and interdisciplinary approaches to realize their full potential for sustainable nanotechnology and environmental solutions.

Fullerenes are spherical or ellipsoidal carbon nanostructures formed through sp^2 hybridization, typically comprising 28–1500 carbon atoms. Their hollow cage-like configuration imparts remarkable mechanical, electrical, and thermal properties, enabling use in photovoltaics, catalysis, biosensing, and drug delivery [40].

Carbon nanotubes (CNTs) consist of rolled graphene sheets forming cylindrical nanostructures, categorized as single-walled (SWCNTs) or multi-walled (MWCNTs). They exhibit exceptional tensile strength, high aspect ratio, and tunable conductivity, making them ideal for applications in electronics, composites, and nanomedicine. However, concerns regarding their toxicity and environmental persistence necessitate further safety evaluation [41].

Graphene is a two-dimensional monolayer of carbon atoms in a hexagonal lattice, noted for its high surface area, superior electron mobility, and mechanical robustness. Its derivatives, including graphene oxide (GO) and reduced graphene oxide (rGO), find applications in water purification, energy storage, and biosensing, although production scalability and safety remain challenges [42].

Carbon-based quantum dots (CQDs)—including carbon and graphene quantum dots—are < 10 nm fluorescent nanostructures exhibiting strong photoluminescence, excellent solubility, and biocompatibility, enabling use in imaging, optoelectronics, and drug delivery [43].

Carbon black (CB), composed of fine carbon spheres from incomplete combustion, serves in catalysis, rubber reinforcement, and as a pigment but also contributes to environmental pollution [44].

4 Synthesis of Nanoparticles

There are two primary types of nanoparticle synthesis: conventional methods and environmentally friendly approaches, also known as green chemistry approaches. These methods can also be broadly categorized into top-down and bottom-up

Table 3 Carbon-based nanoparticles: structures, key properties, synthesis advancements, and applications

Type	Structure/ morphology	Properties	Recent synthesis advancements	Typical applications	References
Carbon nanotubes (CNTs)	Cylindrical tubes (single- or multi-walled)	High tensile strength, electrical and thermal conductivity, chemical stability	Plasma-enhanced CVD for low-temperature CNT growth; pulse arc discharge for high-quality SWCNTs	Drug delivery, biosensors, nanoelectronics, composite materials	[33]
Fullerenes (C_{60}, C_{70})	Spherical (buckyballs)	Antioxidant properties, electron affinity, photoconductivity	Solvothermal and arc-discharge synthesis optimization	Antiviral agents, photovoltaics, drug carriers	[34]
Graphene/graphene oxide	2D single-layer or few-layer sheets	High surface area, mechanical strength, electrical conductivity	Plasma-enhanced CVD for low-temperature graphene growth; N/S doping for enhanced photocatalytic activity	Drug delivery, tissue engineering, water purification, sensors, energy storage	[35]
Carbon quantum dots (CQDs)	Spherical nanoscale carbon particles (< 10 nm)	Fluorescent, biocompatible, low toxicity	Biomass pyrolysis for waste-to-nanomaterial conversion; green synthesis via plant extract mediation	Bioimaging, drug delivery, photodynamic therapy, adsorbents	[36]
Nanodiamonds (NDs)	Crystalline diamond-like nanoparticles	High hardness, chemical inertness, surface modifiability	Detonation synthesis refinement for uniform size distribution	Biomedical imaging, drug delivery, polishing materials	[37]
Carbon black (CB)	Amorphous carbon spheres/aggregates	Conductive, adsorptive, cost-effective	Pyrolysis and combustion-derived optimization	Catalysis, rubber reinforcement, pigments, pollution control	[38, 39]

SWCNTs single-walled carbon nanotubes, *CVD* chemical vapor deposition, *N/S doping* nitrogen and sulfur co-doping (a modification technique to enhance photocatalytic or electronic activity)

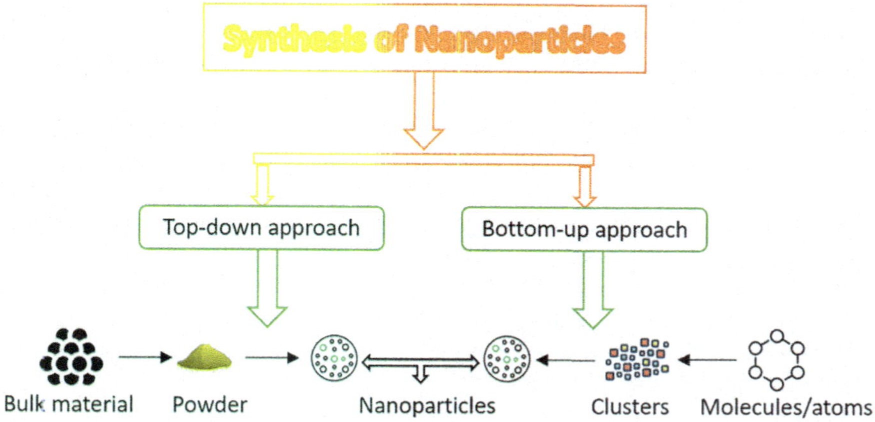

Fig. 1 Diagram illustrating the "top-down technique" and "bottom-up technique" for the production of nanoparticles

techniques [45]. Nanoparticles are synthesized using bottom-up and top-down approaches, including methods like sol–gel, solvothermal processes, mechanical milling, laser ablation, thermal decomposition, and sputtering (Fig. 1). These methods involve chemical reactions and specific conditions to control size, shape, and crystallography [46]. Top-down methods involve breaking down larger materials into nanoparticles through techniques like mechanical milling and physical vapor deposition, while bottom-up methods build nanoparticles from smaller units using processes such as chemical reduction and sol–gel methods. Each approach has distinct advantages and limitations [47, 48]. Nanoparticles can be synthesized using chemical, biological, and physical methods. The biological method, particularly plant-mediated synthesis, is preferred due to its cost-effectiveness, non-toxicity, and use of normal conditions, employing intracellular, extracellular, and phytochemical-mediated routes [49]. Biological synthesis is favored for its eco-friendliness, absence of toxic agents, and ability to control nanoparticle size and shape [50]. Green synthesis using plants is a notable biochemical approach, offering advantages such as simplicity, cost-effectiveness, reproducibility, and stability, utilizing phytochemicals as reducing and capping agents in nanoparticle formation [51]. These techniques enable the creation of nanoparticles with enhanced performance and flexible applications in technology [52].

4.1 Conventional Methods of Nanoparticle Synthesis

Conventional or traditional methods of nanoparticle synthesis primarily include physical and chemical techniques such as chemical vapor deposition (CVD), laser ablation, mechanical milling, sol–gel processing, inert gas condensation (IGC), and physical vapor deposition (PVD). These methods have historically played a crucial role in producing nanoparticles of controlled size, morphology, and composition for diverse industrial and biomedical applications [53]. However, their high-energy demand, use of toxic reagents, and environmental burden have increasingly raised sustainability concerns [54].

In chemical vapor deposition (CVD), gaseous precursors react or decompose on a heated substrate to form thin films or nanostructures. This method provides excellent control over film thickness and composition and is widely applied in semiconductor and coating industries [55]. Similarly, laser ablation employs high-energy laser pulses to vaporize material from a solid target, leading to nanoparticle formation upon condensation [56]. It enables precise size control and high purity but remains limited by high equipment cost and low scalability.

Mechanical milling is a top-down physical method where bulk materials are ground into nanoparticles through mechanical energy transfer [57, 58]. Although it is simple and cost-effective, it often yields non-uniform particles and poses contamination risks. The sol–gel method, on the other hand, is a bottom-up chemical approach involving hydrolysis and polycondensation of metal precursors to form a colloidal sol that transitions into a gel [59]. It allows excellent compositional control but relies heavily on organic solvents and catalysts, raising environmental concerns.

Vapor-phase synthesis techniques such as inert gas condensation (IGC) and physical vapor deposition (PVD) also provide fine control over particle size and purity. In IGC, metal vapors condense in an inert gas atmosphere to form nanoparticles, while in PVD, target materials are vaporized and deposited as thin coatings [60, 61]. Despite their technical versatility, these processes typically require vacuum systems, inert gases, and high temperatures, making them energy-intensive and costly.

As a result of the challenges associated with the traditional synthesis methods, there has been a growing shift toward green synthesis approaches that employ biological materials and eco-friendly reducing and stabilizing agents derived from plants, microorganisms, and natural polymers.

4.2 Green Chemistry Approach

Green synthesis is a clean, safe, cost-effective, and environmentally friendly method of producing nanoparticles [62]. This method avoids harmful chemicals, relying on natural reducing agents, and can incorporate biomass waste, supporting sustainable production practices [63]. The green production of nanoparticles has received a lot of interest in recent years. In this context, plant extracts and natural resources such as microbes and enzymes have been discovered to be effective alternative reagents in nanoparticle synthesis. Using green substances has various advantages, including low energy consumption and moderate operation conditions (e.g., pressure and temperature) without the use of harmful chemicals [64]. Green chemistry for nanoparticle synthesis emphasizes eco-friendly methods using biological resources like plants, bacteria, and fungi. This approach minimizes environmental impact, avoids hazardous chemicals, and utilizes natural compounds as reducing and stabilizing agents, ensuring sustainable and efficient production [65]. It aims to minimize waste and enhance reproducibility while addressing limitations related to raw material availability and geographical variations [66]. These green methods offer advantages like enhanced stability, faster synthesis rates, and milder conditions compared to traditional chemical synthesis methods [67].

The biological method, which is provided as an alternative to chemical and physical methods, offers an environmentally acceptable way to synthesize nanoparticles. Furthermore, this procedure does not require expensive, dangerous, or hazardous chemicals [68]. Several microbes have been examined for their ability to produce metal nanoparticles. Several species, such as bacteria, algae, yeast, and fungi, are capable of lowering metal ions via metalloregulatory mechanisms when exposed to metal ions [69]. Plants and plants extracts are also a green approach for synthesizing nanoparticles, and plant-based nanoparticle synthesis is a green chemistry approach that bridges nanotechnology with plant biotechnology. Plant extracts are utilized in the bio-reduction of metal ions to produce nanoparticles. Plant metabolites such as sugars, terpenoids, polyphenols, alkaloids, phenolic acids, and proteins have been shown to have a key role in the reduction of metal ions into nanoparticles and their subsequent stability [70]. The green chemistry approach for nanoparticle synthesis emphasizes using non-toxic reagents, making it an eco-friendlier alternative to conventional methods [71]. It minimizes toxic reagents, operates under ambient conditions, and enhances stability through biomolecule interactions, addressing environmental concerns associated with traditional synthesis methods [72].

Synthesis Using Microorganisms

The synthesis of nanoparticles using microorganisms represents a promising avenue in green nanotechnology, emphasizing environmentally friendly methods. Microbial synthesis leverages the natural capabilities of various microorganisms, including bacteria, fungi, and algae, to produce nanoparticles with unique properties [73]. Microorganisms play a crucial role in the biosynthesis of metal nanoparticles, offering a novel and promising approach (Fig. 2A). This method leverages biological processes to produce nanoparticles, highlighting diverse synthesis strategies and their potential applications in various fields, including medicine and technology [74]. Prokaryotic bacteria, actinomycetes, fungus, algae, and yeast are widely employed as bioreactors for the manufacture of nanoparticles. Extensive scientific research was conducted to create this approach for creating a range of NPs (Ag, Au, Pd, TiO_2, CdS, etc.). Microorganisms absorb target ions from their surroundings and subsequently convert the metal ions into the element metal using enzymes produced by cellular processes. This synthesis can be characterized as intracellular or extracellular, depending on where the NP is synthesized. The intracellular technique involves transferring metal ions into the microbial cell and forming NPs in the presence of enzymes. The extracellular creation of NPs includes the trapping of metal ions on cell surfaces and reducing ions in the presence of enzymes [75].

The use of bacteria for nanoparticle synthesis represents a promising avenue in green nanotechnology, offering a sustainable and eco-friendly alternative to traditional chemical methods. Bacterial synthesis leverages the unique biochemical pathways of microorganisms to convert metal ions into nanoparticles, which can be utilized in various applications, including antimicrobial agents and environmental remediation [76]. Bacteria, such as *Rhodopseudomonas* palustris and *Bacillus thuringiensis*, contribute to nanoparticle synthesis through mechanisms like lyases and bio-reduction. These microbial processes enhance green nanotechnology by offering environmentally friendly methods for producing nanoparticles with reduced toxicity [73]. Bacteria play a crucial role in the green synthesis of nanoparticles through bioprocesses, utilizing their bio-reduction capacity. This environmentally friendly approach minimizes pollution and energy consumption, making bacterial synthesis a cost-effective alternative to traditional physio-chemical methods [77]. They synthesize metal nanoparticles through mechanisms that involve transporting, regulating, and binding metal ions, leading to nanoparticle formation [78]. Bacteria, specifically *Bacillus* sp., are utilized for

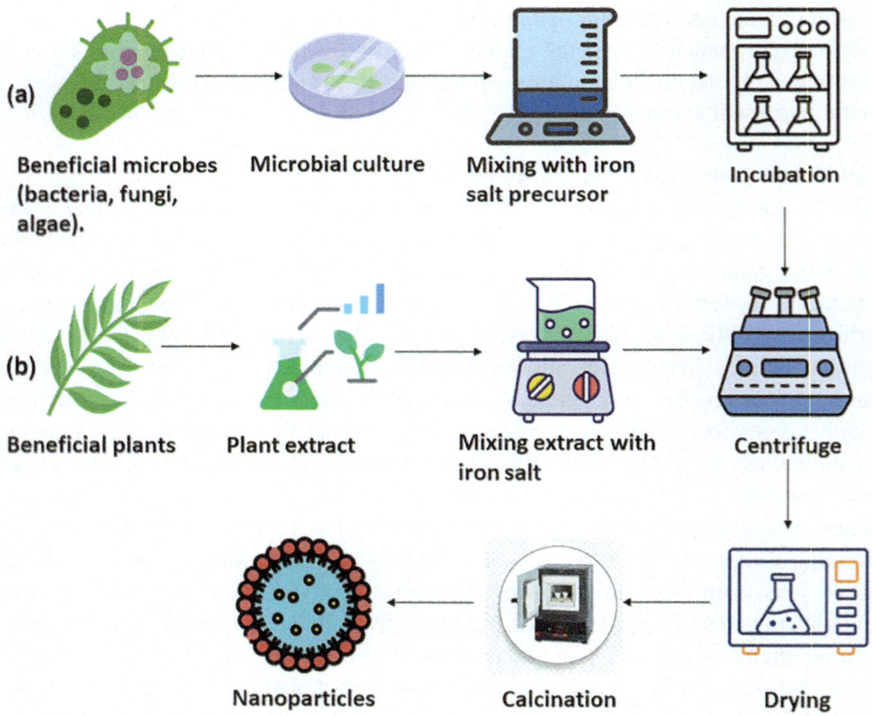

Fig. 2 Synthesis of nanoparticles using beneficial microorganisms (**A**) and beneficial plants (**B**)

eco-friendly synthesis of iron oxide nanoparticles (Fe_3O_4) due to their ability to produce nanoparticles through non-toxic methods, offering advantages like low cost and simplicity in the biosynthesis process [79]. A study by [80] demonstrates the biosynthesis of nickel nanoparticles using *Bacillus* species, optimizing growth duration, pH, and nickel precursor concentration through a microfluidic network. Gangan et al. [81] reported that engineered *Escherichia coli* can synthesize cadmium sulfide nanoparticles by expressing a divalent metal transporter and a nucleating peptide, allowing controlled synthesis at sub-toxic levels within the periplasmic space, resulting in nanoparticles with spherical morphology. A study by Kathiriya et al. [82] investigates the extracellular biosynthesis of silver nanoparticles using the bacteria *Bacillus subtilis*. The formation is confirmed by color change and UV–Vis spectra, indicating potential for eco-friendly synthesis methods and antimicrobial applications against human pathogens.

Nanoparticles can also be synthesized by using fungi as a reducing/stabilizing agent. The synthesis of nanoparticles using fungi, known as mycosynthesis, is an innovative and eco-friendly approach gaining traction in nanotechnology. Fungi possess unique enzymatic systems that facilitate the production of nanoparticles with diverse properties, making them suitable for various applications, including pharmaceuticals and agriculture [83]. Fungi are eukaryotic multicellular organisms that obtain their food by secreting digestive enzymes into their immediate environment, then absorbing the dissolved molecules [84]. Fungi are preferred for biogenic nanoparticle synthesis due to their high biomass production, secretion of stabilizing proteins, and metal tolerance. This method yields nanoparticles with desirable size, morphology, and enhanced physio-chemical characteristics, promoting sustainable applications in agriculture [85]. They facilitate green synthesis of nanoparticles through their enzymatic systems, producing diverse nanoparticles with varying sizes, shapes, and biological activities [86]. Endophytic fungi facilitate the green synthesis of gold (AuNPs) and silver nanoparticles (AgNPs) through various proteins, alkaloids, and phenolics [87]. Some actinomycetes, also referred to as mycelial bacteria, complete the group of fungi that produce NPs because they share certain features with prokaryotes [88]. Actinomycetes are specific microorganisms that produce bioactive compounds, making them effective in the green synthesis of silver nanoparticles. Actinomycetes, particularly *Streptomyces*, facilitate the microbial synthesis of nanoparticles through both extracellular and intracellular methods. They can produce various shapes and sizes of nanoparticles, including silver, gold, and copper, which are utilized across medical, pharmaceutical, environmental, and industrial sectors [89]. The NPs synthesized by actinomycete-mediated approaches appear to be stable, with good polydisperse distribution and high bactericidal activity; however, their potential to be involved in NP production has not been explored at a high capacity. Actinomycetes have been used extensively in the biosynthesis of silver and gold NPs, while a few have also been involved in the formation of zinc and some bimetallic NPs. The most studied actinomycetes in nanobiotechnology belong to the *Nocardia*, *Nocardiopsis*, *Rhodococcus*, *Streptomyces*, *Thermoactinomyces*, and *Thermomonospora* genera [90].

Several algae can also be used as an environmentally benign method for the synthesis of nanoparticles. Algae are sea microorganisms that have been reported not only to absorb heavy metals from various ecosystems, but also to synthesize metallic nanoparticles [91]. Algae-mediated synthesis of nanoparticles utilizes bioactive compounds in algae to reduce and stabilize metal ions, forming nanoparticles. This eco-friendly approach leverages the unique biochemical properties of various algal species, offering sustainable alternatives for applications in medicine, energy, and environmental remediation [92]. Algae are increasingly utilized for synthesizing nanoparticles due to their cost efficiency, sustainability, and minimal secondary pollutants. They offer high stability and surface activity, making them effective in environmental remediation against emerging contaminants like dyes and phenols [93]. Many algae, including brown, red, and blue-green types, utilize their extracellular metabolites to facilitate the synthesis of nanoparticles [94]. Cyanobacteria and eukaryotic green development genera, such as *S. subsalsa*, *L. majuscule*, *R. hieroglyphics*, *C. vulgaris*, *C. prolifera*, *P. pavonica*, *S. Platensis*, and *S. fluitans*, can be employed as economically viable materials for the biorecovery of metal from liquid courses of action [95].

Plants and Plants Extracts

The green synthesis of nanoparticles using plants and their extracts has emerged as a sustainable and eco-friendly alternative to conventional chemical and physical synthesis methods. In this approach, phytochemicals present in plant extracts—such as phenolics, flavonoids, terpenoids, alkaloids, and proteins—act as natural reducing, stabilizing, and capping agents, facilitating the conversion of metal ions into stable nanoparticles under mild reaction conditions [96, 97]. The choice of plant species, extraction method, and reaction parameters such as temperature, pH, and precursor concentration critically influences nanoparticle size, morphology, and stability.

Plant-mediated synthesis (Fig. 2B) offers distinct advantages over other biological routes, particularly due to the abundance, diversity, and renewability of plant biomaterials, which eliminate the need for maintaining microbial cultures. Various plant parts including leaves, roots, stems, fruits, and seeds, have been utilized for the green synthesis of metal and metal oxide nanoparticles, owing to their rich phytochemical composition [98]. For instance, plant extracts from species such as *Artemisia*, *Psylliostachys*, and *Achillea nobilis* have been employed to synthesize nanoparticles with controlled morphology and enhanced functional properties [99].

Moreover, this method enables the large-scale, cost-effective, and rapid production of biocompatible nanoparticles suitable for biomedical, agricultural, and environmental applications [100]. Despite these advantages, the precise biochemical mechanisms underlying plant-mediated nanoparticle synthesis remain inadequately understood. Evidence suggests that phytochemicals play dual roles as reducing and capping agents; however, the kinetics and molecular pathways of metal ion reduction and nanoparticle stabilization require further elucidation [101, 102]. Advancing mechanistic understanding through omics and computational approaches could further optimize this sustainable synthesis route, reinforcing its role in developing green nanotechnologies aligned with global sustainability goals.

5 Characterization of Nanoparticles

The growth of nanotechnology across various research domains has prompted the need for analytical methods to analyze and characterize nanoparticles. Characterization of nanoparticles is essential for understanding their physicochemical properties, which influence their performance and suitability for various applications in medicine, catalysis, electronics, and environmental remediation [103]. Key parameters typically assessed include particle size, shape, crystallinity, surface charge, optical behavior, and chemical composition. A combination of analytical techniques—such as ultraviolet–visible (UV–Vis) spectroscopy, Fourier transform infrared spectroscopy (FTIR), X-ray diffraction (XRD), dynamic light scattering (DLS), scanning electron microscopy (SEM), transmission electron microscopy (TEM), and energy-dispersive X-ray spectroscopy (EDS)—is often employed to provide a comprehensive understanding of nanoparticle morphology, structure, and surface chemistry.

5.1 Ultra-Violet Visible Spectroscopy

UV–Visible spectroscopy (UV–Vis) is one of the most widely used techniques for the preliminary characterization of nanoparticles, providing insights into their optical and electronic properties. It measures the absorption and scattering of light in the ultraviolet and visible regions (typically 300–800 nm), which correspond to electronic transitions within the material [104]. The position and intensity of the absorption peak, known as the surface plasmon resonance (SPR) band for metallic nanoparticles, are directly related to the particle's size, shape, composition, concentration, and degree of aggregation.

From the UV–Vis spectra, several optical parameters such as band gap energy, absorption coefficient, and refractive index can be derived, providing information about the nanoparticles' electronic structure and optical behavior [105]. These parameters are critical for assessing material suitability in optoelectronic, photocatalytic, and biomedical applications. UV–Vis spectroscopy is therefore considered a rapid, non-destructive, and reliable method for monitoring nanoparticle formation, confirming synthesis, and estimating particle stability in colloidal suspensions.

5.2 Fourier Transform Infrared Spectroscopy (FTIR)

FTIR spectroscopy is used to identify the functional groups and chemical bonds associated with nanoparticles and their stabilizing agents. It provides information on surface chemistry and confirms the involvement of biomolecules, such as phenols, flavonoids, proteins, and polysaccharides, in capping or reducing metal ions during green synthesis. Shifts in characteristic absorption peaks indicate chemical interactions between nanoparticles and surface ligands. FTIR analysis therefore helps elucidate nanoparticle–biomolecule interactions, surface modification, and overall stability of the synthesized nanomaterials [106].

5.3 X-Ray Diffraction (XRD)

X-ray diffraction (XRD) is an essential technique for determining the crystalline structure, phase purity, and average crystallite size of nanoparticles. The diffraction peaks correspond to specific lattice planes, allowing identification of crystal phases by comparison with standard Joint Committee on Powder Diffraction Standards (JCPDS) files [107]. Using the Debye–Scherrer equation, crystallite size can be estimated from peak broadening, while lattice strain and interplanar spacing (d-values) reveal information about structural order and defects. XRD analysis thus confirms the successful synthesis of nanocrystalline materials and their degree of crystallinity.

5.4 Scanning and Transmission Electron Microscopy (SEM/TEM)

Scanning Electron Microscopy (SEM) and Transmission Electron Microscopy (TEM) provide direct visualization of nanoparticle morphology, size distribution, and surface texture at high resolution. SEM generates detailed images of surface topography and particle agglomeration, while TEM enables visualization of internal structures, crystallinity, and lattice fringes at the nanoscale [108]. Together, these imaging techniques complement other characterization methods by validating nanoparticle shape, size uniformity, and dispersion. Selected Area Electron Diffraction (SAED) patterns obtained via TEM further confirm crystallographic information, correlating well with XRD results [109].

When coupled with Energy Dispersive X-ray Spectroscopy (EDS), these techniques allow simultaneous elemental analysis of the nanoparticles, confirming their chemical composition and purity. EDS identifies the constituent elements present and verifies successful metal incorporation or reduction in green-synthesized nanoparticles [109]. The combined use of SEM/TEM and EDS therefore provides a comprehensive understanding of nanoparticle morphology, structure, and composition, ensuring accurate correlation between synthesis parameters and material properties.

5.5 Dynamic Light Scattering (DLS)

Dynamic light scattering (DLS) is employed to determine the hydrodynamic diameter and size distribution of nanoparticles in suspension based on fluctuations in scattered light intensity. It also provides the polydispersity index (PDI), which indicates the uniformity of particle dispersion, and the zeta potential, reflecting surface charge and colloidal stability. DLS is particularly useful for monitoring aggregation behavior and stability of green-synthesized nanoparticles in aqueous environments, complementing the size data obtained from microscopy [73].

5.6 Mass Spectrometric Characterization of Nanoparticles

Mass spectrometry (MS) has emerged as a powerful analytical tool for the detailed characterization of nanoparticles, particularly in determining their elemental composition, isotopic distribution, and surface chemistry [110]. Among various MS techniques, Inductively Coupled Plasma Mass Spectrometry (ICP-MS) and its advanced form, Single-Particle ICP-MS (SP-ICP-MS), are widely utilized for nanoparticle analysis. SP-ICP-MS allows for the detection, quantification, and size determination of individual nanoparticles in complex matrices with high sensitivity and precision. This technique operates by introducing nanoparticles into an argon plasma, where they are atomized and ionized, producing a transient signal corresponding to a single-particle event. The signal intensity is proportional to particle mass, enabling estimation of particle size and number concentration [111, 112]. Unlike conventional ensemble techniques, SP-ICP-MS provides information on particle-size distribution, aggregation state, and dissolution kinetics at environmentally relevant concentrations, making it highly valuable for toxicological and environmental studies.

Recent advancements, such as coupling SP-ICP-MS with field-flow fractionation (FFF) or hydrodynamic chromatography (HDC), have enhanced the resolution and differentiation of nanoparticle populations with varying compositions or coatings. Furthermore, integration with isotopic labeling and laser ablation (LA-ICP-MS) extends its applicability to mapping nanoparticle distribution within biological or environmental samples. The high sensitivity, quantitative accuracy, and multi-element detection capability of SP-ICP-MS make it an indispensable technique for comprehensive nanoparticle characterization, particularly in assessing the environmental fate, bioavailability, and safety of both chemically and green-synthesized nanomaterials.

5.7 Characterization and Monitoring of Nanoparticles in Low-Resource Settings

In low-resource countries, limited access to advanced analytical facilities often constrains the identification, monitoring, and management of engineered nanoparticles (ENPs). However, a practical and sustainable approach can be achieved through a tiered analytical framework that integrates low-cost field tools, basic laboratory techniques, and shared regional facilities for confirmatory analyses. At the primary level, portable and low-cost methods such as colorimetric assays, UV–Vis spectrophotometry, and handheld X-ray fluorescence (pXRF) instruments can provide rapid, semi-quantitative screening of nanoparticles and metal contaminants in water, soil, and agricultural samples [113, 114]. These tools are suitable for preliminary assessments and environmental surveillance programs.

At the intermediate level, basic laboratory instruments like dynamic light scattering (DLS), Zeta potential analyzers, and Fourier transform infrared (FTIR) spectroscopy can characterize particle size, stability, and surface chemistry, while Scanning Electron Microscopy (SEM) coupled with Energy Dispersive X-ray Spectroscopy (EDS) can confirm morphology and elemental composition. For definitive and regulatory-grade analyses, regional or national reference laboratories can employ advanced techniques such as Inductively Coupled Plasma Mass Spectrometry (ICP-MS) or Single-Particle ICP-MS (SP-ICP-MS), Transmission Electron Microscopy (TEM), and Field-Flow Fractionation (FFF) coupled with ICP-MS or light scattering detectors [115]. This hierarchical approach allows low-income nations to balance affordability with scientific accuracy by combining routine local screening with selective high-precision confirmatory tests.

To ensure consistency and credibility, standardized sampling protocols, use of field and procedural blanks, and inter-laboratory calibration are recommended. Governments and institutions are further encouraged to establish regional centers of excellence to host advanced facilities, promote training, and facilitate data sharing. Strengthening capacity in nanoparticle monitoring not only supports environmental safety and public health but also aligns with the principles of sustainability and green innovation, ensuring that nanotechnology development proceeds responsibly within local resource constraints.

6 Applications of Nanoparticles

Nanoparticles are at the forefront of the rapidly expanding area of nanotechnology. Their distinctive size-dependent qualities make these materials excellent and indispensable in many areas of human activity [116]. Nanoparticles are a particularly important platform, displaying unique features with potentially broad applications as shown in Fig. 3. NPs have gained considerable attention due to their immense unique properties; they have been used for many applications and in various sectors such as medicine, agriculture, industries, bioremediation, and biological applications [117].

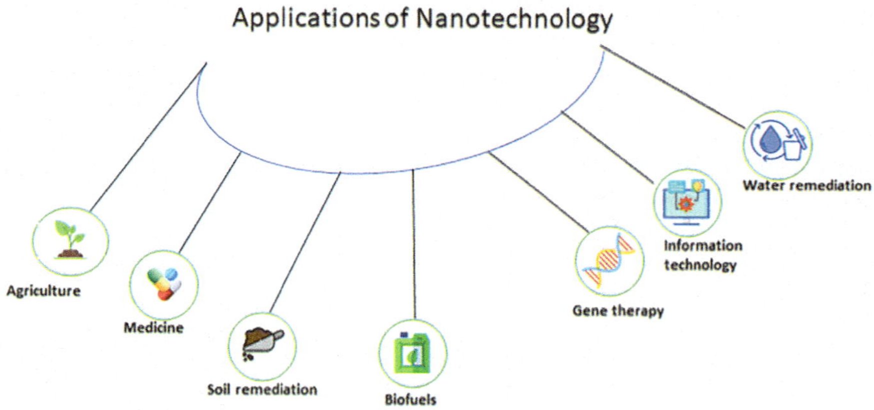

Fig. 3 Various applications of nanoparticles

6.1 Nanoparticles in Medicine

NPs are widely used in medicine for targeted drug delivery; due to their small size, they can easily diffuse through capillaries and accumulate in the active site [118]. The role of nanomedicine is to create more effective instruments for the prevention and treatment of various diseases by interacting nanoparticles with biological molecules. Nanoparticles are particularly important study topics because of their potential applications in biotechnology, pharmacology, optics, but most notably in medicine [119]. Due to the fact that nanoparticles can exist in the same size as that of proteins, it makes nanoparticles suitable for bio-tagging or labeling. Nanoparticles can also be used as an effective and safer method for cancer therapy, by enclosing the hydrophobic version of the dye molecule used for cancer treatment inside a porous nanoparticle. This will prevent the spread of the dye into the patient's body to cause further damage as the dye stays trapped inside the nanoparticle and hence cannot affect the other parts of the body. Proteins are a key aspect of the cell's language, machinery, and structure, and knowing their functions is critical for future advancements in human health. Gold nanoparticles are commonly employed in immunohistochemistry to detect protein–protein interactions [120].

6.2 Nanoparticles for Bio-fuel Production

Biofuels make use of domestically available renewable resources which are biodegradable and non-toxic [121]. Biofuels are fast growing as an alternative renewable energy source due to their cost-effectiveness and nonpolluting features and are environmentally friendly when compared with the fossil fuels. Because of the environmental and economic benefits of biofuels in comparison to petroleum products, they are fast progressing as elective substitutes for sustainable sources

[122]. The usage of nanoparticles in biofuel production processes is gaining traction due to their enhancing effects on metabolic reactions. Several nanomaterials, including nanofibers, nanotubes, and metallic nanoparticles, have been used in biofuel production methods. It has been shown that the activity of microorganisms can be increased by using several nanoparticles in anaerobic conditions because they enhanced the transfer of electrons in metabolic processes. Examples of such biofuels include biohydrogen, biodiesel, biogas and bioethanol [123].

6.3 Nanofertilizers

Nanoparticles have shown great promise in agriculture, particularly as nanofertilizers, where they enhance nutrient availability, improve soil health, and promote sustainable crop production as shown in Table 4. Conventional fertilizers, though vital for modern agriculture, are often inefficient and environmentally damaging—only a small fraction of applied nutrients, such as phosphorus and nitrogen, is taken up by plants, while the remainder is lost through leaching and runoff, leading to eutrophication and soil degradation [124, 125]. In contrast, nanofertilizers offer controlled and targeted nutrient delivery due to their high surface area, nanoscale size, and enhanced reactivity, which improve nutrient uptake efficiency and minimize environmental losses [126].

By acting as slow- or controlled-release nutrient systems, nanofertilizers ensure prolonged nutrient availability, reduce application frequency, and mitigate the adverse effects of excessive chemical fertilizer use. Furthermore, the integration of green-synthesized nanoparticles derived from plant or microbial sources presents a sustainable pathway toward eco-friendly agricultural practices. These bio-based nanofertilizers not only reduce dependence on synthetic chemicals but also align with circular economy principles by utilizing renewable biomaterials. Future research should focus on elucidating the interactions between nanofertilizers, soil microbiota, and plant physiological processes, while assessing their long-term ecological safety. Advancing sustainable nanofertilizer development will play a crucial role in achieving resilient, resource-efficient, and environmentally responsible agricultural systems.

6.4 Soil Remediation

Soil is an important aspect of the environment because it provides crucial ecological benefits for life. However, urbanization and increasing industry have had negative consequences for the soil. In recent decades, soil has been contaminated by hazardous and toxic pollutants produced from anthropogenic sources such as the disposal of chemical waste from industrial, abandoned use of fertilizers and pesticides, and other potential chemical sources that lead to soil contamination [127]. Diverse materials can be used to rehabilitate the environment, providing a variety of approaches to addressing this pressing issue. Among many remediation

Table 4 Applications of nanoparticles

Application area	Nanoparticles	Description of application	References
Medicine	Gold (AuNPs), silver (AgNPs), iron oxide (FeNPs), zinc oxide (ZnONPs), copper (CuNPs), titanium dioxide (TiO_2 NPs)	Drug delivery, diagnostics (MRI contrast agents), tissue engineering (bone growth), antimicrobial agents in wound dressings and medical devices	[132]
Environment	Silver (AgNPs), iron (FeNPs), nanoscale zerovalent iron (nZVI), metal oxides, carbon nanotubes	Bioremediation (removal of heavy metals, organic pollutants from water/soil), water purification, environmental monitoring	[133]
Agriculture	Silver (AgNPs), copper (CuNPs), zinc (ZnNPs), iron (FeNPs), gold (AuNPs), aluminum (AlNPs)	Targeted delivery of pesticides/ herbicides, nanofertilizers for efficient nutrient delivery, plant growth promotion	[134]
Food industry	Silver (AgNPs), zinc oxide (ZnONPs), iron oxide (FeNPs)	Food additives, antimicrobial packaging, preservation, detection of pathogens	[135]
Electronics	Semiconductor NPs (e.g., quantum dots), gold (AuNPs), silver (AgNPs), copper (CuNPs)	Solar cells, LEDs, transistors, sensors, high-density data storage	[136]
Energy	Iron (FeNPs), carbon nanotubes, metal oxides	Batteries, fuel cells, solar panels, energy storage and conversion	[137]
Textiles	Silver (AgNPs), titanium dioxide (TiO_2 NPs), zinc oxide (ZnONPs)	Antimicrobial fabrics, UV-protective clothing, stain-resistant and durable textiles	[132]
Catalysis	Gold (AuNPs), silver (AgNPs), iron (FeNPs), metal oxides	Industrial catalysts for chemical reactions, pollution control	[138]
Cosmetics	Titanium dioxide (TiO_2 NPs), zinc oxide (ZnONPs)	Sunscreens, anti-aging creams, enhanced absorption and stability of active ingredients	[139]

strategies, current efforts to improve the detection and purification of toxins mostly rely on nanotechnologies [128, 129]. Various designed NPs have been shown to have strong reactivity to a wide range of contaminants, including heavy metals and organic pollutants. This is owing to their increased mobility, specific surface area, and reactivity. Green nanoparticles also have these desirable qualities, making them appropriate for soil remediation. Global soil contamination is serious, reducing natural ecosystem functions and harming human health. Excavation followed by disposal or ex situ treatment are two well-established ways for treating polluted soils. However, these systems can be costly to operate, time-consuming, and generate significant amounts of secondary pollution. As a result, low-impact in situ therapy approaches, such as NP injection, are gaining popularity [130].

6.5 Wastewater Purification

Nanotechnology presents innovative and sustainable solutions to address critical challenges in water quality, scarcity, and pollution. The use of nanoparticles in wastewater treatment has demonstrated remarkable efficiency in removing a wide range of contaminants, including heavy metals, organic pollutants, dyes, and microbial pathogens [131]. Metallic and metal oxide nanoparticles such as Ag, CuO, ZnO, and TiO_2, as well as carbon-based nanomaterials like graphene and carbon nanotubes, exhibit superior adsorption, photocatalytic, and antimicrobial properties, enabling efficient purification and disinfection of wastewater [91].

Future progress in this field should focus on the scaling up of green nanomaterials, development of regenerable and reusable nanocomposites, and comprehensive ecotoxicological assessments to ensure environmental safety. Integrating green nanotechnology into wastewater treatment systems thus represents a pivotal step toward achieving sustainable water management, circular economy objectives, and cleaner production practices globally.

7 Sectors of Nanomaterial Application in Ecosystem Sustainability

Nanotechnology plays a pivotal role in advancing ecosystem sustainability, with its most significant applications found in the agricultural, environmental, and energy sectors. These areas directly contribute to sustainable development goals by promoting resource efficiency, environmental protection, and green innovation.

In agriculture, nanomaterials are utilized as nanofertilizers, nanopesticides, and nanosensors, offering controlled nutrient release, enhanced crop productivity, and reduced chemical pollution. Such innovations improve nutrient uptake and soil health while minimizing fertilizer runoff, aligning agricultural practices with sustainable land and water management [140]. The environmental sector employs nanoparticles for wastewater treatment, air purification, and heavy metal remediation. Metal and metal oxide nanoparticles such as TiO_2, ZnO, and Fe_3O_4 are particularly effective in degrading organic pollutants, adsorbing toxic ions, and disinfecting pathogens [4]. These nanomaterials enable cleaner water production, reduced contaminant load, and sustainable waste management systems.

In the energy sector, nanomaterials support sustainability through their integration into solar cells, fuel cells, and energy storage systems, enhancing efficiency and durability while reducing material and energy consumption. For example, TiO_2 and graphene-based nanostructures improve photocatalytic and electrochemical performance, contributing to the global transition toward renewable and low-carbon energy technologies [141].

Collectively, these sectors exemplify how nanotechnology can drive ecosystem sustainability by promoting resource conservation, pollution control, and circular economy principles, ultimately fostering a more resilient and environmentally responsible technological future. However, to fully harness this potential, future research should prioritize scalable, low-cost green synthesis methods, standardized life-cycle assessments, and regulatory frameworks that ensure both performance and ecological safety. In the context of developing and resource-limited countries, emphasis should be placed on accessible characterization techniques, capacity building, and local resource utilization—such as plant-based nanomaterial synthesis, to promote affordability and environmental compatibility. Strengthening interdisciplinary and international collaborations among scientists, engineers, and policymakers will be key to establishing sustainable nanotechnology systems that enhance productivity, energy efficiency, and environmental protection while supporting equitable global development.

8 Conclusion and Future Perspectives

Nanotechnology has emerged as a pivotal area of modern science, offering innovative solutions through the development and application of nanoparticles with distinctive physicochemical and functional properties. This review has summarized the advancements in nanotechnology, highlighting physical, chemical, and biological synthesis methods, each presenting specific advantages and limitations. Special attention has been given to biological and green synthesis methods by several studies, due to their eco-friendly nature and growing relevance in sustainable nanotechnology.

Future perspectives suggest that upcoming research should prioritize the refinement of environmentally friendly, cost-effective, and scalable synthesis methods. Integration of plant-based, microbial, and enzymatic synthesis routes, alongside computational modeling and machine learning for nanoparticle design and property prediction, is anticipated to enhance synthesis control and reproducibility. The development of multifunctional, targeted, and biocompatible nanoparticles will open new avenues, particularly in fields such as nanomedicine, precision agriculture, environmental remediation, and renewable energy.

Additionally, comprehensive studies on nanoparticle toxicity, long-term stability, and environmental impact are imperative to ensure safe and responsible applications. Strengthened interdisciplinary collaborations between nanotechnology, artificial intelligence, materials science, biotechnology, and data science are expected to further accelerate innovations and facilitate the translation of laboratory-scale findings to industrial and clinical applications.

Acknowledgements We thank INTERACT—Africa project under ACE—II, phytochemicals, textile and renewable energy (PTRE), Moi University, Kenya, for supporting this work. The authors express their sincere gratitude to editors and the anonymous reviewers for their constructive feedback and recommendations to enhance the quality of this paper.

Author Contributions The authors made the following contributions: S.B was responsible for conceptualization and initial writing; R.R handled conceptualization, review, and editing; J.M contributed to the review and editing; and N.N.P engaged in review and editing.

Declaration of Conflict of Interest The authors state that they have no identifiable financial conflicts or personal connections that might have seemed to affect the research presented in this paper.

References

1. Hughes KJ, Ganesan M, Tenchov R, Iyer KA, Ralhan K, Diaz LL et al (2025) Nanoscience in action: unveiling emerging trends in materials and applications. ACS Omega 10(8):7530–7548. https://doi.org/10.1021/acsomega.4c10929
2. Kurul F, Turkmen H, Cetin AE, Topkaya SN (2025) Nanomedicine: how nanomaterials are transforming drug delivery, bio-imaging, and diagnosis. Next Nanotechnol 7:100129. https://www.sciencedirect.com/science/article/pii/S2949829524000901
3. Kumah EA, Fopa RD, Harati S, Boadu P, Zohoori FV, Pak T (2023) Human and environmental impacts of nanoparticles: a scoping review of the current literature. BMC Public Health 23(1):1–28. https://doi.org/10.1186/s12889-023-15958-4
4. Li K, De Rancourt De Mimérand Y, Jin X, Yi J, Guo J (2020) Metal oxide (ZnO and TiO_2) and Fe-based metal–organic-framework nanoparticles on 3D-printed fractal polymer surfaces for photocatalytic degradation of organic pollutants. ACS Appl Nano Mater 3(3):2830–2845. https://doi.org/10.1021/acsanm.0c00096
5. Hamda AS, Areti HA, Gudeta RL, Abo LD, Jayakumar M (2025) Carbon-based nanomaterials for water treatment: a comprehensive review of recent advances and mechanisms. Chem Eng J Adv 23:100834. https://www.sciencedirect.com/science/article/pii/S2666821125001310
6. Qamer S, Romli MH, Che-Hamzah F, Misni N, Joseph NMS, Al-Haj NA et al (2021) Systematic review on biosynthesis of silver nanoparticles and antibacterial activities: application and theoretical perspectives. Molecules 26(16):5057. https://pmc.ncbi.nlm.nih.gov/articles/PMC8398138/
7. Sune PR, Jumde KS, Hatwar PR, Bakal RL, More SD, Korde AV (2024) Nanoparticles: classification, types and applications: a comprehensive review. GSC Biol Pharm Sci 29(3):190–197. https://scispace.com/papers/nanoparticles-classification-types-and-applications-a-3fe3q1pf9nd2
8. Singh AR, Maniraj M, Jain S (2021) Classification and synthesis of nanoparticles: a review, pp 1113–1125. https://scispace.com/papers/classification-and-synthesis-of-nanoparticles-a-review-18ag45zp4t
9. Habib B, Shabbir K, Ahmad A, Batool S, Liaqat I, Saddiqua A et al (2024) Nanoparticles unveiled: a comprehensive review of classification, properties, synthesis, and applications. Int J Adv Res 12(08):861–875. https://scispace.com/papers/nanoparticles-unveiled-a-comprehensive-review-of-48y84a4lol5x
10. Awan MSUD, Anwar MT, Khan HI, Asghar MR, Raza MR, Husnain N et al (2024) Nanoparticles' classification, synthesis, characterization and applications—a review. Charact Appl Nanomater 8(1):8899–8899. https://scispace.com/papers/nanoparticles-classification-synthesis-characterization-and-6kiaad6klq42
11. Vijeata A, Chaudhary GR, Chaudhary S, Ibrahim AA, Umar A (2024) Recent advancements and prospects in carbon-based nanomaterials derived from biomass for environmental remediation applications. Chemosphere 357:141935. https://scispace.com/papers/recent-advancements-and-prospects-in-carbon-based-jjb9z8uhvs
12. Abdulrahman HH, Ahmed KM, Fathulla Qader A, Anwar Omer R, Abdulkareem EI (2024) An overview of the preparation methodologies for the synthesis of organic and bioorganic nanoparticles. J Phys Chem Funct Mater 7(2):227–235. https://dergipark.org.tr/jphcfum
13. Young H, He Y, Joo B, Ferguson S, Demko A, Butterfield SK et al (2024) Toward the scalable, rapid, reproducible, and cost-effective synthesis of personalized nanomedicines at the point

of care. Nano Lett 24(3):920–928. https://scispace.com/papers/toward-the-scalable-rapid-reproducible-and-cost-effective-2rasj55m25
14. Niknam Z, Hossein Zadeh F, Toosi S, Haji Shoreh NK, Rasmi Y, Saleem I (2024) Organic-based nanomaterials for regenerative medicine. In: Handbook of nanomaterials, volume 2: biomedicine, environment, food, and agriculture, pp 359–400. https://scispace.com/papers/organic-based-nanomaterials-for-regenerative-medicine-4kz7bu9rqd
15. Ijaz I, Gilani E, Nazir A, Bukhari A (2020) Detail review on chemical, physical and green synthesis, classification, characterizations and applications of nanoparticles. Green Chem Lett Rev
16. Rao JP, Geckeler KE (2011) Polymer nanoparticles: preparation techniques and size-control parameters. Prog Polym Sci 36(7):887–913
17. Srivastava A, Singh A, Srivastava P, Wal A, Wal P, Yadav AK et al (2024) Liposomes as transformative tools in drug delivery, therapeutics, and beyond. Curr Drug Ther 19. https://scispace.com/papers/liposomes-as-transformative-tools-in-drug-delivery-3yf6avre5i
18. Arias JL, Barretta R, Nguyen TTL, Duong VA (2022) Solid lipid nanoparticles. https://doi.org/10.3390/encyclopedia2020063
19. Chauhan I, Yasir M, Verma M, Singh AP (2020) Nanostructured lipid carriers: a groundbreaking approach for transdermal drug delivery. Adv Pharm Bull 10(2):150–165. https://scispace.com/papers/nanostructured-lipid-carriers-a-groundbreaking-approach-for-4utl6fw8k2
20. Gillella S, Divyanjali M, Rishitha S, Amzad S, Reddy TU, Girish C et al (2024) Polymeric nanoparticles—a review. J Innov Appl Pharm Sci 25. https://doi.org/10.37022/jiaps.v9i1.575
21. Kumari SC, Mohammed D, Anuradha K (2024) Dextran-mediated nanoparticles and its various applications. In: Microbial essentialism: an industrial prospective: developments in applied microbiology and biotechnology, pp 113–123. https://scispace.com/papers/dextran-mediated-nanoparticles-and-its-various-applications-4eikli8dfx
22. Verma D, Gulati N, Kaul S, Mukherjee S, Nagaich U (2018) Protein based nanostructures for drug delivery. https://doi.org/10.1155/2018/9285854
23. Negi P, Chardra V, Kumar A, Kumar P, Priya, Yashmin K et al (2025) Plant-mediated synthesis of metal oxide nanocomposites: emerging prospects in electrochemical corrosion inhibition and biosensing. Sustain Chem Clim Act 7:100107. https://www.sciencedirect.com/science/article/pii/S2772826925000525
24. Pechyen C, Tangnorawich B, Toommee S, Marks R, Parcharoen Y (2024) Green synthesis of metal nanoparticles, characterization, and biosensing applications. Sens Int 1:5
25. Ghamarpoor R, Fallah A, Jamshidi M (2024) A review of synthesis methods, modifications, and mechanisms of ZnO/TiO_2-based photocatalysts for photodegradation of contaminants. ACS Omega 9(24):25457–25492. https://doi.org/10.1021/acsomega.3c08717
26. Burlec AF, Corciova A, Boev M, Batir-Marin D, Mircea C, Cioanca O et al (2023) Current overview of metal nanoparticles' synthesis, characterization, and biomedical applications, with a focus on silver and gold nanoparticles. Pharmaceuticals 16(10):1410. https://pmc.ncbi.nlm.nih.gov/articles/PMC10610223/
27. Malathi S, Pakrudheen I, Kalkura SN, Webster TJ, Balasubramanian S (2022) Disposable biosensors based on metal nanoparticles. Sens Int 3:100169. https://pmc.ncbi.nlm.nih.gov/articles/PMC8889882/
28. Alhalili Z (2023) Metal oxides nanoparticles: general structural description, chemical, physical, and biological synthesis methods, role in pesticides and heavy metal removal through wastewater treatment. Molecules 28(7):3086. https://pmc.ncbi.nlm.nih.gov/articles/PMC10096196/
29. Li K, De Rancourt De Mimérand Y, Jin X, Yi J, Guo J (2020) Metal oxide (ZnO and TiO_2) and Fe-based metal-organic-framework nanoparticles on 3D-printed fractal polymer surfaces for photocatalytic degradation of organic pollutants. ACS Appl Nano Mater 3(3):2830–2845
30. Larionov KB, Tsibulskiy SA (2017) A review on the classification, characterisation, synthesis of nanoparticles and their application

31. Zhao C, Kang J, Li Y, Wang Y, Tang X, Jiang Z (2023) Carbon-based stimuli-responsive nanomaterials: classification and application. Cyborg Bionic Syst 4. https://doi.org/10.34133/cbsystems.0022
32. Katubi KMM, Alsaiari NS, Alzahrani FM, Siddeeg SM, Tahoon MA (2021) Synthesis of manganese ferrite/graphene oxide magnetic nanocomposite for pollutants removal from water. Processes 9(4)
33. Devi N, Kumar R, Chen YS, Singh RK (2023) Carbon-based nanomaterials: carbon nanotube, fullerene, and carbon dots. In: Nanomaterials: advances and applications, pp 27–57. https://doi.org/10.1007/978-981-19-7963-7_2
34. Yu H, Feng W (2024) Carbon-based multifunctional nanomaterials. Nanomaterials 14(19):1600. https://pmc.ncbi.nlm.nih.gov/articles/PMC11478235/
35. Ealias AM, Saravanakumar MP (2017) A review on the classification, characterisation, synthesis of nanoparticles and their application. IOP Conf Ser Mater Sci Eng 263(3)
36. Farmand M, Jahanpeyma F, Gholaminejad A, Azimzadeh M, Malaei F, Shoaie N (2022) Carbon nanostructures: a comprehensive review of potential applications and toxic effects. 3 Biotech 12(8):159. https://pmc.ncbi.nlm.nih.gov/articles/PMC9259781/
37. Sonkar PK, Narvdeshwar, Gupta PK (2021) Characteristics of carbon nanotubes and their nanocomposites. In: Fundamentals and properties of multifunctional nanomaterials, pp 99–118
38. Hooda K, Singh A (2024) Carbon-based nanomaterial: properties and applications. In: Futuristic trends in chemical material sciences & nano technology, vol 3, book 9, pp 1–15
39. Pathak K, Kumar M, Misra SK, Kumari B, Kaushal N (2022) Carbon-based nanomaterials: potential therapeutic applications. In: Functionalized carbon nanomaterials for theranostic applications, pp 263–285
40. Madkour LH (2024) Fullerene-based modifier: biosensing and nanobiotechnological industrial applications. In: Emerging engineering technologies and industrial applications, pp 59–96
41. Gacem A, Modi S, Yadav VK, Islam S, Patel A, Dawane V et al (2022) Recent advances in methods for synthesis of carbon nanotubes and carbon nanocomposite and their emerging applications: a descriptive review. J Nanomater 2022(1):7238602. https://doi.org/10.1155/2022/7238602
42. Biswas HS, Kundu AK, Poddar S (2024) Harnessing carbon-based nanomaterials for sustainable environmental solutions, pp 57–77
43. Guan X, Li Z, Geng X, Lei Z, Karakoti A, Wu T et al (2023) Emerging trends of carbon-based quantum dots: nanoarchitectonics and applications. Small 19(17)
44. Gómez-Hernández R, Panecatl-Bernal Y, Méndez-Rojas MÁ (2019) High yield and simple one-step production of carbon black nanoparticles from waste tires. Heliyon 5(7)
45. Parthasarathy S, Behera S, Das DK, Patnaik P, Das N, Mallick P et al (2024) Nanoparticle synthesis methods. In: Advances in chemical and materials engineering book series, pp 33–68. https://scispace.com/papers/nanoparticle-synthesis-methods-31ml977z0gh7
46. Torvi AI, Sangeetha J, Shettar AK, Thangadurai D, Rajole P (2022) Nanoparticles and nanomaterials: an update. In: Biogenic nanomaterials: structural properties and functional applications, pp 1–17. https://scispace.com/papers/nanoparticles-and-nanomaterials-an-update-1oxmd8ym
47. Fatemi M, Bahrami Z (2023) Synthesis of different types, shapes, and sizes of nanocarriers using synthetic and biological approaches. In: Nano drug delivery for cancer therapy, pp 25–52. https://scispace.com/papers/synthesis-of-different-types-shapes-and-sizes-of-bb9508cwc5
48. Triphati A, Pirzadah TB (2023) Synthesis methods of nanoparticles and their key applications. In: Synthesis of bionanomaterials for biomedical applications, pp 57–76. https://scispace.com/papers/synthesis-methods-of-nanoparticles-and-their-key-1najzpoq

49. Ankamwar B, Kirtiwar S, Shukla AC (2020) Plant-mediated green synthesis of nanoparticles. In: Advances in pharmaceutical biotechnology: recent progress and future applications, pp 221–234. https://scispace.com/papers/plant-mediated-green-synthesis-of-nanoparticles-49tukbqeih
50. Malinoshevska M, Shydlovska O (2022) Silver and cerium nanoparticles synthesis methods. Біологія та екологія 8(1):73–79. https://scispace.com/papers/silver-and-cerium-nanoparticles-synthesis-methods-4xzlnkjzqr
51. Soroodi F, Jamal P, Noorbatcha IA (2020) Experimental methods for the phytochemical production of nanoparticles. In: Advances in nanotechnology and its applications, pp 83–95. https://scispace.com/papers/experimental-methods-for-the-phytochemical-production-of-3pnbsz1l8c
52. Mohil R, Kaushik N (2024) Overview on nanoparticle classes, design and perspectives. In: Innovations in science and technology: shaping a sustainable future, vol 1, pp 47–54. https://scispace.com/papers/overview-on-nanoparticle-classes-design-and-perspectives-1fozugy7qv
53. Wegner K, Schimmoeller B, Thiebaut B (2011) Pilot plants for industrial nanoparticle production by flame spray pyrolysis. KONA Powder Part J 29(29):251–265
54. Kumari S, Raturi S, Kulshrestha S, Chauhan K, Dhingra S, András K et al (2023) A comprehensive review on various techniques used for synthesizing nanoparticles. J Mater Res Technol 27:1739–1763. https://www.sciencedirect.com/science/article/pii/S223878542302416X
55. Pedersen H, Elliott SD (2014) Studying chemical vapor deposition processes with theoretical chemistry. Theor Chem Acc 133(5):1–10
56. Saeed M, Alshammari Y, Majeed SA, Al-Nasrallah E (2020) Chemical vapour deposition of graphene—synthesis, characterisation, and applications: a review. Molecules 25(17):3856. https://pmc.ncbi.nlm.nih.gov/articles/PMC7503287/
57. Ijaz I, Gilani E, Nazir A, Bukhari A (2020) Detail review on chemical, physical and green synthesis, classification, characterizations and applications of nanoparticles. Green Chem Lett Rev. https://doi.org/10.1080/17518253.2020.1802517
58. Sherif El-Eskandarany M, Al-Hazza A, Al-Hajji LA, Ali N, Al-Duweesh AA, Banyan M et al (2021) Mechanical milling: a superior nanotechnological tool for fabrication of nanocrystalline and nanocomposite materials. Nanomaterials 11(10):2484. https://pmc.ncbi.nlm.nih.gov/articles/PMC8539264/
59. Abid N, Khan AM, Shujait S, Chaudhary K, Ikram M, Imran M et al (2022) Synthesis of nanomaterials using various top-down and bottom-up approaches, influencing factors, advantages, and disadvantages: a review. Adv Colloid Interface Sci 300:102597. https://www.sciencedirect.com/science/article/abs/pii/S0001868621002384
60. Malekzadeh M, Swihart MT (2021) Vapor-phase production of nanomaterials. Chem Soc Rev 50(12):7132–7249
61. Escorcia-Díaz D, García-Mora S, Rendón-Castrillón L, Ramírez-Carmona M, Ocampo-López C (2023) Advancements in nanoparticle deposition techniques for diverse substrates: a review. Nanomaterials 13(18):2586. https://pmc.ncbi.nlm.nih.gov/articles/PMC10537803/
62. Huston M, Debella M, Dibella M, Gupta A (2021) Green synthesis of nanomaterials, pp 1–29
63. Miu BA, Dinischiotu A (2022) New green approaches in nanoparticles synthesis: an overview. Molecules 27(19)
64. Wang X, Feng Y, Dong P, Huang J (2019) A mini review on carbon quantum dots: preparation, properties, and electrocatalytic application. Front Chem 7:1–9
65. Álvarez-Chimal R, Ángel Arenas-Alatorre J (2023) Green synthesis of nanoparticles: a biological approach. green chemistry for environmental sustainability—prevention-assurance-sustainability (P-A-S) approach
66. Jain K, Takuli A, Gupta TK, Gupta D (2024) Rethinking nanoparticle synthesis: a sustainable approach vs. traditional methods. Chem Asian J
67. Girotra V, Kaushik P, Vaya D (2024) Exploring sustainable synthesis paths: a comprehensive review of environmentally friendly methods for fabricating nanomaterials through green chemistry approaches. Turk J Chem 48(5):703–725

68. Nadaroğlu H, Alayli Güngör A, İnce S (2017) Synthesis of nanoparticles by green synthesis method. Int J Innov Res Rev 1(1):6–9
69. Park TJ, Lee KG, Lee SY (2016) Advances in microbial biosynthesis of metal nanoparticles. Appl Microbiol Biotechnol 100(2):521–534
70. Parveen K, Banse V, Ledwani L (2016) Green synthesis of nanoparticles: their advantages and disadvantages. AIP Conf Proc 1724
71. Nair GM, Sajini T, Mathew B (2022) Advanced green approaches for metal and metal oxide nanoparticles synthesis and their environmental applications. Talanta Open 5
72. Niveditha M, Shelar VR, Bhalerao BM (2024) Green synthesis: an alternative sustainable route for nanotechnology. J Adv Biol Biotechnol 27(7):1054–1069
73. Sharma C, Tripathi P, Wadhwa N (2024) Nanoparticles emerging through microbial routes. In: Futuristic trends in biotechnology, vol 3, book 1, pp 127–139
74. Desai PP, Radha MJ, Savitha G, Boregowda R (2023) Versatile strategies for multifaceted nanoparticle synthesis—an overview. In: Nanotechnology and in silico tools: natural remedies and drug discovery, pp 155–168
75. Dhand C, Dwivedi N, Loh XJ, Jie Ying AN, Verma NK, Beuerman RW et al (2015) Methods and strategies for the synthesis of diverse nanoparticles and their applications: a comprehensive overview. RSC Adv 5(127):105003–105037
76. Arora A, Lashani E, Turner RJ (2024) Bacterial synthesis of metal nanoparticles as antimicrobials. Microb Biotechnol 17(8)
77. Singhal A, Baweja P, Gupta S, Chopra H, Gandhi PB (2023) Biogenesis of nanoparticles using microorganisms: a review. Plant Sci Today 10(4):97–104
78. Campaña AL, Saragliadis A, Mikheenko P, Linke D (2023) Insights into the bacterial synthesis of metal nanoparticles. Front Nanotechnol 5
79. Jubran AS, Al-Zamely OM, Al-Ammar MH (2020) A study of iron oxide nanoparticles synthesis by using bacteria. Int J Pharm Qual Assur 11(1):88–92
80. Yusuf HA, Redha Zainab M, Thawadi S, Maki H, Abbas H, Mohsin J (2024) Optimization of *Bacillus* bacteria mediated biosynthesis of nickel nanoparticles using a precise gradient generated through microfluidic network. https://www.researchsquare.com/article/rs-4977308/v1
81. Gangan MS, Naughton KL, Boedicker JQ (2023) Utilizing a divalent metal ion transporter to control biogenic nanoparticle synthesis. J Ind Microbiol Biotechnol 50(1)
82. Kathiriya N, Dudhagara D, Sojitra R (2024) Synthesis of silver nanoparticles from bacteria and fungi for evaluating efficacy against human pathogens. In: Futuristic trends in biotechnology, vol 3, book 8, pp 219–239
83. Rezghi Rami M, Meskini M, Ebadi Sharafabad B (2024) Fungal-mediated nanoparticles for industrial applications: synthesis and mechanism of action. J Infect Public Health 17(10)
84. Shelke D, Chambhare M, Sonawane H (2022) Mycosynthesis of nanoparticles and their potential application in pharmaceutical bioprocessing. In: Nanotechnology in the life sciences, pp 425–442
85. Chauhan A, Anand J, Parkash V, Rai N (2023) Biogenic synthesis: a sustainable approach for nanoparticles synthesis mediated by fungi. Inorg Nano-Met Chem 53(5):460–473
86. Loshchinina EA, Vetchinkina EP, Kupryashina MA (2023) Diversity of biogenic nanoparticles obtained by the fungi-mediated synthesis: a review. Biomimetics 8(1)
87. Bloch K, Ghosh S (2022) Endophytic fungi-mediated synthesis of gold and silver nanoparticles. In: Endophytic association: what, why and how, pp 19–36
88. Abdelhafez ZN, Moubarak AA, El-Shora HM, Hamed AA, Hassan MG (2024) Screening of soil actinomycetes ability to biosynthesis zinc oxide nanoparticles. Benha J Appl Sci 9(3):133–138
89. Venkateswaran SP, Palaniswamy VK, Vishvanand R, Periakaruppan R (2021) Actinomycetes-assisted nanoparticles: synthesis and applications. In: Agri-waste and microbes for production of sustainable nanomaterials, pp 375–395
90. Miu BA, Dinischiotu A (2022) New green approaches in nanoparticles synthesis: an overview

91. Debnath S, Kumar RS, Babu MN (2024) Green synthesis of metallic nanoparticles and their potential applications. Pharm Times 56(1bbb):6–11
92. Venkateswarlu Y, Supraja N (2024) A review on algal mediated synthesized metallic nanoparticles: an eco-friendly approach for sustainable nanotechnology. Curr J Appl Sci Technol 43(6):1–10
93. Premarathna KSD, Lau SY, Chiong T, Show PL, Vithanage M, Lam MK (2024) Greening up the fight against emerging contaminants: algae-based nanoparticles for water remediation. Clean Technol Environ Policy
94. Fayyad RJ, Ali ANM, Ali TR, Faris SA (2024) Extracellular algae based nanoparticles, overview, mechanisms and some medical applications. South Asian Res J Biol Appl Biosci 6(04):114–119. https://scispace.com/papers/extracellular-algae-based-nanoparticles-overview-mechanisms-2ywl5m5r0ztv
95. Journal AI, Gour A, Jain NK (2019) Advances in green synthesis of nanoparticles. Artif Cells Nanomed Biotechnol 47(1):844–851
96. Kumari M, Sadhu P, Talele C, Shah N (2024) Green nanotechnology: how plants can help synthesise nanoparticles for biomedical and environmental purposes. J Nat Remedies 24(5):1021–1034
97. Tanwar SN, Parauha YR, There Y, Swart HC, Dhoble SJ (2024) Plant-based biosynthesis of metal and metal oxide nanoparticles: an update on antimicrobial and anticancer activity. ChemBioEng Rev
98. Jadoun S, Arif R, Jangid NK, Meena RK (2021) Green synthesis of nanoparticles using plant extracts: a review. Environ Chem Lett 19(1):355–374. https://doi.org/10.1007/s10311-020-01074-x
99. Lakhani KG, Hamid R, Motamedi E, Marviya GV (2025) A review on plant metabolite-mediated nanoparticle synthesis: sustainable applications in horticultural crops. Front Nanotechnol 9(7):1545413
100. Ali BH, Baghdadi M (2024) A review of plant-derived metallic nanoparticles synthesized by biosynthesis: synthesis, characterization, and applications. In: Green and sustainable approaches using wastes for the production of multifunctional nanomaterials, pp 251–272. https://scispace.com/papers/a-review-of-plant-derived-metallic-nanoparticles-synthesized-142ws3g9s8
101. Escárcega-González CE, Garza-Cervantes JA, Vázquez-Rodríguez A, Montelongo-Peralta LZ, Treviño-González MT, Díaz Barriga Castro E et al (2018) In vivo antimicrobial activity of silver nanoparticles produced via a green chemistry synthesis using acacia rigidula as a reducing and capping agent. Int J Nanomed 13:2349–2363. https://doi.org/10.2147/IJN.S160605
102. El Shafey AM (2020) Green synthesis of metal and metal oxide nanoparticles from plant leaf extracts and their applications: a review. Green Process Synth 9(1):304–339
103. Hong SJ, Mun HJ, Kim BJ, Kim YS (2021) Characterization of nickel oxide nanoparticles synthesized under low temperature. Micromachines 12(10):1168. https://www.mdpi.com/2072-666X/12/10/1168/htm
104. Calderón-Jiménez B, Sarmanho GF, Murphy KE, Bustos ARM, Vega-Baudrit JR (2017) NanoUV-VIS: an interactive visualization tool for monitoring the evolution of optical properties of nanoparticles throughout synthesis reactions. J Res Natl Inst Stand Technol 122
105. Kafle BP (2020) Introduction to nanomaterials and application of UV–visible spectroscopy for their characterization. In: Chemical analysis and material characterization by spectrophotometry, pp 147–198. https://scispace.com/papers/introduction-to-nanomaterials-and-application-of-uv-visible-57d6r6qzjb
106. Jyoti K, Baunthiyal M, Singh A (2016) Characterization of silver nanoparticles synthesized using *Urtica dioica* Linn. leaves and their synergistic effects with antibiotics. J Radiat Res Appl Sci 9(3):217–227. https://www.sciencedirect.com/science/article/pii/S1687850715001132?via%3Dihub

107. Muhammad W, Ullah N, Haroon M, Abbasi BH (2019) Optical, morphological and biological analysis of zinc oxide nanoparticles (ZnO NPs) using *Papaver somniferum* L. RSC Adv 9(51):29541–29548. https://pubs.rsc.org/en/content/articlehtml/2019/ra/c9ra04424h
108. Gowtham Sanjai S, Thejaraju R, Pradhann S, Kumar A, Abdul Zubair S, Kathuria R et al (2023) Synthesis of titanium dioxide (TiO_2) nanoparticles by sol-gel method and its characterization. Mater Sci Forum 1109:3–9
109. Velásquez AA, Urquijo JP (2021) Synthesis and characterization of magnetite-maghemite nanoparticles in presence of polyethylene glycol obtained by mechanical milling. Mater Sci Eng B 263
110. Bustos ARM, Encinar JR, Sanz-Medel A (2013) Mass spectrometry for the characterisation of nanoparticles. Anal Bioanal Chem 405(17):5637–5643. https://doi.org/10.1007/s00216-013-7014-y
111. Comby-Zerbino C, Dagany X, Chirot F, Dugourd P, Antoine R, Antoine R (2021) The emergence of mass spectrometry for characterizing nanomaterials. Atomically precise nanoclusters and beyond. Mater Adv 2(15):4896–4913. https://pubs.rsc.org/en/content/articlehtml/2021/ma/d1ma00261a
112. Costa-Fernández JM, Menéndez-Miranda M, Bouzas-Ramos D, Encinar JR, Sanz-Medel A (2016) Mass spectrometry for the characterization and quantification of engineered inorganic nanoparticles. TrAC Trends Anal Chem 84:139–148. https://www.sciencedirect.com/science/article/abs/pii/S0165993616300759
113. Petersen EJ, Henry TB, Zhao J, MacCuspie RI, Kirschling TL, Dobrovolskaia MA et al (2014) Identification and avoidance of potential artifacts and misinterpretations in nanomaterial ecotoxicity measurements. Environ Sci Technol 48(8):4226–4246. https://doi.org/10.1021/es4052999
114. McQueen A, Kimble A, Krupa P, Longwell A, Calomeni-Eck A, Moore D (2025) Review of emerging and nonconventional analytical techniques for per- and polyfluoroalkyl substances (PFAS): application for risk assessment. Water 2025 17(3):303. https://www.mdpi.com/2073-4441/17/3/303/htm
115. Firoozi AA, Firoozi AA, Maghami MR (2025) Transformative impacts of nanotechnology on sustainable construction: a comprehensive review. Results Eng 26:104973. https://www.sciencedirect.com/science/article/pii/S2590123025010497
116. Serhan M, Sprowls M, Jackemeyer D, Long M, Perez ID, Maret W et al (2019) Total iron measurement in human serum with a smartphone. In: AIChE annual meeting, conference proceedings, Nov 2019
117. De M, Ghosh PS, Rotello VM (2008) Applications of nanoparticles in biology. Adv Mater 20(22):4225–4241
118. Musielak E, Feliczak-Guzik A (2022) Synthesis and potential applications of lipid nanoparticles in medicine
119. Selvamuthukumar S, Velmurugan R (2012) Nanostructured lipid carriers: a potential drug carrier for cancer chemotherapy, pp 1–8
120. Salata OV (2004) J Nanobiotechnol 6(3):1–6
121. Palaniappan K (2017) An overview of applications of nanotechnology in biofuel production. World Appl Sci J 35(8):1305–1311
122. Zhao D, Jiaqiang E, Pham M, Zhao D, Deng Y, Le D et al (2021) Effect of different technologies on combustion and emissions of the diesel engine fueled with biodiesel: a review. Renew Sustain Energy Rev 80:620–647
123. Sekoai PT, Ouma CNM, du Preez SP, Modisha P, Engelbrecht N, Bessarabov DG et al (2018) Application of nanoparticles in biofuels: an overview. Fuel 237:380–397
124. Negi S, Batoye S, Singh K (2021) New frontiers of nanomaterials in environmental science
125. Negi R, Yadav N, Yadav AN (2025) Microbial biofertilizers: a paradigm shift towards agricultural sustainability. Biologia 80(2):389–414. https://www.researchgate.net/publication/387808681_Microbial_biofertilizers_A_paradigm_shift_towards_agricultural_sustainability
126. Pestovsky YS, Martínez-Antonio A (2017) The use of nanoparticles and nanoformulations in agriculture. J Nanosci Nanotechnol 17(12):8699–8730

127. Kristanti RA, Mei R, Liong Y, Hadibarata T (2021) Soil remediation applications of nanotechnology. Trop Aquat Soil Pollut 1(1):35–45
128. Shipley HJ, Engates KE, Guettner AM (2011) Study of iron oxide nanoparticles in soil for remediation of arsenic, pp 2387–2397
129. New WX, Ogbezode JE, Gani P (2023) Nanoparticles in soil remediation: challenges and opportunities. Ind Domest Waste Manag 3(2):127–140
130. Wang Y, Connor DO, Shen Z, Lo IMC, Tsang DCW (2019) Green synthesis of nanoparticles for the remediation of contaminated waters and soils: constituents, synthesizing methods, and influencing factors. J Clean Prod
131. Araújo R, Meira AC, Fiúza A (2015) The use of nanoparticles in soil and water remediation processes. Mater Today Proc 2(1):315–320. https://doi.org/10.1016/j.matpr.2015.04.055
132. Altammar KA (2023) A review on nanoparticles: characteristics, synthesis, applications, and challenges. Front Microbiol 14:1155622. https://pmc.ncbi.nlm.nih.gov/articles/PMC10168541/
133. Jain S, Kumar A, Dhewa T, Panwar S, Majumdar RS (2018) Applications of nanotechnology in agriculture—a review: concepts, aspects, prospects, and constraints. In: Sustainable biological systems for agriculture, pp 3–20
134. Periakaruppan R, Romanovski V, Thirumalaisamy SK, Palanimuthu V, Sampath MP, Anilkumar A et al (2023) Innovations in modern nanotechnology for the sustainable production of agriculture. ChemEngineering 7(4)
135. Singh BV, Girase IP, Sharma M, Vidhya CS, Tiwari AK, Baral K et al (2024) Nanoparticle-enhanced approaches for sustainable agriculture and innovations in food science. Int J Environ Clim Change 14(1):293–313
136. Mekuye B, Abera B (2023) Nanomaterials: an overview of synthesis, classification, characterization, and applications. Nano Select 4(8):486–501. https://doi.org/10.1002/nano.202300038
137. Khan Y, Sadia H, Ali Shah SZ, Khan MN, Shah AA, Ullah N et al (2022) Classification, synthetic, and characterization approaches to nanoparticles, and their applications in various fields of nanotechnology: a review. Catalysts 12(11)
138. Bishnoi A, Patel P, Ambegaonkar N (2019) Synthesis and applications of nanomaterials in various industries. J Emerg Technol Innov Res 6(4):757–770
139. Eker F, Duman H, Akdaşçi E, Bolat E, Sarıtaş S, Karav S et al (2024) A comprehensive review of nanoparticles: from classification to application and toxicity. Molecules 29(15):3482. https://www.mdpi.com/1420-3049/29/15/3482/htm
140. Mridha D, Lamsal B, Antonangelo JA (2025) Nanotechnology in agriculture: innovations for sustainability and greenhouse gas mitigation—a review. Sci Total Environ 995:180065. https://www.sciencedirect.com/science/article/pii/S004896972501705X
141. Sead FF, Jain V, Roopashree R, Devi A, Kashyap A, Sharma GC et al (2025) Engineering TiO_2-based nanostructures for enhanced electrocatalytic and photocatalytic redox reactions. Results Chem 17:102544. https://www.sciencedirect.com/science/article/pii/S2211715625005272

Intelligent Optimization of PI-Controlled Grid-Connected Inverters for Enhanced Power Quality

Tiku Fidelis Etanya, Pierre Tsafack, and Divine Khan Ngwashi

Abstract

The rapid integration of distributed renewable generators, such as solar PV and wind turbines, into contemporary power networks presents notable challenges in maintaining power quality. These include voltage and frequency fluctuations arising from the variable nature of renewable resources, as well as harmonics introduced by inverter-based interfaces. Effective mitigation of these issues demands an advanced control strategy that improves the reliability and operational performance of the grid. In this work, a proportional-integral current controller is developed and fine-tuned using Particle Swarm Optimization (PSO) and Genetic Algorithm (GA), with conventional Ziegler-Nichols tuning serving as a benchmark. The effectiveness of the proposed control approach is quantified through the Integral of Time-weighted Absolute Error (ITAE) criterion. Simulation results show that PSO achieved ITAE values of 0.1605 (direct-axis component) and 0.1491 (quadrature-axis component), representing reductions of 90.91% and 91.02% compared to Z-N values of 1.7658 and 1.6595, and further reductions of 21.82% and 1.84% compared to GA values of 0.2053 and 0.1519, respectively, confirming its superior control accuracy and significant performance advantage. In terms of power quality, the total voltage harmonic distortion (THD_V) remains at 0.0% due to effective filtering, while the total current harmonic distortion (THD_i) is minimized to 0.17% with PSO, compared to 0.22% for GA and 1.98% for Z-N. The optimized PI control strategy ensures

T. F. Etanya (✉) · P. Tsafack · D. K. Ngwashi
Department of Electrical and Electronic Engineering, Faculty of Engineering and Technology, University of Buea, Buea, Cameroon
e-mail: tiku.fidelis@ubuea.cm

D. K. Ngwashi
e-mail: ngwashi.divine@ubuea.cm

R. Ramkat et al. (eds.), *Towards the Sustainability of our Ecosystems*, Springer Proceedings in Geography and Sustainability,
https://doi.org/10.1007/978-3-032-30076-8_16

that THD values comply with IEEE 519 - 2014 and IEC 61000 - 3 - 6 Standards demonstrating its effectiveness in mitigating power fluctuations, reducing harmonics, and improving grid stability. These findings emphasize the critical role of intelligent optimization in enhancing power quality for grid-connected renewable energy systems.

Keywords

Power quality • Grid-connected inverter • PI control strategy • Total harmonic distortion • Intelligent optimization

1 Introduction

The growing deployment of renewable energy sources, especially solar photovoltaic and wind systems, within modern power networks has brought about notable power quality (PQ) issues, as illustrated in Fig. 1 [1]. These sources, while environmentally beneficial, are inherently intermittent and require power electronic interface devices such as grid-connected inverters for efficient grid integration [2]. However, these inverters contribute to power fluctuations and increased total harmonic distortion, posing a threat to grid stability and system reliability. Ensuring compliance with international PQ standards such as IEEE 519–2014 and IEC 61000–3-6 requires the implementation of effective inverter control strategies [3, 4]. PI controllers using traditional tuning methods are commonly employed due to their structural simplicity and effective steady-state performance [5]. However, these methods are limited by slow transient response and poor adaptability under dynamic conditions, revealing a gap in achieving optimal controller performance [6]. To address this challenge, this study investigates the application of intelligent optimization algorithms, specifically GA and PSO, for tuning PI controller gains in a grid-connected inverter [7]. The primary objective of this work is to enhance current regulation while minimizing harmonic distortion, utilizing ITAE. By addressing the limitations of traditional tuning methods, such as Ziegler-Nichols (Z-N) and trial-and-error approaches, which often result in poor dynamic response, high overshoot, and inadequate mitigation of harmonics and power fluctuations, this work contributes to the development of more robust and adaptive control strategies that support the reliable and scalable integration of renewable energy into modern power grids.

2 Materials and Methods

This work improves the quality of power delivered by grid-connected inverters through an intelligently tuned PI-based current regulation approach. As shown in Fig. 2, the system models a renewable energy source as an infinite bus for a stable voltage reference. PI controllers regulate D- and Q-axis currents, with gains

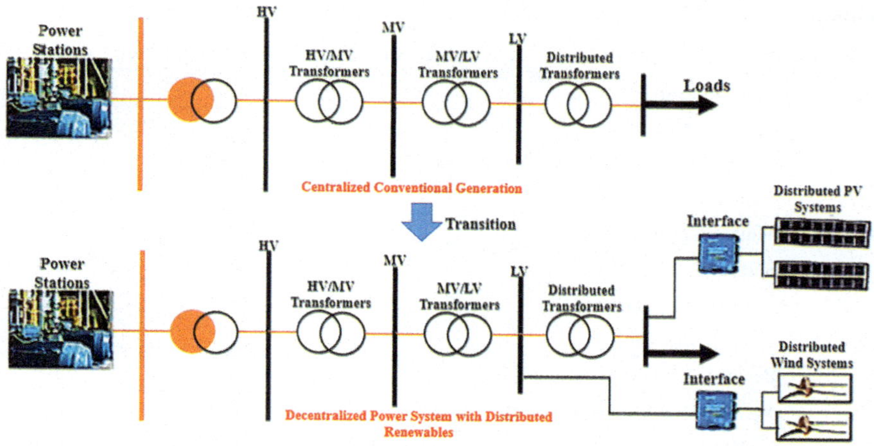

Fig. 1 Depiction of power system evolution [3]

tuned via ZN, GA, and PSO, and evaluated using the ITAE metric. Simulated in MATLAB/Simulink, the inverter interfaces the source with the grid and a three-phase load, operating under a current control scheme driven by active and reactive power references. Table 1 presents the tuning parameters and optimization settings for GA and PSO used in the PI controller gain optimization process.

Table 2 depicts system design specifications.

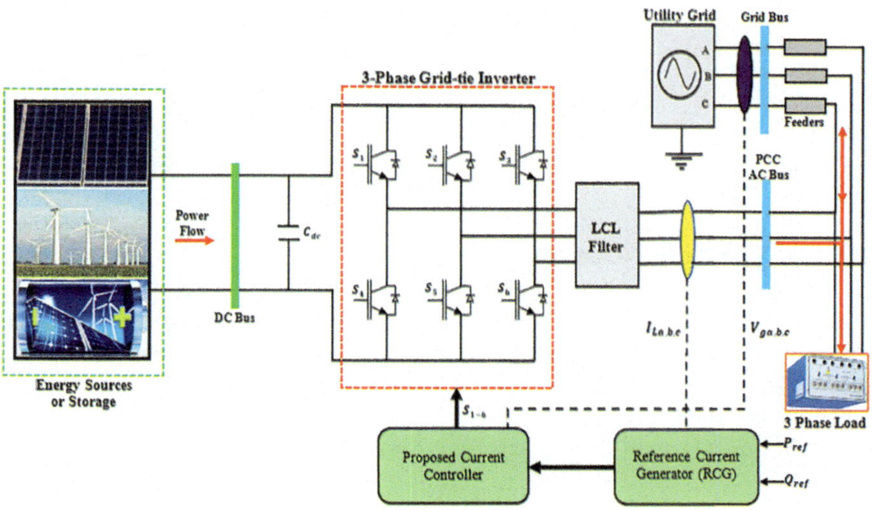

Fig. 2 Overall system design (own contribution)

Table 1 GA and PSO configuration for optimal PI gain selection (own contribution)

Description		Symbol	Value/method
GA	Population size	n	100
	Crossover operator	–	Uniform crossover
	Crossover probability	pC	1
	Gain	PI	4
	Mutation probability	μ	0.02
	Selection method	–	Roulette wheel selection
	Number of iterations	k	50
PSO	Inertia coefficient	w	1
	Personal acceleration coefficient	c_1	2
	Social acceleration coefficient	c_2	2
	Controller gains	k_{pd}, kid and kpq k_{iq}	4
	Search space	–	1–10,000

Note that, n and k were the same for both GA and PSO

Table 2 System design specifications (own contribution)

Nominal title		Design values
DC-link voltage	V_{DC}	800 V
Three-phase inverter voltage	$V_{3\emptyset LL}$	400 V
DG capacity	S_G	100 kVA
Switching frequency	f_{sw}	10 kHz
Grid frequency	f_G	50 Hz
Modulation index	M_I	0.817
DG	L_i	0.5093 mH
	L_g	0.5093 mH
	C_f	99.472 μF
	R_d	0.5333 Ω

2.1 Controller Design

The PI controller design begins with the conventional Z-N trial-and-error method for initial tuning. To improve performance, a cost function is formulated as a minimization problem for GA and PSO to optimize the gains [8]. The objective functions are defined separately for both the direct-axis and the quadrature-axis as shown in Eqs. (1) and (2), respectively.

$$\beta_1 = \int_{t=0}^{t=t_{\mathrm{sim}}} \left|I'_D - I_{D\mathrm{cal}}\right|.t\mathrm{d}t, \tag{1}$$

$$\beta_2 = \int_{t=0}^{t=t_{\mathrm{sim}}} \left|I'_Q - I_{Q\mathrm{cal}}\right|.t\mathrm{d}t, \tag{2}$$

where I'_D and $I_{D\text{cal}}$ represent the reference and computed currents along the direct-axis, while I'_Q and $I_{Q\text{cal}}$, correspond to the reference, and calculated currents along the quadrature-axis. The overall cost function is then formulated as a multi-objective function, as shown in Eq. (3).

$$\beta = \text{Min}[W_1\beta_1 + W_2\beta_2], \tag{3}$$

where W_1 and W_2 are weighting factors for the direct-axis and quadrature-axis errors, respectively. The optimization process is subjected to predefined constraints that ensure feasible PI gain selection as shown in Eqs. (4)–(7).

$$K_{Pd}{}^{CC_{\text{Min}}} \leq K_{Pd}{}^{CC} \leq K_{Pd}{}^{CC_{\text{Max}}}, \tag{4}$$

$$K_{id}{}^{CC_{\text{Min}}} \leq K_{id}{}^{CC} \leq K_{id}{}^{CC_{\text{Max}}}, \tag{5}$$

$$K_{Pq}{}^{CC_{\text{Min}}} \leq K_{Pq}{}^{CC} \leq K_{Pq}{}^{CC_{\text{Max}}}, \tag{6}$$

$$K_{iq}{}^{CC_{\text{Min}}} \leq K_{iq}{}^{CC} \leq K_{iq}{}^{CC_{\text{Max}}}, \tag{7}$$

where $K_{Pd}{}^{CC}$, $K_{id}{}^{CC}$, $K_{pq}{}^{CC}$ and, $K_{iq}{}^{CC}$ are the proposed intelligent controller gains. The optimization problem expressed in Eq. (3) is solved subject to the feasibility constraints outlined in Eqs. (4)–(7). In Fig. 2, the reference current generator (RCG) computes DQ current references from the active P_{ref} and reactive Q_{ref} power setpoints using V_D and V_Q obtained from V_{abc}, as expressed in Eqs. (8) and (9).

$$I_{oD\text{ref}} = \frac{V_D P_{\text{ref}} + V_Q Q_{\text{ref}}}{V_D{}^2 + V_Q{}^2}, \tag{8}$$

$$I_{oq\text{ref}} = \frac{V_Q P_{\text{ref}} - V_D Q_{\text{ref}}}{V_D{}^2 + V_Q{}^2}. \tag{9}$$

The proposed current controller, shown in Fig. 3, utilizes two proportional-integral controllers integrated with feed-forward compensation and cross-coupling decoupling to achieve precise regulation. Normalization is achieved by dividing V_{ref} by the DC-link voltage [9].

A closed-loop Synchronous Reference Frame Phase-Locked Loop (SRF-PLL) is employed to extract the grid voltage phase angle (θ), enabling precise synchronization and ensuring the inverter output remains frequency and phase aligned with the grid [10].

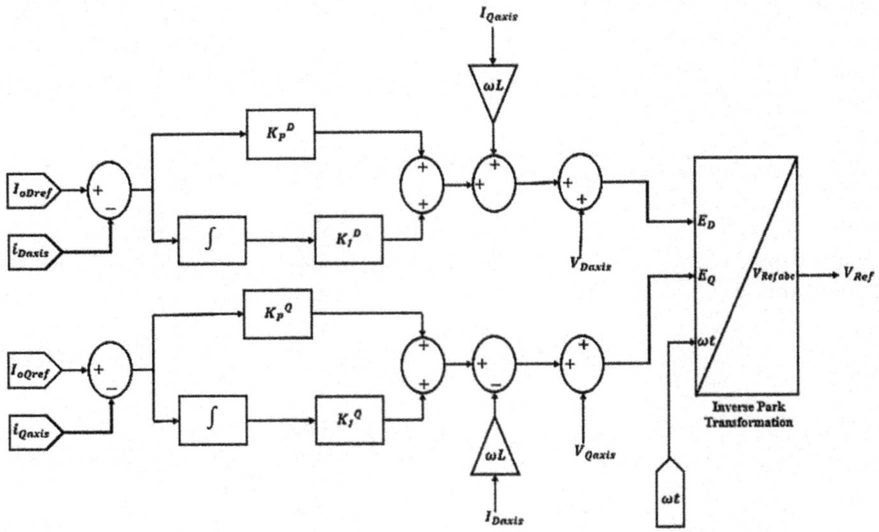

Fig. 3 Proposed current controller design (own contribution)

3 Results and Discussions

The proposed controller was tested in MATLAB Simulink on a 100 kVA grid-connected inverter, delivering balanced power to the grid once the load demand was satisfied. Initial controller gains were set using the traditional Z-N method and subsequently refined through GA and PSO optimization. Table 3 presents the tuning results, with dynamic stability assessed under severe disturbances.

3.1 Control Performance Criterion

Table 4 confirms the superior performance of intelligent optimization over Ziegler-Nichols (ZN) tuning. GA and PSO achieved much lower ITAE values for both direct-axis and quadrature-axis controllers, with PSO slightly outperforming GA. In contrast, ZN showed significantly higher errors, indicating slower response

Table 3 Controller gains tuned using Z-N, GA, and PSO optimization methods

Controller	Parameter	Z-N gains	GA gains	PSO gains
D-axis	K_{PD}	2.25	2598.52259	1825.732
	K_{ID}	2692.46	441.070005	5457.313
Q-axis	K_{PQ}	3.6	1994.121442	4267.707
	K_{IQ}	1313.07	2203.872723	6108.578

Table 4 Comparative ITAE performance using Z-N, GA, and PSO

Z-N results		
Direct-axis controller	ITAE	1.76570
Quadrature-axis controller	ITAE	1.65950
GA results		
Direct-axis controller	ITAE	0.20530
Quadrature-axis controller	ITAE	0.15190
PSO results		
Direct-axis controller	ITAE	0.16050
Quadrature-axis controller	ITAE	0.14910

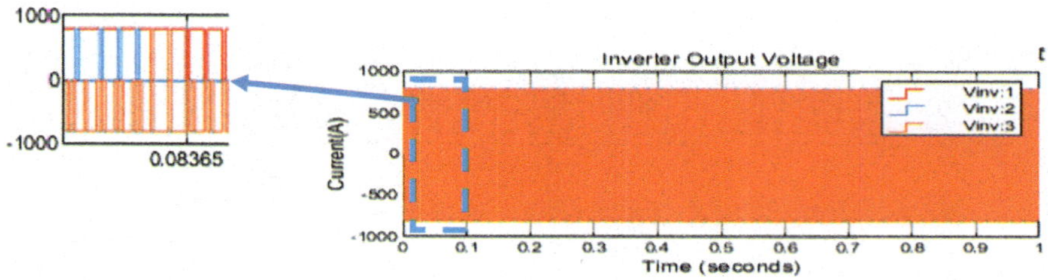

Fig. 4 Three-phase inverter output voltage

and less accurate tracking. Intelligent methods offer more precise and responsive control.

Fast Fourier Transform (FFT) analysis assessed power quality, focusing on harmonics. Figures 4 and 5 show the inverter output voltage and its filtered output for load voltage, current, grid voltage, and PCC current, all approximating a pure sine wave. The analysis reveals that THD_v is 0.00% for Z-N, GA, and PSO, as the filter design primarily impacted voltage distortion. THD_i was lowest for PSO at 0.17%, outperforming GA (0.22%) and Z-N (1.98%), as shown in Figs. 6, 7, and 8 for THD_v and THD_i respectively. These results comply with IEEE 519–2014 and IEC 61000-3-6 standards. Filter design improved TVHD, while PSO tuning proved superior in reducing TCHD.

3.2 Power Injection Response

The dynamic performance of the proposed PSO-based three-phase inverter was evaluated under variable renewable generation and a constant 50 kW resistive load. Among the tested control strategies (Z-N, GA, and PSO), PSO exhibited the best tracking performance as evidenced by the lowest ITAE values (Table 4). Consequently, only the results obtained using PSO are presented. With reactive power $Q = 0$, the inverter functioned as a grid-feeding unit, supplying only active power to the strong grid, as is typical in PV grid-tied applications. As shown in Fig. 9,

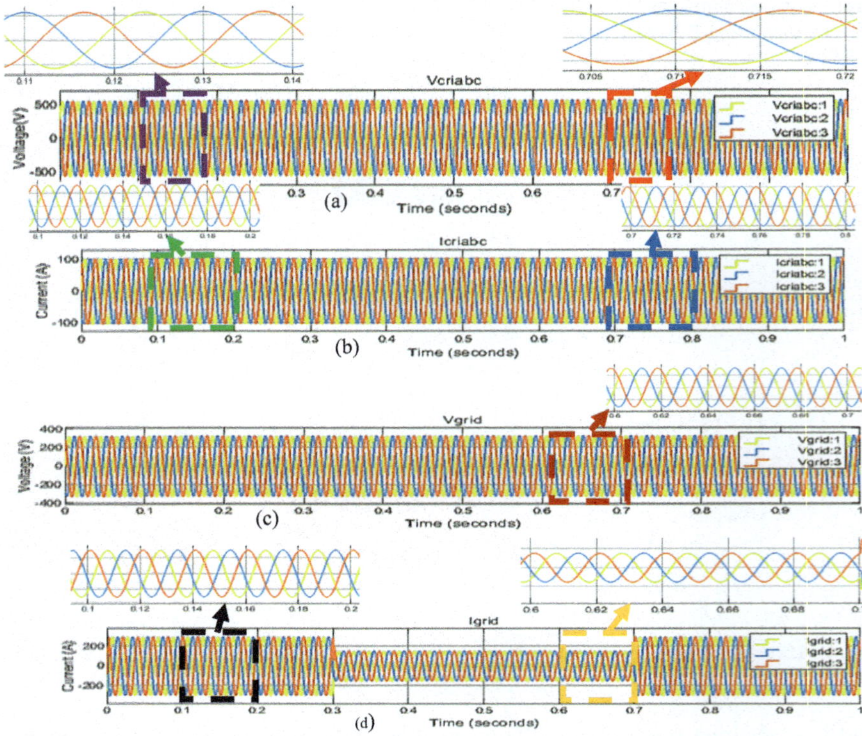

Fig. 5 PCC voltage and current waveforms: **a** load demand voltage, **b** load demand current, **c** utility grid voltage, **d** utility grid current

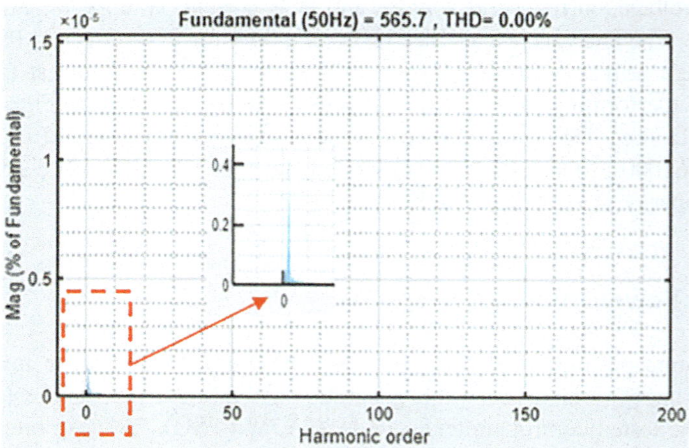

Fig. 6 Overall harmonic distortion of voltage at PCC

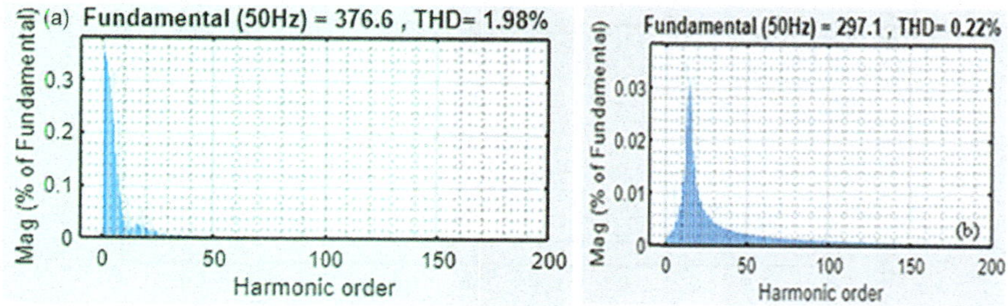

Fig. 7 Overall harmonic distortion of current at PCC: **a** applying Z-N; **b** applying GA

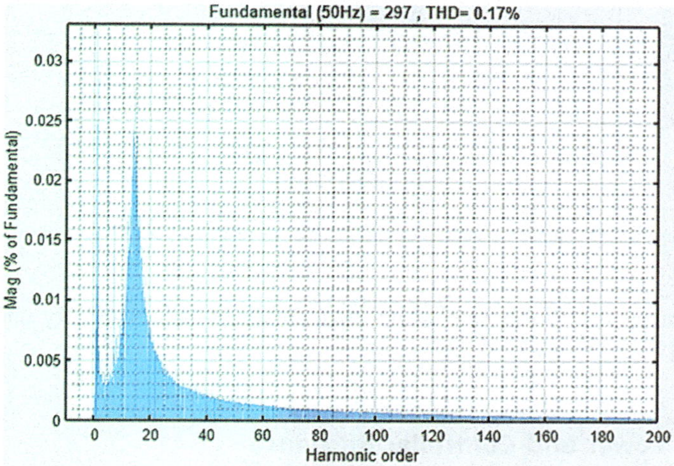

Fig. 8 Overall harmonic distortion of current at PCC using PSO optimization

the system injected a 50 kW surplus into the grid during the periods 0 – 0.2 s and 0.7 – 1 s when the generation was 100 kW. These results demonstrate the PSO-optimized controller's superior capability in sustaining active power balance and upholding system stability under dynamic operating conditions [11].

After evaluating inverter performance under a resistive load, further tests involved a PQ load to assess dynamic behavior under both active (P) and reactive (Q) power demand. As shown in Fig. 10, the system responded to load disturbances with accurate power tracking. From 0 to 0.4 s, the generator produced 100 kW and 100 kVar, meeting an 80 kW and 80 kVar load while injecting a 20 kW and 20 kVar surplus into the grid. At 0.4 to 0.8 s, the generation and load were balanced at 50 kW and 50 kVar, with no injection. Between 0.8 to 1.2 s, the 100 kW and 100 kVar load exceeded generation, drawing −50 kW and −50 kVar from the grid. Finally, from 1.2 to 2 s, a 50 kW and a 50 kVar surplus was injected. These results confirm the inverter's grid-supporting capability, delivering

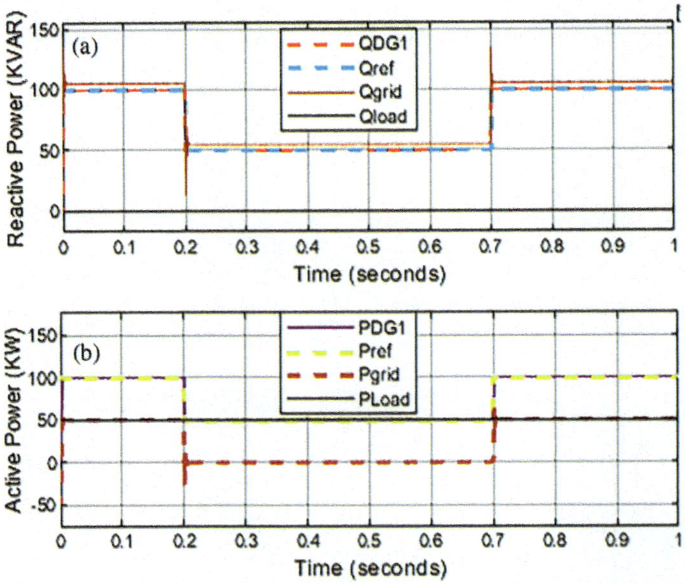

Fig. 9 Constant active power injection: **a** reactive power response; **b** active power response

both *P* and *Q* to a weak grid while maintaining dynamic stability and tracking accuracy [12].

3.3 Power and Controller Response

Figure 11 illustrates the system's behavior in response to a step change in both active and reactive power output. The power response in Fig. 11a and b demonstrates low tracking error, minimal overshoot, and fast settling, confirming the controller's robustness under dynamic generation conditions. The DQ-axis current controller, optimized using the PSO algorithm, exhibited enhanced transient and steady-state performance. As depicted in Fig. 11c and d, the controller achieved a settling time of 0.00352 s, with steady-state errors of 0.5 (direct-axis) and 0.4 (quadrature-axis), ensuring precise current tracking. Furthermore, the ITAE values of 0.1605 and 0.1491 for the direct and quadrature axes, respectively, highlight reduced cumulative error. These findings demonstrate the PSO-optimized controller's capability to enhance control precision and dynamic performance, establishing it as an effective approach for real-time management of inverter currents in grid-connected distributed energy systems.

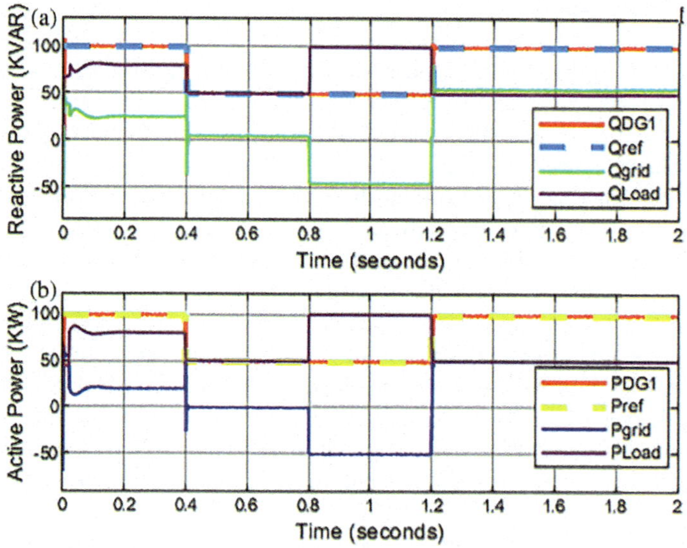

Fig. 10 Variable power injection: **a** reactive power output; **b** active power output

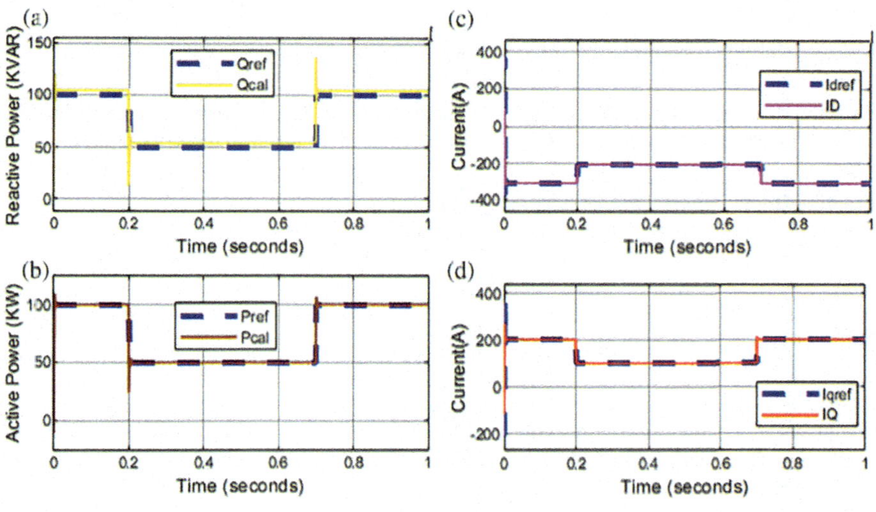

Fig. 11 Step change active/reactive power disturbances: **a** reactive power response; **b** active power response; **c** direct-axis controller response; **d** quadrature-axis controller response

4 Conclusion

This work developed an optimized PI-based current control approach for a grid-connected inverter to improve power quality in renewable energy distribution systems. Using PSO and GA, with Z-N tuning as a benchmark, the controller achieved superior tracking accuracy, reduced ITAE values, and minimized harmonic distortion. PSO outperformed the other methods under both resistive and PQ load conditions, maintaining dynamic power balance, injecting surplus active and reactive power, and effectively supporting both strong and weak grid conditions. These results confirm the robustness and effectiveness of the proposed strategy in mitigating power fluctuations and harmonics, in compliance with IEEE and IEC standards, which is critical since distributed energy generator sources, such as solar PV and wind energy, are already widely used in Sub-Saharan Africa but often deliver unstable and poor-quality power. Future work will explore the implementation of a hybrid GA-PSO-optimized PI controller to further improve performance and support real-time deployment in three-phase inverter systems.

Danksagung **Authors Contribution** Tiku Fidelis Etanya: Conceptualization, Methodology, Software, Formal analysis, Visualization, Investigation, Analysis, Validation, original draft writing, Writing—review and editing. Pierre Tsafack: Supervision, Writing—review, and editing. Divine Khan Ngwashi, Supervision, Writing—review and editing.

Data Availability Statement No data was used for the research described in the article.

Conflict of Interest Statement The authors declare that they have no known competing financial interests or personal relationships that could have appeared to influence the work reported in this paper.

Funding Information This research received no external funding.

References

1. Razmi D, Lu T, Papari B, Akbari E, Fathi G, Ghadamyari M (2023) An overview on power quality issues and control strategies for distribution networks with the presence of distributed generation resources. IEEE Access 11:10308–10325. https://doi.org/10.1109/ACCESS.2023.3238685
2. Kumar SS et al (2024) Advanced power electronics for efficient integration of renewable energy into the grid. In: E3S Web of conferences. EDP Sciences, p 5005. https://doi.org/10.1051/e3sconf/202459105005
3. Etanya TF, Tsafack P, Ngwashi DK (2025) Grid-connected distributed renewable energy generation system: power quality issues, and mitigation techniques—a review. Energy Rep 13(2):3181–3203. https://doi.org/10.1016/j.egyr.2025.02.050
4. Nkambule MS, Hasan AN, Shongwe T (2024) Advanced control strategies for photovoltaic power quality and maximum power point tracking optimization. IEEE Access 12:74456–74481. https://doi.org/10.1109/ACCESS.2024.3404497

5. Yessef M et al (2024) Experimental validation of feedback PI controllers for multi-rotor wind energy conversion systems. IEEE Access 12:7071–7088. https://doi.org/10.1109/ACCESS.2024.3351355
6. Rahli C et al (2024) PV systems operating in dynamic climatic circumstances using a PSO-based SMC and PID controller. In: 2024 2nd international conference on electrical engineering and automatic control (ICEEAC). IEEE, pp 1–4. https://doi.org/10.1109/ICEEAC61226.2024.10576441
7. Boutaghane K, Bennecib N, Benidir M, Benbouhenni H, Colak I (2024) Performance enhancement of a three-phase grid-connected PV inverter system using fractional-order integral sliding mode controls. Energy Rep 11:3976–3994. https://doi.org/10.1016/j.egyr.2024.03.049
8. Rodríguez-Abreo O, Rodríguez-Reséndiz J, García-Cerezo A, García-Martínez JR (2024) Fuzzy logic controller for UAV with gains optimized via genetic algorithm. Heliyon 10(4). https://doi.org/10.1016/j.heliyon.2024.e26363
9. Kumar P, Arya SR, Kumar S (2024) Adaptive fuzzy integrated LMS-based control strategy for voltage power quality enhancement. E-Prime-Adv Electr Eng Electron Energy 9:100635. https://doi.org/10.1016/j.prime.2024.100635
10. Priyanka B, Lavanya MC (2024) Phase-locked loop (PLL) techniques for grid synchronization: a comprehensive review. In: 2024 second international conference on emerging trends in information technology and engineering (ICETITE). IEEE, pp 1–9. https://doi.org/10.1109/ic-ETITE58242.2024.10493448
11. Sreenu C, Mallesham G, Shekar TC, Salkuti SR (2024) Pairing voltage-source converters with PI tuning controller based on PSO for grid-connected wind-solar cogeneration. Franklin Open 8:100138. https://doi.org/10.1016/j.fraope.2024.100138
12. Radwan A, Elshenawy MA, Mohammed YA-RI, El-Saadany EF (2024) Grid-forming voltage-source converter for hybrid wind-solar systems interfacing weak grids. IEEE Open J Power Electron 5:956–975. https://doi.org/10.1109/OJPEL.2024.3410908

Assessment of Nutrient Retention in Yala Wetland Ecosystem, Kenya

Kelvin Kariuki Muli, Nzula Kitaka, Julius Kipkemboi, and Risper Ajwang' Ondiek

Abstract

The Yala Wetland is a significant wetland in Kenya that offers ecosystem services and, as a result, a source of revenue for the local population. This study investigates the nutrient retention capacity at the Yala wetland by using a black box approach and focusing on the selected parameters at the inlet and outflow over a given temporal range that captures high and low flow in the River Yala. To assess nutrient retention and temporal changes, concentrations of dissolved oxygen (DO), electrical conductivity (EC), water temperature, pH, total suspended solids (TSS), nitrogen (N), and phosphorus (P) were selected for monitoring. There was no significant change in the mean DO concentration at the outlet ($F = 5.657$, $p = 0.063$), but there was a significant temporal variation at the inlet ($F = 7.315$, $p = 0.051$). Significant variation was observed in the EC values at both sites ($F = 28.387$, $p = 0.004$), with the inlet peaking at 341.1 µS/cm in March. Ammonium concentrations at the inlet ($F = 8.257$, $p = 0.035$) and outlet ($F = 7.463$, $p = 0.028$) varied significantly over months, according to nutrient analyses. Over the course of the study, there was little variation in the levels of nitrate and total nitrogen. Remarkably, the marsh retained 28.57% total phosphorus, 30.52% nitrate, and 40% ammonium, acting as a nutrient sink. Maintaining the lake's buffer zone in terms of water quality and fostering biodiversity require an understanding of these connections. Understanding the nutrient retention dynamics and water quality indicators of the Yala Wetland is essential to its continued function as a natural buffer that filters pollutants before they reach downstream ecosystems like Lake Victoria.

K. K. Muli (✉) · N. Kitaka · J. Kipkemboi · R. A. Ondiek
Department of Biological Sciences, Egerton University, Egerton, Njoro, Kenya

K. K. Muli
Department of Climate Change, Pan African Network for Climate Action, Nairobi, Kenya

R. Ramkat et al. (eds.), *Towards the Sustainability of our Ecosystems*, Springer Proceedings in Geography and Sustainability,
https://doi.org/10.1007/978-3-032-30076-8_17

The sustainability of ecosystem services and local livelihoods depend on biodiversity conservation and water quality maintenance, both of which are aided by this information. In conclusion, to preserve Yala Wetland's integrity and strengthen its resistance to impending environmental threats, ongoing research and proactive management techniques are required.

Keywords

Buffering capacity • Ecological monitoring • Water quality • Nitrogen retention • Phosphorus retention

1 Introduction

Wetlands are unique and vital habitats that are characterized by the seasonal or continuous presence of water [1]. They are essential to preserving biodiversity and ecological balance. Among the ecological benefits, they offer are carbon sequestration, flood control, water purification, and habitat for a range of plants and animals [2]. Among these wetlands, Yala Wetland in western Kenya is particularly significant both ecologically and economically. The Yala River provides water for the approximately 17,500-hectare Yala Wetland, which is home to a variety of aquatic and terrestrial species and supports communities through agriculture, tourism, and fishing [3]. Management and conservation of Yala Wetland depend on an understanding of its ecological processes, especially its ability to buffer nutrients.

The physico-chemical properties of wetland ecosystems and the health of aquatic life are greatly influenced by a number of factors, including pH, dissolved oxygen (DO), electrical conductivity (EC), total suspended solids (TSS), and temperature [4]. These characteristics may influence the biological processes taking place in wetland systems and function as indicators of environmental conditions [1]. For instance, changes in pH can impact the solubility and availability of nutrients, and dissolved oxygen levels are necessary for aquatic life to survive [5]. Furthermore, the growth of aquatic plants and microorganisms may be impacted by the concentration of dissolved salts, which can be ascertained by electrical conductivity [6].

Temperature, dissolved oxygen (DO), electrical conductivity (EC), pH, and total suspended solids (TSS) are some of the factors that have a major impact on the physico-chemical features of wetland ecosystems and the health of aquatic species [4]. These traits can influence the biological activities that take place in wetland systems and act as environmental indicators [1]. For example, changes in pH may impact the availability and solubility of nutrients, and aquatic life depends on dissolved oxygen levels to thrive [5]. Furthermore, the concentration of dissolved salts that may have an impact on the growth of aquatic plants and microorganisms can be determined using electrical conductivity [6].

Yala Wetland receives runoff from adjacent agricultural areas, which can encourage nutrient loading and eutrophication; it is particularly relevant to studies on nutrient retention [7]. The wetland is an essential site for maintaining the biodiversity and water quality of the area. Although nutrient retention mechanisms may improve water quality in outlet areas, runoff from agricultural lands frequently leads to higher nutrient concentrations in wetland inlet areas [8]. This work is significant because it uses a black-box approach to empirically demonstrate nutrient retention in the Yala Wetland and calculates precise retention rates for phosphorus, nitrate, and ammonium over time scales. Unlike previous general observations, it takes seasonal variations into account and offers a data-driven basis for wetland management and policy development. However, ongoing human pressures like agricultural expansion and land use changes pose a serious threat to this wetland's health [9]. Understanding the physico-chemical properties and nutrient dynamics at the inlet and outlet is necessary to develop strategies to lessen these impacts and increase Yala Wetland's resilience.

Wetland physico-chemical properties and nutrient retention have been the subject of numerous studies; however, quantitative information unique to Yala Wetland is scarce, particularly with regard to seasonal retention efficiency and temporal nutrient fluxes using a standardized approach [8]. This gap limits site-specific conservation planning and evidence-based wetland management. By comparing the physico-chemical properties and nutrient retention of the entrance and outlet, a comprehensive understanding of this wetland's functioning and the ecological processes at play can be gained. Plans for sustainable land and water management that benefit the local community and ecosystem can also be developed with the aid of these assessments [10].

In addition to being helpful in evaluating the ecosystem regulating services of Yala Wetland, this study will give knowledge that will influence policies geared at maintaining and developing wetland ecosystems in Kenya and beyond.

2 Materials and Methods

2.1 Site Description

Yala Wetland in the Kenyan counties of Siaya and Busia served as the study's site. Yala Wetland's coordinates are latitude 0° to 0° 45 north and longitude 33°55″ to 34°25 north (Fig. 1). With 17,500 hectares, it is the largest wetland ecosystem in the Lake Victoria basin. The wetland is located between 1140 and 1150 meters above sea level [11]. The climate of Yala Wetland is mostly tropical and humid, with average highs of 24 °C and highs of 16 to 28 °C. Between 600 and 2700 mm of precipitation fall on average each year [12].

The wetland was formed by the backflow of water from Lake Victoria and the flooding of the Yala and Nzoia rivers [13]. The Yala River serves as the wetland's main water source. The wetland contains four lakes: Lake Nyamboyo (1 ha), Lake

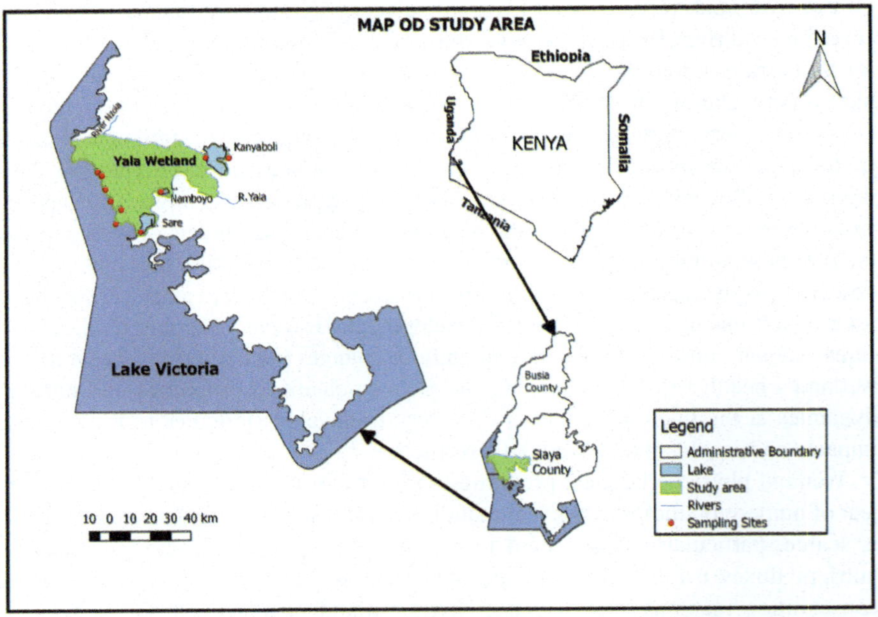

Fig. 1 Map of Yala wetland showing sampling sites. *Source* Author

Sare (500 ha), Lake Bob (70 ha), and Lake Kanyaboli (1000 ha) [14]. The predominant vegetation is papyrus. *Cyperus papyrus* is mostly found in permanently flooded areas, whereas Phragmites mauritianus is found in seasonally flooded areas along the perimeter [13]. The satellite lakes are home to endangered species such as the Sitatunga antelope (*Tragelaphus spekei*) and several rare fish species. Furthermore, fertilizers and pollutants from the nearby catchment are naturally filtered by the wetland. Before entering Lake Victoria, the river water also undergoes sedimentation, which removes silt and sediments from the water [15].

The biological functions and ecosystem services that wetlands provide have been under pressure from a range of anthropogenic activities in the wetland, surrounding area, and upstream area over the past 50 years. These elements include population expansion, changes in land use, and how these elements interact with lake and water level dynamics. As in many other regions of the world, these ecosystems have deteriorated due to a variety of stresses brought on by the ensuing pressures [16].

2.2 Methods

Sampling and Processing

As shown in Fig. 1, a number of physico-chemical parameters were measured in situ at the outlets adjacent to Lake Victoria and at the inlets upstream of the main river that flow into the floodplain wetland. These parameters included electrical conductivity (EC), pH, water temperature, and dissolved oxygen (DO). Measurements of physico-chemical parameters and the collection of water samples for nutrient assessment in a laboratory were accumulated over a five-month period, from November 2023 to March 2024. These tests were conducted using a calibrated HQ 40d (HACH) multimeter. During the sampling period, 500 ml plastic bottles that had been acid-washed were used to collect water samples in triplicate.

A variety of environmental and human-caused factors affecting Yala Wetland was identified during the five-month monitoring period, which spanned November 2023 to March 2024. The main seasonal weather patterns of the catchment area, including distinct wet and dry spells, occur during this time and directly affect the wetland's hydrological behavior and water quality. The sampling period includes both wet and dry seasons to adequately understand how changes in rainfall, runoff, and human activity during these months affect the physico-chemical characteristics of the wetland, such as dissolved oxygen (DO), pH, water temperature, and electrical conductivity (EC). This approach can be used to evaluate the wetland's ecological health, buffering capacity, and responsiveness to changing environmental conditions.

A number of nitrogen forms, such as ammonium-nitrogen (NH_4^--N), nitrate-nitrogen (NO_3^--N), nitrite-nitrogen (NO_2^--N), and total nitrogen (TN), were evaluated using established protocols [17]. The sodium-salicylate method was used to test NH_4^--N by mixing 2.5 ml of sodium-salicylate and hypochloride solutions with 25 ml of filtered wetland water. After 90 min of dark incubation, absorbance at 665 nm was measured using a GENESYS 30 10uv spectrophotometer. With the exception of samples that were dissolved in H_2SO_4 and then evaporated at 95 °C, the outcomes for NO_3^--N were the same. At 420 nm, the absorbance was measured. NO_2^--N was found by reacting with sulfanilamide and N-Naphthyl-(1) ethylenediamine-dihydrochloride, and the absorbance at 543 nm was noted.

Using persulfate digestion, which involves adding 1 milliliter of warm potassium persulfate to 25 milliliters of unfiltered water to change different forms of nitrogen into ammonium, total nitrogen (TN) was calculated. The samples were autoclaved at 120 °C and 1.2 atm for 90 min. After digestion, the reduced nitrogen forms were examined using the sodium-salicylate method. The final concentrations of NH_4^--N, NO_3^--N, NO_2^--N, and TN were calculated using the corresponding equations obtained from standard calibration curves [17].

Soluble reactive phosphorus (SRP) was assessed using the ascorbic acid technique [17]. Sulfuric acid (B), ascorbic acid (C), potassium antimonyl tartrate solution (D), and ammonium molybdate solution (A) were mixed in a 2:5:2:1 (ml) ratio. After applying this mixed solution at a 1:10 ratio to the filtered water sample for 15 min, the absorbance at 885 nm was measured using a GENESYS 30 10uv spectrophotometer. The concentration was determined using standard solutions. Total phosphorus (TP) was measured using persulfate digestion of unfiltered water, which transformed all forms of phosphorus into SRP. The SRP was then examined using the ascorbic acid method, with concentrations measured using standard solutions [17].

A known volume of water was filtered through pre-weighed Whatman GF/C filters with a pore size of 0.45 μm in order to measure the total suspended solids (TSS) gravimetrically. After that, the filter papers were dried for three hours at 95 °C until their weight remained constant. A modified version of the [17] method formula was used to determine the TSS's weight;

$$\mathrm{TSS} = \left(\left(W_c\text{--}W_f\right) \times 106\right)V^{-1}, \tag{1}$$

where TSS = Total suspended solids (mg L^{-1}), W_f = Weight of dried filter paper in grams, W_c = Constant weight of filter paper + residue in grams, and V = Volume of water filtered (ml).

Determination of nutrient retention

The nutrient retention of Yala wetland was determined by comparing the nutrient levels between the inlets of Yala wetland and the outlets across the wetland downstream.

$$\text{Nutrients retention} = \text{Nutrients levels at the outlets} - \text{Nutrients levels at the inlets} \tag{2}$$

Data Analysis

The data were managed using Microsoft Office Excel 2007, and the statistical analysis was conducted using IBM SPSS Statistics 21 (USA). The data were subjected to tests for homogeneity of variance and normality, each with a significance level of $p < 0.05$. Levene's test for homogeneity of variance and the Shapiro-Wilk test for normality both verified that the in situ measurement data had a normal distribution. Values were normalized, and the variables were log-transformed before parametric tests were performed. The Bartlett test was used to assess if the variance was homogeneous. A one-way ANOVA was used to determine whether the nutrient concentration at the inlets and sampling sites varied significantly over the transects. At each sampling location, the association between nutrient levels and changes in land use was ascertained using Pearson correlation. The F-test and adjusted R-squared value were used to assess each regression model's fit to the data. Less than 0.05 was considered a significant p-value.

3 Results

3.1 Variation of in Situ Physico-Chemical Parameters at the Inlets and Outlets of Yala Wetland

Table 1 shows the average water temperature, pH, dissolved oxygen (DO), and electrical conductivity (EC) values over the sampling months, as well as any notable differences between them as shown in Fig. 2. Over the course of the experiment, there was a significant change in the mean DO concentration at the Yala wetland's inlet (ANOVA, $F = 7.315$, d.f $= 7$, $p = 0.051$). However, Tukey's post hoc test revealed no significant differences between the individual sampling months ($p > 0.05$). On the other hand, the DO content at the wetland's exit did not significantly change over time (ANOVA, $F = 5.657$, d.f $= 7$, $p = 0.063$). At the exit, the average DO content ranged from 0.37 to 2.37 mg/l, while at the wetland inlet, it ranged from 4.56 to 6.56 mg/l.

During the sampling period, there was a significant variation in the electrical conductivity (EC) at the wetland inlet (ANOVA, $F = 28.387$, d.f $= 7$, $p = 0.004$). March saw the highest EC value of 341.1 µS/cm, while January saw the lowest value of 206.33 µS/cm. Tukey's post hoc analysis revealed a significant difference ($p > 0.05$) between the EC values from November and March. There was a

Table 1 Mean values and ranges of physico-chemical variables in Yala wetland, upstream and downstream of the wetland. ± represents standard error of the mean

Physico-chemical parameters	Mean values and range at inlets	Mean values and range at outlets
DO (mg/l)	5.46 ± 0.3	1.37 ± 0.3
	4.56–6.56	0.37–2.37
Temp°C	26.54 ± 1.1	23.43 ± 1.1
	25.44–27.64	22.33–24.53
pH	8.87 ± 1.4	7.9 ± 0.9
EC (µS/cm)	228.1 ± 4.6	126.01 ± 5.4
	223.5–232.7	120.61–131.41

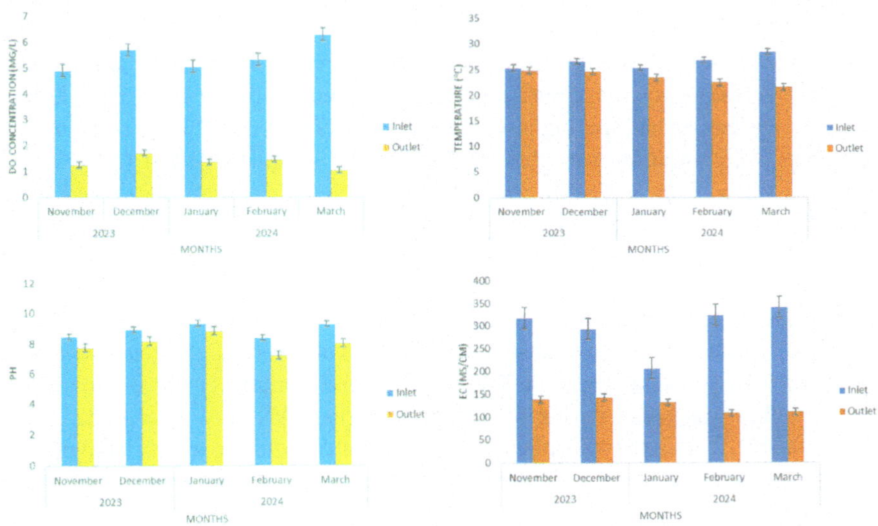

Fig. 2 Temporal variation of in situ measurements of DO, temperature, pH and EC at the inlet and outlet of Yala wetland

comparable pattern at the wetland outlet, where EC varied significantly between sampling months (ANOVA, $F = 27.876$, d.f $= 7$, $p = 0.004$).

The water temperature at the wetland inlet did not significantly change during the sampling period (ANOVA, $F = 0.275$, d.f $= 7$, $p = 0.851$). Additionally, there were no appreciable changes in the outflow's trend over time (ANOVA, $F = 0.234$, d.f $= 7$, $p = 0.872$). During the sampling period, the water temperatures at the intake and exit ranged from 25.44 °C to 27.64 °C and 22.33 °C to 24.53 °C, respectively.

The pH levels at the wetland inlet did not significantly change during the observation period (ANOVA, $F = 0.258$, d.f $= 7$, $p = 0.849$). Similarly, no appreciable changes in pH over time were observed at the outlet (ANOVA, $F = 0.229$, d.f $= 7$, $p = 0.859$). Throughout the sample period, the pH levels at the wetland's inflow and outlet ranged from 7.47 to 10.27 and 7.0 to 8.8, respectively.

3.2 Temporal Variation in TSS, N, and P Concentration in Yala Wetland

The results for the temporal change in nitrogen (N), phosphorus (P), and total suspended solids (TSS) at the entrance and exit of the wetland between November 2023 and March 2024 are shown in Fig. 3. The mean ammonium content at the wetland's inflow varied significantly over the sampling period (ANOVA, $F = 8.257$, d.f $= 7$, $p = 0.035$), with an increase observed between December and February. Tukey's post hoc analysis revealed that ammonium concentrations at the inlet and exit were significantly different in November, December, February, and March ($p < 0.05$), but comparable in January ($p > 0.05$). The ammonium content at the wetland outlet also varied significantly over the sampling period (ANOVA; $F = 7.463$, d.f $= 7$, $p = 0.028$).

The mean nitrate content displayed a clear trend in contrast to ammonium. January had the lowest concentration at the wetland's inflow (41.15 ± 3.00 mg/l), while December had the highest concentration (71.22 ± 6.26 mg/l). However, no appreciable variation was observed during the sampling period (ANOVA, $F = 0.941$, d.f $= 7$, $p = 0.478$). At the outlet, a similar pattern was noted, with December having the highest nitrate content (45 ± 2.54 mg/l) and February having the lowest (19.3 ± 2.26 mg/l). Again, no significant variance was observed over the whole sampling period (ANOVA, $F = 1.416$, d.f $= 7$, $p = 0.327$).

The mean concentration of total nitrogen (TN) followed a similar pattern to that of nitrate. At the wetland's inlet, TN levels were highest in December (84.03 ± 6.5 mg/l) and lowest in March (18.69 ± 2.9 mg/l). However, no significant variation was found during the sampling period (ANOVA, $F = 1.032$, d.f $= 7$, $p = 0.456$). The outflow displayed the same pattern with no appreciable change in TN concentration over the sampling period (ANOVA, $F = 1.263$, d.f $= 7$, $p = 0.398$).

The highest mean concentrations of soluble reactive phosphorus (SRP) were found in November and February at the inlet and in November, December, and

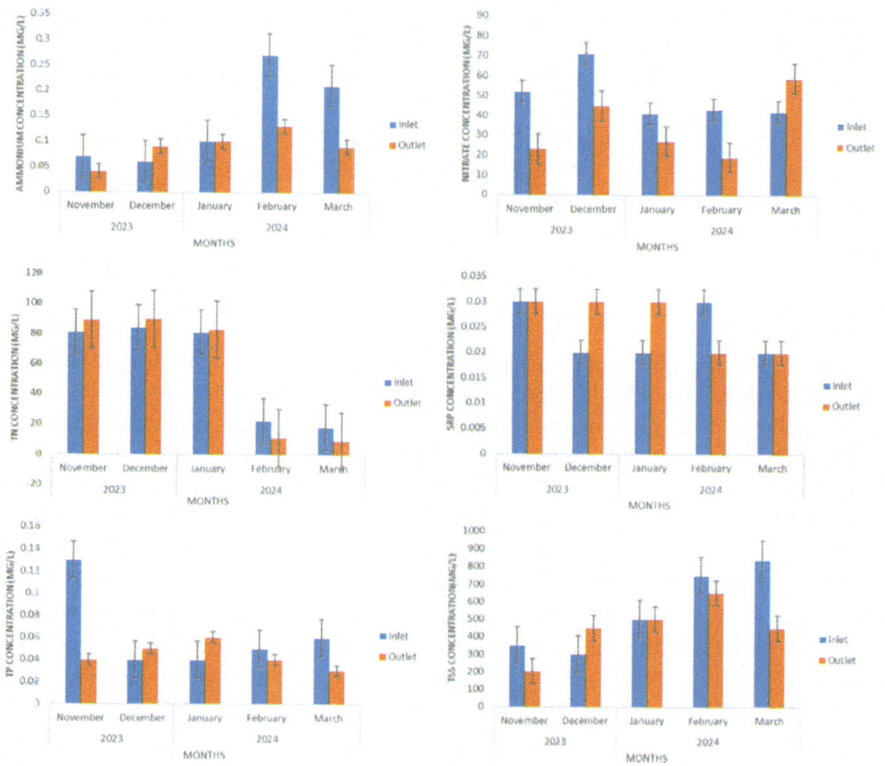

Fig. 3 Temporal variation in nutrient concentrations at the inlet and outlet of Yala wetland

January at the wetland's exit. The intake and outflow SRP concentrations were 0.03 ± 0.01 mg/l. The SRP concentration at intake varied significantly over the sampling period (ANOVA, $F = 11.510$, d.f $= 7$, $p = 0.018$). Tukey's post hoc analysis revealed that although the concentration in November was similar ($p > 0.05$), it differed significantly from that in December and March. The average SRP concentration at the outlet did not significantly change from December to March (ANOVA, $F = 1.808$, d.f $= 7$, $p = 0.284$).

Total phosphorus (TP) levels increased at the inlet from December to March and at the exit from November to January. The inlet's TP concentration varied significantly between November and March (ANOVA, $F = 10.729$, d.f $= 7$, $p = 0.021$). Tukey's post hoc test revealed that while the concentrations in December and January were similar ($p > 0.05$), those in November and March were significantly different ($p < 0.05$). There was no appreciable change during the sampling period, although the TP concentration at the outlet increased between November and January and then decreased by March (ANOVA, $F = 6.024$, d.f $= 7$, $p = 0.057$).

Table 2 Mean nutrient concentrations and retention rates at Yala Wetland's inlet and outlet

Variables	Concentration (mg/l)		Nutrients retention (%)
	Inlet	Outlet	
TSS	548	450	17.88
NH_4–N	0.15	0.09	40
NO_3–N	49.8	34.6	30.52
TN	57.2	56.4	1.4
SRP	0.03	0.03	0
TP	0.07	0.05	28.57

The concentration of total suspended solids (TSS) was at its highest at the inlet in March (840.50 ± 170.40 mg/l) and at the outlet in February (650 ± 94 mg/l), after declining from November to December and then gradually increasing through March. Over the sampling period, the TSS concentration at the inlet changed significantly (ANOVA, $F = 16.721$, d.f = 7, $p = 0.009$). TSS levels in November and December were comparable (p > 0.05), according to Tukey's post hoc analysis, but January and March had a significant difference. TSS levels at the outlet did not significantly change over the course of the sampling period (ANOVA, $F = 2.261$, d.f = 7, $p = 0.223$). The results of the Pearson correlation analysis showed that TSS and TP at the inlet and outlet had a significant positive correlation ($r = 0.960$).

3.3 Nutrients Retention (%) in Yala Wetland

After decreasing from November to December and then gradually increasing through March, the total suspended solids (TSS) concentration peaked at the inlet in March (840.50 ± 170.40 mg/l) and at the outlet in February (650 ± 94 mg/l), as shown in Table 2. The TSS concentration at intake varied significantly over the sampling period (ANOVA, $F = 16.721$, d.f = 7, $p = 0.009$). Tukey's post hoc analysis revealed that TSS levels in January and March differed significantly from those in November and December, which were comparable ($p > 0.05$). During the sampling period, there was no significant change in TSS levels at the outflow (ANOVA, $F = 2.261$, d.f = 7, $p = 0.223$). According to the findings of the Pearson correlation analysis, there was a significant positive correlation between TSS and TP at the inlet and outlet ($r = 0.960$).

4 Discussion

The study provides a thorough biogeochemical evaluation of Yala Wetland's ability to regulate water quality and hold onto nutrients. The analysis of physico-chemical parameters across time and space gradients offers important new insights into the fluxes and retention dynamics of important components, including suspended particles, nutrients, and dissolved oxygen (DO). In addition to assessing inflow and outflow concentrations, it is crucial to compute fluxes (mass per unit area per unit time), retention efficiency, and storage potentials in order to comprehend wetland performance as a biogeochemical reactor.

From a range of 4.56 to 6.56 mg/L at the inlet, the dissolved oxygen (DO) concentrations at the exit fell precipitously to 0.37 to 2.37 mg/L. The breakdown of organic matter and microbial respiration in anoxic zones are most likely the causes of this regional decline, which suggests a high oxygen demand in the wetland (1). With an estimated surface area of 15 km^2, an average flow of 0.5 m^3/s, and a mean DO drop of 4 mg/L, the wetland system is removing about 172.8 kg of oxygen per day (0.5 $m^3/s \times 4$ mg/L $\times$ 86,400 s/day). A high biochemical oxygen demand (BOD) suggests possible hypoxic hazards to aquatic life, particularly in the vicinity of the outlet, and validates the function of microbial activity in carbon metabolism [18].

The temperature fluctuations (inlet: 25.44–27.64 °C; outlet: 22.33–24.53 °C) stayed within the ideal ranges for aquatic life, highlighting the wetland's ability to act as a thermal buffer. Enzymatic and microbial metabolic activities that are essential to biogeochemical transformations are supported by this thermal stability.[4] Similarly, pH values stayed neutral to moderately alkaline (7.47–10.27 inlet; 7.0–8.8 outlet), suggesting a balanced aquatic environment and stable carbonate equilibrium. PH stability is very important in fostering ecological resilience, and this is consistent with their findings [19].

Significant variation in electrical conductivity (EC) readings (206.33–341.1 μS/cm) indicates that the dissolved ions are influenced by evaporation, agricultural runoff, and seasonal inflows. EC fluxes averaged 267 μS/cm per acre, corresponding to ionic concentrations of approximately 180–220 mg/L TDS. The filtration function of the wetland and its ability to retain and alter ionic species through sedimentation and adsorption are reflected in these changes [20]. Nutrient loading and landscape connectivity can be substituted with EC monitoring.

The mineralization of organic materials and inflow pulses was linked to the temporal variability of ammonium-nitrogen (NH_4^+-N). The NH_4^+-N loading into the wetland was roughly 19.4 kg/day (0.45 mg/L $\times$ 0.5 $m^3/s \times$ 86,400 s) at a flow rate of 0.5 m^3/s and an average input concentration of 0.45 mg/L. Nitrification, ammonification, and either microbial uptake or subsequent assimilation by aquatic plants were indicated by decreased amounts in the discharge. The total nitrogen (TN) concentrations in a 15 km^2 wetland peaked in December (mean 2.4 mg/L) and then declined by March (mean 1.1 mg/L). This translates to a daily net retention of about 119 kg, or 7.9 kg/ha. These figures demonstrate the wetland's capacity to serve as a nitrogen sink [19].

While Total Phosphorus (TP) showed noticeable increases at the intake, reaching 0.68 mg/L, Soluble Reactive Phosphorus (SRP) levels stayed relatively constant at approximately 0.03 mg/L. The calculated phosphorus input is approximately 29.4 kg/day, but the measured outflow decreases show a retention efficacy of 42%. Adsorption to sediments and co-precipitation with iron and aluminum oxides are the main processes that propel this capability. The rise in total suspended solids (TSS) in March, which had a positive correlation with TP, implies that sediment resuspension affects P dynamics [20]. The estimated ~ 0.82 kg/ha/day of phosphorus retained per acre indicates that sediment control is essential for improving phosphorus sequestration.

These results demonstrate how crucial Yala Wetland is for controlling nutrient flows and water quality. The wetland reduces the risk of eutrophication and preserves aquatic biodiversity by efficiently absorbing and holding onto large amounts of nitrogen and phosphorus as a biogeochemical buffer. This supports the findings on the contribution of wetland ecosystems to the global carbon and nitrogen budgets [21]. Due to its low retention capacity, the wetland is susceptible to external factors such as saturation, silt obstruction, and changes in land use.

Therefore, it is essential to combine wetland restoration initiatives, continuous monitoring, and land-use management [22]. The quantification of nutrient budgets using fluxes per unit area (e.g., kg/ha/day) allows for a more accurate assessment of ecosystem service delivery and facilitates the development of adaptive management strategies in response to climatic and anthropogenic causes [23].

5 Conclusion

This study emphasizes the importance of Yala Wetland for maintaining biological stability, managing water quality, and conserving nutrients. The significant temporal variations in dissolved oxygen, ammonium, and total phosphorus concentrations, particularly between the entrance and outflow, show the wetland's capacity to hold onto aquatic life and buffer pollutants. The necessity of integrated sediment-nutrient management is highlighted by the positive correlation between TSS and TP, which further demonstrates the interconnected dynamics of sediment transport and nutrient loading. The wetland shows indications of stress from upstream anthropogenic forces, including eutrophication and seasonal nutrient surges, despite its resilience. Declining DO levels at the outlet point to possible threats to aquatic biodiversity and ecosystem function, even though other metrics like temperature, pH, and EC stayed within acceptable ranges, indicating ecosystem stability.

To better understand the nutrient cycling processes in the wetland, future research should look into microbial and hydrological interactions and employ longer time scales. Furthermore, spatial modeling of nutrient pathways and sediment dynamics can help direct more targeted conservation efforts. A multidisciplinary approach integrating land-use planning, community-based monitoring, and policy enforcement will be crucial to enhancing Yala Wetland's resilience to climate change and human-caused stresses. Beyond mere conservation, safeguarding this ecological resource is essential to ensuring water quality, biodiversity, and livelihoods for future generations.

Acknowledgements Our gratitude goes to the LWM program, Department of Biological Sciences and Department of Soil Science Egerton University. We are also grateful to Mr. Erick Owino of Egerton University for field assistance.

Conflict of Interest The authors affirm that this research was carried out without any commercial or financial ties that could be perceived as a potential conflict of interest.

Data Availability Statement The data that support the findings of this study are available from Kelvin Kariuki Muli but restrictions apply to the availability of these data, which were used under license for the current study, and so are not publicly available. Data are however available from the authors upon reasonable request and with permission of Kelvin Kariuki Muli.

Ethics Statement In compliance with Kenya's national research regulations, this study was approved by the National Commission for Science, Technology & Innovation (NACOSTI). Local officials, such as chiefs and village elders, were notified about the study prior to data collection and, following a thorough briefing, gave their consent as well.

Authorship Contribution Declaration Conceptualization, Author 1.; Methodology, Author 1, Author 2, Author 3 and Author 4.; Software, Author 1, Author 3.; Validation, Author 1, Author 2, Author 3 and Author 4.

Funding Declaration This research was funded by Rotary Club of Vienna.

References

1. Mitsch WJ, Gosselink JG (2015) Wetlands, 5th edn. Wiley
2. Maltby E, Barker T, Smith C (2017) The role of water temperature in wetland ecosystems: implications for biodiversity and management. Aquat Conserv: Mar Freshw Ecosyst 27(4):819–831
3. Oginde PO, Mbogo G, Karanja E (2021) Assessment of the hydrological and ecological functions of Yala Wetland. Afr J Aquat Sci 46(1):1–13
4. Wetzel RG (2001) Limnology: lake and river ecosystems. Academic Press
5. Reckhow KH, Van Drecht G, Regan K (2015) Phosphorus retention and transport in wetland ecosystems: a review of processes and management implications. Water Res 68:97–109
6. Chowdhury SK, Islam MN, Mukul SA (2020) Assessing the role of electrical conductivity in wetland ecosystems: a comprehensive review. Aquat Ecol 54(3):579–596
7. Njeru RW, Muthoni JW, Muthama A (2019) Role of Yala Wetland in nutrient management and its implications for local livelihoods. Int J Wetl Res 9(2):105–115
8. Ego JO, Mugo RW, Wairia LW (2022) Nutrient dynamics in wetland ecosystems: a comparative analysis of inlet and outlet water quality. Wetl Ecol Manage 30(4):473–487
9. Bishop TD, Scharfenberger A, Gill G (2019) The impacts of agricultural expansion on wetland ecosystems: a case study from Yala Wetland in Kenya. Environ Sci Policy 94:191–198
10. Chakraborty SK, Sanyal P, Ray R (2023) Ecology and history of wetland research: operating scientific principles of eco-dynamics of wetland ecosystem with special reference to east Kolkata wetland, India. In: Wetlands ecology: eco-biological uniqueness of a Ramsar site (East Kolkata Wetlands, India). Springer International Publishing, Cham, pp 39–165
11. Kenya Wetlands Forum (2006) Report of a rapid assessment of the Yala Swamp Wetlands., Wetlands Forum, Nairobi, Kenya
12. Ondere LA (2016) Spatial-temporal changes of landcover types in response to anthropogenic dynamics in Yala Swamp, Siaya County. Masters Desertation, Moi University, Kenya, Kenya

13. Kinaro, Z. O. (2008). Wetland conversion to large-scale agricultural production; implications on the livelihoods of rural communities, Yala Swamp, Lake Victoria basin, Kenya (Doctoral dissertation, Linköping University, Sweden)
14. BirdLife International (2012) Kenya's important bird areas: status and trends 2012. Nature Kenya—The East Africa Natural History Society
15. BirdLife International. (2008). Saving Lake Victoria by conserving the Yala Swamp. BirdLife International Africa Partnership Secretariat, Nairobi, Kenya
16. Davis AJ, Koop K (2006) The role of dissolved oxygen in aquatic ecosystems. Aquatic Ecology 40(4):501–510
17. American Public Health Association (APHA) (2004) Standard Methods for the Examination of Water and Wastewater, 21st edn. American Public Health Association, American Water Works Association, and Water Environment Federation, Washington, DC
18. Smith VH, Tilman GD, Nekola JC (1999) Eutrophication: impacts of excess nutrient inputs on freshwater, marine, and terrestrial ecosystems. Environ Pollut 100(1–3):179–196
19. Horne AJ, Goldman CR (1994) Limnology, 2nd edn. McGraw-Hill
20. Chale FMM (2004) Trace metal and major ion concentrations in algae and water in the rivers and wetlands of Lake Victoria basin, Tanzania. Tanzan J Sci 30(2):51–62
21. Raburu PO, Okeyo-Owuor JB, Kwena F (2012) Community-based approach to the management of Nyando Wetland, Lake Victoria Basin, Kenya. Afr J Environ Sci Technol 6(12):475–483
22. Kadlec RH, Wallace SD (2009) Treatment Wetlands, 2nd edn. CRC Press
23. Reddy KR, DeLaune RD (2008) Biogeochemistry of wetlands: science and applications. Wetl Ecol Manage 16(4):193–204

Backmatter

This section recognizes the key stakeholder input and participation in the International Conference on Climate Change and Ecosystem Restoration (ICRECC 2025). The section begins with a sample of the keynotes presented in the conference, followed by an article from some of the key climate financiers, who were also major ICRECC 2025 supporters. This section concludes with a dedicatory page for each of the keys stakeholders that also sponsored the conference.

Keynotes

Valorization of Agricultural Waste: Preliminary Results of the Proposal to Establish a Spinning Unit for Banana Fibers in Eldoret, Kenya

The management of agricultural waste is critical in addressing environmental concerns and promoting health working and living environments. This case study focuses on the valorization of banana stems through value addition processes that align with global sustainable industrialization strategies. Eldoret town in Kenya, home to several textile factories including the revitalized Rivatex facility, presents a unique opportunity for integrating banana stem fiber into textile production. Eldoret's strategic location, surrounded by banana-producing counties such as Kisii, Kakamega, Siaya, and Bungoma and in close proximity to Eastern Uganda, positions it as a potential processing and innovation hub. The study explores the potential of converting banana pseudo-stems, which are often discarded post-harvest, into spun yarn suitable for sustainable fashion products (jeans, jackets, socks, etc.). By analyzing the Kenya Textile and garment sector (its potential and hotspots), the production of banana in Kenya and Eastern Uganda and mapping this with the unique potential of Eldoret as a regional manufacturing, the study demonstrates the feasibility of establishing a banana fiber value chain anchored in Eldoret. The proposed model highlights rural-industrial linkages, job creation, use of renewable energy and resource efficiency in both

R. Ramkat et al. (eds.), *Towards the Sustainability of our Ecosystems*, Springer Proceedings in Geography and Sustainability,
https://doi.org/10.1007/978-3-032-30076-8

agriculture and manufacturing. By leveraging existing infrastructure and promoting cross-border agro-industrial collaboration, this initiative will contribute to Kenya's textile revival and address other pressing issues, which includes waste management and responsible production and consumption.

Josphat Igadwa Mwasiagi, Moi University, Eldoret, Kenya,
Email: igadwa@gmail.com

Contextualizing, Localizing and Actualizing Ecosystem Restoration

Ecosystem restoration entails deliberate efforts to assist a degraded ecosystem toward recovery pathway. Restoration encompasses a wide continuum of practices, depending on local conditions and societal choice. It also extends to conserving the ecosystems that are still intact. Premised on the fact that healthier ecosystems provide us with better ecosystem services (ES) it will be beneficial to start by reflecting on what goes wrong during degradation. Conceptualization of restoration entails understanding what is to be transformed, how, by whom and to what extent. Climate change is expected to exacerbate the effects of degradation while aggravating the implications on humanity, nature and businesses and hence compromise the attainment of Sustainable Development Goals (SDGs). Contextualization ecosystem restoration should be viewed from the lens of global and local frameworks. The United Nations Decade on Ecosystem Restoration (2021–2030) challenges everyone to scale up restoration efforts that breathe new life into our degraded ecosystems. In the Kenyan context such guidelines and regulations buttress the Environmental Management and Coordination Act (EMCA). Localizing restoration efforts should focus on sustainable management of ecosystems. Such efforts should include but not limited to capacity strengthening, promoting sustainable cities and green businesses and Agrifood Systems within the landscapes in question. Actualizing restoration entails initiating projects and initiatives at local level. Such initiatives could be in the form of transformative pathways whereby stakeholders are brought together through a T-Lab approach in diagnosing and visioning how to deal with environmental and societal change through practices and structures that can be used as one of the approaches for sustainable natural resource management. Another recent notable restoration effort in Kenya is that of Nairobi river in which a commission has been formed to engage in rehabilitation, protection, restoration of the riparian and from a basin approach. Ecosystem restoration is not without challenges. Some of the challenges that can slow down ecosystem restoration efforts include limited scientific knowledge and community capacity, inadequate government support and financial resources. Nevertheless ecosystem restoration is an essential component of regaining the lost ecosystem service in any landscape and pathway toward sustainability of our ecosystems.

By Prof. Julius Kipkemboi, Kaimosi Friends University
Email: jkipkemboi@kafu.ac.ke

Harnessing AI for Resilience: Addressing Climate-Induced Health, Food Security, and Livelihood Challenges in Africa

Africa stands at a critical juncture where climate change is intensifying vulnerabilities across health systems, food security, and livelihoods. From droughts and floods to the spread of vector-borne diseases, the impacts are widespread and disproportionally affect vulnerable populations. Amidst these growing challenges, there is a powerful opportunity to harness Artificial Intelligence (AI) to develop adaptive strategies and strengthen resilience across sectors. Here, we explore the transformative potential of AI in addressing climate-induced challenges across key sectors. AI can enable more accurate climate modeling, predictive analytics that forecast disease outbreaks and optimize agricultural practices. AI-driven powered solutions such as satellite-based monitoring and machine learning algorithms, are helping detect early signs of food shortage, health risks and disruptions to livelihoods, providing a critical window for timely interventions. However, realizing this potential requires more than just technology. Key challenges include data scarcity, limited technological infrastructure, and the need for capacity at the local level. Additionally, ethical concerns around data privacy, algorithmic bias, and equitable access to AI tools must be addressed to ensure that solutions are inclusive and beneficial to all communities. As we move toward a future shaped by both climate uncertainty and digital transformation, Africa has the chance to leapfrog with contextually relevant AI applications that are equitable, scalable, and sustainable. This session aims to inspire cross-sector dialogue and collaboration, urging all stakeholders—researchers, policymakers, technologists, and citizens—to co-create AI solutions that build resilience, protect human health, secure food systems, and safeguard livelihoods across the continent.

Prof. Ann Mwangi, Moi University,
Email: annwsum@gmail.com

Sponsors and Partners

Stima Sacco's Green Financing and Digital Transformation Drive Sector Leadership in Climate Action

Stima DT Sacco 50+ years of empowering members for life has solidified its position as a leader and pacesetter in Kenya's cooperative sector thanks to its strategic leadership, innovation, and prudent financial management practices.

In serving its over 220,000 members with an asset base of over KES 66.4 billion (USD 510 million) and investing over KES 50 billion (USD 384.6 million) in impact lending to its members, Stima Sacco has led all deposit-taking SACCOs with KES 10.26 billion (USD 78.919 million) in total income, contributing to the Sacco sector's historic KES 1 trillion (USD 7.692 billion) asset milestone (Sacco Sector Supervision Report, 2024).

The 2024 sustainability report of Stima Sacco demonstrates its focus on the 7th cooperative principle on concern for the community through climate action. The successful integration of green financing with its digital innovation by investing KES 1.74 billion (USD 13.384 million) through green lending toward affordable housing, water, and electricity projects to improve community livelihoods is remarkable. Sacco's green loan products—including solar energy, eco-housing, and electric mobility financing—are accessible via mobile and online platforms, thereby reducing carbon footprints and expanding reach.

The digital transformation at Stima DT Sacco enables faster loan processing, improves member engagement, enhances sustainability outcomes and contributes to climate-smart strategy. All these initiatives have made positive contributions to its compliance with SASRA's (Sacco Societies Regulatory Authority) regulatory framework as well as Kenya's Vision 2030 and UN-SDGs positions in addition to being a model for cooperative-led green growth.

At Stima DT Sacco, sustainability is embedded in the cooperative values that guide its operations, thereby showcasing its commitment to economic resilience, environmental stewardship, and social impact.

By integrating Environmental, Social, and Governance (ESG) principles in its 2025–2029 strategic plan, there is no doubt that the Sacco is well positioned to create long-term value for the members, stakeholders, and the wider community thus aligning with SDG-13 on climate action through responsible business practices, digital banking initiatives, and sustainable financing that contribute to environmental conservation and climate resilience.

By Amos Mungo, Manager, Stima Sacco, Kisumu Branch
Email: mungoamos@gmail.com/amungo@stima-sacco.com

Kingdom Bank: Empowering a Low-Carbon Future Through Climate-Conscious Banking in Kenya

Kingdom Bank has emerged as a key driver of Kenya's transition toward a low-carbon, climate-resilient economy. Through a deliberate strategy grounded in sustainability, innovation, and inclusion, the Bank has implemented a suite of initiatives, products, and services that directly address climate change and accelerate renewable energy adoption across Kenya's social and economic sectors. These actions are guided by both national and global regulatory frameworks for green finance, positioning Kingdom Bank as a leader in responsible banking and climate stewardship.

Regulatory and Policy Alignment
Kingdom Bank operates fully in line with Kenya's strengthened green-finance framework. Its programs are anchored in the Climate Change Act (2016) and successive National Climate Change Action Plans, which promote low-carbon, climate-resilient growth. The Bank has also integrated the Central Bank of Kenya (CBK) Green Finance Taxonomy and Climate-Risk Disclosure Framework (2025) into its operations, enabling accurate classification of green assets, transparent disclosure of climate risk, and consistent reporting to regulators and investors. Globally, Kingdom Bank's governance and reporting practices are aligned with leading standards, including IFRS S2 climate-related disclosures, Task Force on Climate-Related Financial Disclosures (TCFD) principles, and the UN Principles

for Responsible Banking. These frameworks ensure accountability, transparency, and strategic alignment with the global net-zero agenda.

Implemented Initiatives
Kingdom Bank has successfully mainstreamed climate considerations into all aspects of its strategy and lending operations. Key initiatives include:

Governance and risk integration: Climate and environmental risk assessment integrated into enterprise risk management, board oversight, and credit decision processes.

Green finance policy: Internal taxonomy and exclusion list developed to guide lending toward renewable energy, clean technology, and climate-resilient sectors.

Staff capacity building: Comprehensive training for credit officers and relationship managers on evaluating renewable energy and climate-smart agriculture projects.

Strategic partnerships: Collaboration with development finance institutions, fintech innovators, and donor agencies to de-risk lending and scale blended-finance mechanisms.

Operational sustainability: Bank-wide greening of branches and offices through rooftop solar installations, energy-efficient equipment, e-banking platforms, and paper-light operations.

Green Products and Services
Kingdom Bank offers a diverse portfolio of products that directly mobilize capital toward climate-positive investments, to mention but a few in summary:

I. Green asset finance: Loans for solar home and educational institutions use, commercial solar PV, mini-grids, and energy-efficient machinery for SMEs.
II. Eco-mortgages and retrofit loans: Financing for energy-efficient housing, building retrofits, and clean-cooking solutions to homes and schools.
III. Green bonds and sustainability-linked loans: Innovative instruments tied to measurable environmental performance indicators.
IV. Climate-smart agri finance: Working-capital facilities, microinsurance, and technical support for agribusinesses adopting sustainable, low-emission practices.
V. Advisory and carbon services: Project-bankability assessments and transaction advisory, and facilitation of access to carbon markets and credit mechanisms to the extent possible currently.

Measured Impact
Kingdom Bank's green-finance initiatives are already generating measurable outcomes across Kenya:

I. Reduced emissions: Financing solar energy and clean-cooking technologies has lowered carbon intensity in households and enterprises.
II. Expanded energy access: Upcoming systems such as Pay-as-you-go (PAYG) and mini-grid financing have improved energy access for off-grid communities and boosted productivity among micro-enterprises.
III. Enhanced climate resilience: Lending to climate-smart agriculture and water-efficient systems strengthens the adaptive capacity of smallholder farmers.
IV. Job creation: Green projects have generated employment opportunities in equipment manufacturing, installation, and maintenance.
V. Financial system stability: Integration of climate risk into credit assessment has improved portfolio quality and reduced exposure to climate-related defaults.

Challenges and Continuous Improvement
While success has been significant, Kingdom Bank continues to address challenges common to the green-finance sector, including data gaps for climate-risk modeling and the need to scale concessional capital for early-stage technologies. The Bank is pursuing continuous improvement through:

I. Simplified MSME product templates: Streamlined documentation to enhance accessibility for small borrowers.
II. Blended-finance and guarantee partnerships: Leveraging concessional funds to reduce risk and attract private capital.
III. Enhanced technical capacity: Ongoing staff development in climate analytics and project due diligence.
IV. Transparent reporting: Annual publication of climate-risk disclosures aligned to IFRS S2 and CBK standards.

Through deliberate strategy, innovation, and strong governance, Kingdom Bank has positioned itself as a national leader in sustainable finance, directly contributing to Kenya's renewable energy transition and climate-resilient growth. Its proactive approach demonstrates how local banks can align profitability with environmental responsibility—proving that banking for the planet is banking for the future.

By Jane Wangari Gathuya, Branch Manager, Kingdom Bank, Eldoret Branch
Email: Jane.Gathuya@kingdombankltd.co.ke/jgathuya@gmail.com

Sponsors and Partners Logos

Index

R. Ramkat et al. (eds.), *Towards the Sustainability of our Ecosystems*, Springer Proceedings in Geography and Sustainability,
https://doi.org/10.1007/978-3-032-30076-8

The manufacturer's authorised representative in the EU is Springer Nature Customer Service Centre GmbH, Europaplatz 3, 69115 Heidelberg, Germany. If you have any concerns regarding our products, please contact ProductSafety@springernature.com

Printed and bound by CPI Group (UK) Ltd, Croydon, CR0 4YY
17/09/2026
02218437-0008